Forces and Fields

Forces and Fields

The Concept of Action at a Distance in the History of Physics

Mary B. Hesse

DOVER PUBLICATIONS, INC.
Mineola, New York

Copyright

Bibliographical Note

This Dover edition, first published in 2005, is an unabridged republication of the first edition originally published in London and New York by T. Nelson in 1961.

Library of Congress Cataloging-in-Publication Data

Hesse, Mary B.
Forces and fields : the concept of action at a distance in the history of physics / Mary B. Hesse.
p. cm.
Originally published: London ; New York : T. Nelson, 1961.
Includes bibliographical references and indexes.
ISBN 0-486-44240-3 (pbk.)
1. Physics—History. 2. Physics—Philosophy. 3. Force and energy. I. Title.

QC7.H62 2005
530.1'09—dc22

2005043259

Manufactured in the United States of America
Dover Publications, Inc., 31 East 2nd Street, Mineola, N.Y. 11501

Preface

In this book I have traced through the history of physics some of the problems clustering round the question : 'How do bodies act on one another across space ?', and I have used the various answers to this question to illustrate the role of fundamental analogies or models in physics, and the ways in which so-called unobservable entities are introduced into it. It has not been my purpose to write an exhaustive history of the subject—this is in any case an unrealisable goal, and the attempt to reach it has had the result too often in the historiography of science of obscuring the principles of historical selection which are always inevitably present. I have selected for most detailed discussion those periods of transition in fundamental physics in which new concepts and ideas have been introduced and made scientifically testable. I have also thought it desirable to make a certain philosophical interpretation of science explicit at the beginning, for this interpretation has no doubt affected both my selection of historical material and my comments upon it. This is after all an acceptably scientific procedure : to state a theory about the nature of science, which is tested by reference to specific historical situations, and then if necessary to modify the theory in the light of the tests. But one thesis which will be argued here with respect to science itself is also assumed to hold in respect to the history of science, namely, that there are no bare and uninterpreted facts ; all facts, whether experimental or historical, are interpreted in the light of some theory. In writing the history of science there will always be present, either implicitly or explicitly, some philosophical view of the nature of science. The first chapter, in which the philosophical view adopted here is developed, is however somewhat more technical than the rest of the book, and a reader whose main interest is historical will find that he can omit the details without affecting his understanding of what follows.

The history of a particular theoretical idea, in which the emphasis is rather upon theory-constructing than upon theory-testing, fortunately lends itself to non-technical presentation. The detailed examination of how and why a certain theory is confirmed or refuted by experimental tests is inevitably somewhat technical, and becomes more so in later historical periods for the more advanced sciences. The fact that it has not been possible or necessary here to enter into details about either mathematical or experimental technicalities has inevitably distorted the over-all picture of physics presented, but there are after all many books devoted to this aspect of the history of science, and still too few which trace the history of scientific ideas and their relations with philosophy. There is therefore some justification for trying to redress the balance even at the risk of over-compensation.

It may be felt that, for a historical study of physics, a disproportionate amount of space has been given to pre-seventeenth-century work. Aristotle may have written a book called the *Physics*, but many histories of physics devote all but a brief introductory chapter to the period beginning with Gilbert and Galileo, and do not appear to lose much thereby. But here again the periods selected depend on one's purpose, and where different fundamental theories and the relations between them are to be discussed, those periods which were failures and cul-de-sacs from the point of view of twentieth-century physics may be as important as the cumulative successes of the past three hundred years. There are now signs in physics that the time is ripe for another major transition of fundamental ideas, and if this occurs it may be that the confident physicalism of the modern period will also be seen by future generations to have led in the end to a cul-de-sac, and this in spite of the numerous technical achievements accomplished on the way. In that case we may hope that twentieth-century physics will be spared the judgments of historians whose criteria are wholly in terms of present success.

There is another reason why I have devoted a considerable amount of space to the Greeks, and have made my account of their science somewhat more comprehensive than the accounts of later periods. There are not yet many general histories of science which incorporate those studies of the Pre-Socratics initiated by Cornford and continued by contemporary classical historians. Burnet's interpretation of the early Greek philosophers as forerunners of later science was no doubt sympathetic to the inductivist view of science then current, in which science was seen as a linear progress towards the 'correct' ideas, but many of Burnet's interpretations are now discredited among classicists, and it seems right to try to put this whole most significant period in new perspective in relation to a different view of the nature of scientific advance. I have also tried to bring out the debt which the seventeenth-century natural philosophers owed to Greek ideas, for in spite of their emotionally charged polemics, they did not in fact erase everything and start again with a clean slate, but proceeded in the fashion of all later science, by criticism, testing, and modification of received ideas.

No-one could claim a specialist knowledge of the history, philosophy and science of all the periods which I have discussed in this book. But some of the effects of my shortcomings in this respect have certainly been mitigated by the generous help I have received from colleagues and friends, and I should like to acknowledge my debt to them, but without saddling them with any responsibility for the views the book now contains. I should like in particular to thank Professor H. Dingle for his encouragement, and also Dr A. Armitage, Mr D. J. Furley, Dr N. H. de V. Heathcote, Dr H. R. Post, Dr J. R. Ravetz, and Dr G. J. Whitrow, all of whom read individual chapters in manuscript. I should also like to express my debt to Dr J. Agassi for conversations and for his unpublished thesis 'The Function of Interpretations in Science' from which I have learnt to disentangle myself from some stereotyped and false interpretations of the history of science, and to Dr P. W. Higgs for advice relating to quantum field theory. The English edition of Professor Popper's *Logic of Scientific Discovery* appeared too late for me to make detailed reference to it, but my debt to the point of view initiated by the first German edition of 1934, and since expanded in Professor Popper's other writings, will be obvious. In order to avoid a multiplicity of footnotes, I have not made detailed reference to secondary historical sources in the text. I hope, however, that my great debt to them is adequately acknowledged in the Bibliography.

Acknowledgments are due to the publishers and Editor of the *British Journal for the Philosophy of Science* and to those of *Isis* for permission to reproduce certain material in Chapters I and VIII respectively.

MARY B. HESSE
M.Sc. Ph.D.

Contents

I THE LOGICAL STATUS OF THEORIES

Physics and epistemology	1
A realist view of theories	3
Operationalism	6
The hypothetico-deductive method and falsifiability	8
The dictionary theory	13
Models	21

II THE PRIMITIVE ANALOGIES

Analogies in primitive scientific explanation	29
Nature as 'Thou' : indefiniteness of the problem of interaction	34
The Pre-Socratics : distinction between matter and force	39
Atomism	42
Immaterial causes	46

III MECHANISM IN GREEK SCIENCE

Mechanical analogies among the cosmologists and medical writers	51
The *horror vacui* and *antiperistasis*	54
Aristotle : matter and form and the primary qualities	60
Aristotle : theory of motion and change	63
Aristotle : the principle of action by contact	67
Aristotle : the unmoved first mover	70

IV THE GREEK INHERITANCE

The primitive analogies in medieval belief	74
The emanation theories	77
Multiplication of species	79
The theory of vacuum-suction	82
Gilbert's *De Magnete*	86
Francis Bacon's classification of actions at a distance	91

V THE CORPUSCULAR PHILOSOPHY

Falsifiability as a seventeenth-century criterion for theories	98
Descartes's mechanical continuum	102
Descartes's method	108
Corpuscular and medium theories	112
Locke on the mechanical philosophy	121

VI THE THEORY OF GRAVITATION

Gravity as internal tendency or external attraction	126
The planetary orbits	129
An analysis of Newton's laws of motion	134
Universal gravitation as a mathematical law	144
Universal gravitation as a physical hypothesis	148
Newton's atomism and active principles	153

VII ACTION AT A DISTANCE

Leibniz's attack on action at a distance 157
Philosophical justifications of action at a distance 163
Kant : *The Metaphysical Foundations of Natural Science* 170
Kant : infinite divisibility of matter as a regulative principle 173
Kant : attraction and repulsion 176
Elastic fluid theories in physics and chemistry 180

VIII THE FIELD THEORIES

Euler's hydrodynamics 189
Criteria for continuous contact-action 195
Faraday : the physical nature of lines of force 198
Faraday : criteria for action at a distance 203
Maxwell : mechanical and field theories of continuous action 206
Hertz : interpretations of Maxwell's equations 212
The Continental action-at-a-distance school 216
Gravitation 222

IX THE THEORY OF RELATIVITY

Interpretations of the Michelson-Morley experiment 226
Consequences of the Lorentz transformation 235
Milne's action-at-a-distance theory 239
Einstein's theory of gravitation 245
Conventional and factual aspects of geometry 253

X MODERN PHYSICS

The empirical basis of quantum mechanics 259
The wave-particle duality and uncertainty principle 263
Sub-quantum theories 267
Modes of action in the quantum field 270
The meson and Maxwell fields 275
The action-at-a-distance theory of Wheeler and Feynman 279
Reversibility of cause and effect 285

XI THE METAPHYSICAL FRAMEWORK OF PHYSICS

Some heuristic and metaphysical considerations 290
Theoretical aspects of extrasensory perception 295

APPENDIX I 305

APPENDIX II 306

BIBLIOGRAPHY 308

INDEX OF PROPER NAMES 313

INDEX OF SUBJECTS 316

Chapter I

THE LOGICAL STATUS OF THEORIES

Physics and Epistemology

DURING the last hundred years, the question of the meaning and logical status of scientific theories has become acute. Whereas it had become the fashion since the disputes of the seventeenth century for scientists to regard philosophy as at best irrelevant and at worst a hindrance to their work, it has now become common to find physicists making philosophical judgments about their theories, and terms like ' reality ', ' epistemology ', ' mental construction ' appear in serious expositions of modern physics, and even invade the antiseptic atmosphere of scientific journals.

The immediate cause of this ferment has been the need to understand and interpret the revolutionary developments in physics since the end of the nineteenth century, and the problem has been to make sense of the apparently paradoxical statements which physicists have been led to make in formulating the new theories. But the deeper problem has been concerned with the nature of scientific theories themselves. This is by no means a new problem, and it is relevant to all kinds of theories, whether they appear paradoxical or not, and whether they are controversial or well-established. It is, however, a problem which is liable to break out with renewed vigour during periods of radical scientific change, and when scientific results appear to contradict traditional ideas.

Throughout the history of science, one can trace the influence of two contrasting accounts of what scientific theories are, what kind of information they give about the world, and, closely connected with this, what is the best procedure to follow in developing them. The well-known razor of William of Ockham is a classic example of the empirical, positivist approach, which forbids the postulation of entities without necessity, and the occasion of it was a realist natural philosophy deriving from Aristotle, in which were postulated hidden causes such as undetectable forces causing motion, and invisible *species* conveying radiation. Another example occurs at the beginning of the modern scientific movement, when Copernicus and Galileo were both urged by well-meaning interpreters to adopt a positivist position regarding the motion of the earth : to claim, that is, only the convenience of a calculating device for the heliocentric system, not that it represented a true account of the structure of the world. Here, as with Ockham earlier and Kant later, part of the purpose of distinguishing sharply between the immediate empirical data and theories which might be built upon them, and of denying that such theories have any descriptive value, was to confine the authority of science to the immediately observed, and leave the province of the unobserved to theology, or metaphysics, or whatever else was believed to have access to non-empirical facts.

Newton's view of science, at least in his non-speculative writings and as expressed in the famous 'hypotheses non fingo', is also in the positivist tradition, since he wished to confine theories to what could be *deduced* from phenomena : 'whatever is not deduced from the phenomena is to be called an hypothesis ; and hypotheses, whether metaphysical or physical, whether of occult qualities or mechanical, have no place in experimental philosophy '.[1] Newton did not however draw the conclusion that the heliocentric system was a calculating device ; for him it was the 'true system of the world', because he regarded Kepler's laws as being deduced from phenomena and 'rendered general by induction', and his own gravitational theory as being a further inference of the same kind. And he, unlike most positivists, was not afraid to speculate as to possible 'hidden causes', although he always did so with apologies and with the implied hope that such causes would eventually be found to be deducible from phenomena, and hence not hypothetical.

The general characteristic of a positivist view of science is suspicion and caution with regard to the claims of theories, and this view therefore always assumes importance as a possible solution of difficulties which arise when the theories to which science seems to be committed are in conflict with common-sense (as with Newton's apparent action at a distance), or traditional belief (as with the motion of the earth), or are unduly complex or apparently self-contradictory (as with the incompatible properties of the nineteenth-century aether or twentieth-century fundamental particles). The suggested solution along positivist lines is then to assert that scientific theories which go beyond immediate experience are mere mental constructions, tools for correlating and predicting the results of possible experiments, and not descriptions of physical reality.

[1] *Philosophiae Naturalis Principia Mathematica* (2nd ed., London, 1713), III, General Scholium to Prop. XLII ; ed. of F. Cajori, Berkeley, 1947, p. 547

This was the line taken by Berkeley and Kant in relation to Newtonian science, by Mach, Pearson and Duhem in relation to nineteenth-century physics, and by many leading exponents of modern physics, where, as we shall see, the issue is confused by special difficulties.

The two traditional views of scientific theories must now be described more precisely. I shall continue to refer to them as the 'realist' and 'positivist' views respectively, and after discussing their characteristics I shall go on to try to show that neither is adequate as an account of science, and to develop with the aid of recent discussions in the philosophy of science a modified realist view which accords more closely with the actual procedures of scientists.[1]

A Realist View of Theories

The realist view holds that scientific theories are in a straightforward sense literal descriptions of nature. This view is easily and almost unconsciously adopted in the early stages of a science, when the statements that appear in it are hardly more than direct descriptions of observations and the results of experiments, using a minimum of technical expression. Examples would be the description of observed positions of a planet, or the optical properties of mirrors or refracting media. When theories become more complex, however, it seems that some statements are being made which are not subject to immediate test in this way, that is, they are not 'observation statements', because, for example, they may be about events inaccessible in space or time, or on too small a scale to be observed with existing scientific apparatus. The realist account asserts that these difficulties in the way of observational tests are merely accidental, and that such theories are still to be regarded as literal descriptions of entities existing in nature, and to be understood in exactly the same way as they would be understood if they were describing observations. In support of this, the

[1] In 'Three Views concerning Human Knowledge' (*Contemporary British Philosophy*, ed. H. D. Lewis (London, 1956, p. 357) Professor Popper has discussed two views of theories which he names 'essentialist' and 'instrumentalist'. The former does not correspond to what is meant here by 'realist', for, according to Popper, essentialists believe that the purpose of science is to discover essences which are final and not subject to correction by future experience. By instrumentalism he means the thesis that theories are computation rules, and in criticising this view he is interested not so much in the problem of meaning, as in examining what scientists in fact do with theories, and showing that this is incompatible with the belief that they are merely rules. His conclusion, that theories are intended to describe the real, although the scientist never knows whether they are in fact true descriptions, is similar to that we shall reach below on somewhat different grounds.

realist might point out that we *never* restrict the meaning of our statements to what is actually observed ; we are confident, for instance, that physical objects remain in existence when we are not looking at them, and it may be argued that the difference between this and the assumption on indirect grounds that there are entities too small or too distant to be observed is one of degree and not of kind.

Consider for example the dynamical theory of gases. This describes gases as being made up of particles moving at random in a containing vessel, and from the mechanics of such particles the properties of gases are deduced and shown to correspond with their actual properties as observed. If these particles are identified with chemical molecules, further correlations between the theory and observations of chemical reactions are possible. Now in the absence of indications to the contrary, it is natural to assume that the theory has shown gases to be literally composed of minute particles, too small to be detected by microscopes, but nevertheless particles of the same kind and obeying the same mechanical laws as marbles or tennis balls.

It is already clear, however, that this similarity between marbles and molecules is not complete, and therefore the use of the word ' particle ' is not entirely literal, since while it makes sense, for example, to ask about the size of both marbles and molecules, other qualities such as colour can be predicated only of marbles. A molecule cannot, consistently with the physical theory in which it appears, have a colour, because colour is described in that theory as a function of collections of molecules.

When theoretical science becomes more complex than in this example, the naïve realist view just described breaks down irreparably. Even in the nineteenth century this can be seen to be happening, and the leading theoretical physicists were not unaware of it. It is well known that most physicists then regarded mechanics as the basic physical theory, and explanation in other fields, namely, heat, light, electricity and magnetism, was thought to be incomplete unless it could be given in terms of the mechanics of rigid, elastic, or fluid bodies. But what is not so often realised is that these mechanical models were hardly ever regarded as literal descriptions of entities existing in nature. In many cases it would have been fantastic to regard them as such. Here for instance is a description of Kelvin's model of the luminiferous aether :

' Suppose . . . that a structure is formed of spheres, each sphere

> being in the centre of a tetrahedron formed by its four nearest neighbours. Let each sphere be joined to these four neighbours by rigid bars, which have spherical caps at their ends so as to slide freely on the spheres. . . . Now attach to each bar a pair of gyroscopically mounted flywheels, rotating with equal and opposite angular velocities, and having their axes in the line of the bar ; . . . the structure as a whole will possess that kind of quasi-elasticity which was first imagined by MacCullagh.' [1]

This aether was supposed to pervade all space from the interior of molecules to the furthest star, but of course neither Kelvin [2] nor anyone else believed that indefinitely small and indefinitely numerous replicas of this model extended throughout the whole of space and pervaded all matter. It had already become clear that the language which described the model was not being used literally to describe the world. Maxwell, whose incidental remarks in scientific papers show him to have been an exceptionally clear thinker about the logical significance of theories, says that such mechanical models must not be taken as modes of connection existing in nature, but are only intended to show that a mechanism can be imagined which is equivalent to the electromagnetic connections of the aether.[3] He goes on to remark that there is generally an infinite number of mechanisms corresponding to a given electromagnetic system. Any such mechanism is neither postulated to exist nor is it unique. It is merely that some of its properties correspond with the observed features of nature, but beyond the observations it does not necessarily correspond with nature in any literal sense.

With the advent of modern theories of matter and radiation it became impossible to maintain a realist view in the old mechanical sense. In quantum theory, for example, it is impossible to find a single mechanical model to represent either atomic structure or radiation. The theory itself will be described in more detail in a later chapter ; its relevance here is that the difficulties to which it led gave support to the positivist view of theories, and this view is still quite widely thought to be a necessary epistemological consequence of quantum physics.

[1] E. T. Whittaker, *History of the Theories of Aether and Electricity*, I, London, 1951, p. 145

[2] Kelvin did indeed remark, 'I never satisfy myself until I can make a mechanical model of a thing. If I can make a mechanical model I can understand it' (quoted from the original edition of the *Baltimore Lectures*, 1884, by S. P. Thompson, *Life of William Thomson*, II, London, 1910, p. 835), but he nowhere claimed that the model was necessarily identical with the thing.

[3] *Treatise on Electricity and Magnetism*, II, Oxford, 1873, p. 416

Operationalism

Parallel with these developments in physics went a good deal of detailed philosophical discussion which sought to interpret physics from a positivist standpoint. Ernst Mach put forward a phenomenalist theory of science, according to which theories are not attempts to explain phenomena by describing a real world which somehow causes the phenomena, but are merely shorthand accounts of the phenomena themselves, effecting an ' economy of thought ' in dealing with them. Russell at one stage of his thought tried to eliminate theoretical concepts from science altogether by defining them explicitly in terms of direct observations, or sense-data, so that their status would be simply that of a shorthand, replaceable in principle by longer descriptions of the sense-data themselves. Thus, according to Russell,[1] when we speak about light waves, we are really using the term as shorthand for a series of observation statements about measurements of reflections, refractions, spectral shifts and so on. Various operational theories have had the same programme of defining theoretical concepts explicitly in terms of the laboratory operations required to measure their numerical values, and according to this view of science, all concepts which cannot be so defined are meaningless, leading to the asking of unanswerable questions, and should therefore be eliminated from scientific theory. The view gained some plausibility from two developments of nineteenth-century physics which culminated in Einstein's special theory of relativity : first, the impossibility of finding any experimental procedure for measuring the absolute velocity of the earth through the aether, and second, the absence of any clear meaning of the concept of simultaneous times at distant points of space, a concept which had been tacitly presupposed to be meaningful in Newtonian mechanics. In the first case the lack of possible operations for measuring absolute velocity led to its abandonment as a meaningful concept within the framework of special relativity. In the second case an explicit ' operational ' definition for simultaneity was provided in terms of observers equipped with clocks and light-signalling apparatus, and it was then found that the acceptable definition differed from Newtonian assumptions in that two times at spatially separated points might be simultaneous for some observers but not for others. Absolute simultaneity therefore lost its meaning within special relativity.

At the time when he was beginning to develop relativity theory, Einstein himself, influenced by Mach, was inclined to interpret

[1] ' The Relation of Sense-Data to Physics ', *Mysticism and Logic*, London, 1918

scientific theories according to the positivist view. Later, however, he modified his opinion considerably, and spoke of theories as free constructions of the mind, not restricted by any necessity of *explicit* definition in terms of observations.[1] But many of his scientific disciples did not follow him in his later philosophy of science. It was Eddington, who gave the first detailed account of relativity theory in English in his *Mathematical Theory of Relativity* in 1923, who also gave in the Introduction to that book one of the first clear statements of operationalism : ' A physical quantity is defined by the series of operations and calculations of which it is the result '.[2] In 1927 P. W. Bridgman gave what is still the classical account of this view of science in his *Logic of Modern Physics*, and although Bridgman himself has since considerably modified his earlier views, this book has remained a typical expression of the approach of many scientists. Lip-service was also paid to the principle of operationalism by some philosophers, including those of the Vienna Circle in their early writings, although it was immediately clear that strict adherence to it would require very drastic reformulations of existing physical theories, including, for example, that of the succeeding chapters of Eddington's own *Mathematical Theory of Relativity*. Bridgman did in fact give some examples of the kind of reformulation required, and claimed that certain purely technical puzzles in classical physics were cleared up by the operational technique,[3] but it was quantum theory that provided the worst conceptual puzzles, and here leading physicists began to use the epistemological language of operationalism, and to claim that certain formulations of quantum theory did correspond to the criterion of operational definition, and thereby eliminated paradoxical and meaningless talk about quantities which could in principle not be measured.

However, the programme of eliminating non-operational concepts from physics was subjected to damaging criticism and has now been abandoned in its original form by practically all philosophers of science including Bridgman and the former members of the Vienna Circle. The programme broke down for three main reasons. First it was shown by F. P. Ramsey and others [4] that it is not possible either in existing scientific theories, or in artificially constructed examples, to define all the concepts in terms of operations.

[1] See, for example, his ' Reply to Criticisms ' in *Albert Einstein : Philosopher-Scientist*, ed. P. A. Schilpp, Evanston, 1949, p. 665

[2] *Mathematical Theory of Relativity*, Cambridge, 1923, p. 3

[3] cf. *The Nature of some of our Physical Concepts*, New York, 1952

[4] F. P. Ramsey, *Foundations of Mathematics*, London, 1931, p. 212 ; R. B. Braithwaite, *Scientific Explanation*, Cambridge, 1953, Chap. iii ; M. B. Hesse, ' Operational Definition and Analogy in Physical Theories ', *B.J.P.S.*, II, 1952, p. 281

The notion of wave-function in quantum physics for example cannot be defined in terms of observational concepts, although the argument can proceed in the reverse direction : if wave-functions are postulated, certain properties of matter and radiation can be deduced, and these are the properties which are observed. No-one would now claim that explicit definitions of such concepts as wave-function are carried out in quantum theory. The second objection to the operational programme is that the possibility of explicit definitions is not generally one of the considerations which weigh with scientists in judging a good theory, although it may sometimes be useful to pay attention to such possibilities when considering what concepts may safely be abandoned, as in the case of absolute velocity and simultaneity. Absence of any possible operational definition *permits* but does not *compel* elimination of a concept, and it sometimes happens that a concept has been retained unnecessarily because it has been assumed to have a direct operational meaning, although this is later found not to be the case. Thus it was not clear before the Michelson-Morley and similar experiments that the ' unobservable ' properties ascribed to the aether by nineteenth-century physics were in contradiction with the empirical facts, and when the experiments had been performed, the aether was abandoned, not primarily because it was unobservable, but because of these contradictions. The fact, however, that the aether was in some senses unobservable, meant that it could be abandoned without conflict with other empirical facts. The technique of operationalism certainly played a useful part in clarifying cases like these. But the third and most damaging objection to its general application [1] is that if explicit definitions of theoretical concepts in terms of observations were possible, the theory would become useless because incapable of growth. Theories must have ' open texture ', in Waismann's phrase, that is, a fringe of meaning not defined by observation, otherwise the whole meaning of the theory would change whenever it was desired to incorporate into it observations of a novel kind, and it is precisely the function of theories to assimilate such new observations without the meaning of the theories being radically altered. We shall return to this point below.

The Hypothetico-Deductive Method and Falsifiability

In dealing with these difficulties, the positivist view of theories passed into a second phase. In this theories are described as

[1] cf. N. R. Campbell, *Physics, the Elements*, Cambridge, 1920, Chap. vi ; F. P. Ramsey, loc. cit. ; F. Waismann, ' Verifiability ', *Logic and Language* (1st series), ed. A. G. N. Flew, Oxford, 1952, p. 117 ; R. B. Braithwaite, loc. cit.

hypothetico-deductive systems, that is, as consisting of hypotheses [1] in the form of postulates and deductions from the postulates in which some, but not all, of the statements can be interpreted as observation statements and confirmed or refuted by experiment. Since it is now admitted that not all the statements of the hypothesis can be given direct empirical meaning by being translated into observation statements, some further conditions have to be satisfied by the non-observational, or theoretical, statements if they are to qualify as meaningful parts of science.

Clearly, such theoretical statements cannot be conclusively verified, for the logical form of the hypothetico-deductive system is : '*A* (hypothesis) implies *B* (observation), and *B* is true', and these premisses do not permit the inference : '*A* is true'. If *B* is false, however, it does follow that *A* is false, and it may be said that theories are conclusively falsified if their consequences are contradicted by observation. So instead of a criterion of empirical verification which the early logical positivists had demanded as the condition for the meaning of a statement, a search began for criteria of falsifiability which would admit genuine scientific theories to be meaningful, but would eliminate non-empirical metaphysics and pseudo-science.[2] The criteria are intended to ensure that a theoretical statement, though not directly verifiable by observation, does nevertheless have empirical consequences, in the sense that some prediction about the world can be deduced from it, together with other theoretical statements, in such a way that the non-fulfilment of the prediction would show one or more of the theoretical statements to be false. The condition would thus eliminate statements which could never be shown to be false by any empirical happening whatever. For example, if an opponent of Newton's gravitational theory had wished to maintain that it is really the intelligent souls of the planets which direct them in their orbits, no future observation of the orbits could show this statement to be false, since it would always be open to anyone who asserted it to say that the planetary souls had willed whatever orbits happened to be observed. The

[1] The word 'hypothesis' is not used here, or in what follows, to denote a doubtful and tentative theory, but to mean that part of *any* theory, however well established, which is not immediate description of observation, nor deduced from such description, and which may logically therefore be false. The notion of 'immediate description of observation' will be examined further below.

[2] Popper has pointed out ('Philosophy of Science : a personal report', *British Philosophy in the Mid-Century*, ed. C. A. Mace, London, 1957, p. 155) that when, in the 1920s, he introduced the notion of falsifiability as characteristic of science, he saw it, not as a criterion of *meaning*, in the fashion of the positivists, but as a criterion of *demarcation* between science on the one hand, and myth, metaphysics and pseudo-science on the other. It does not follow for Popper that non-falsifiable statements are meaningless.

statement is consistent with any empirical happening whatever, it therefore gives no empirical information, and does not state any scientific fact. Positivists wish to claim in general that most of the statements of metaphysics are vacuous in this way, and if they do not then go on to assert that they are meaningless, they do assert at least that they are statements of a different kind from those of theoretical or experimental science.

The criterion of falsifiability in science certainly represents a condition which scientists will recognise. It lies behind the discomfort felt about *ad hoc* postulates which are invented simply to accommodate otherwise troublesome pieces of empirical information which do not fit easily into established theories. If such an *ad hoc* theory has no empirical consequences other than those it was invented to explain, then it is unfalsifiable, and therefore unsatisfactory. Again, if a theory is worded so vaguely or expressed so generally that it can be moulded to accommodate any possible empirical facts, it is an unsatisfactory explanation because unfalsifiable. Descartes's vortex theory is an example of vagueness: almost any physical happening could be described in terms of combinations of vortices in the aether, and Descartes himself showed considerable ingenuity in inventing several such descriptions. But this procedure is not ultimately satisfactory because independent tests of the theory are not sought, and in any case no empirical facts are allowed to count against it, because all can be described in terms of it. It may provide a satisfying picture of facts already known, but it cannot make predictions unless it becomes more precise and submits therefore to the possibility of falsification. Again, the criterion of falsifiability can be used to eliminate redundant hypotheses, for if a theoretical statement is not necessary to, or necessarily implied in, the deductive system leading to the observation statements, then no actual observation can falsify it.

Some such criterion is clearly implicit in the practice of scientific research, and may in fact be said, as we shall see later, to be a major distinction between the scientific movement of the seventeenth century and much previous theorising about the natural world. But in science the criterion is not precisely formulated, and when logicians attempt to formulate it, various difficulties arise. Suggested conditions for a theoretical statement H are usually of the following kind:

1. H in conjunction with other premisses entails one or more observation statements $O_1 \ldots O_n$. This ensures that the theory including H has empirical consequences.

2. $O_1 \ldots O_n$ are not entailed by the other premisses alone. This ensures that H is not redundant.
3. The other premisses are either logical or mathematical, or observation statements (for example, initial conditions specifying the experimental situation of which $O_1 \ldots O_n$ state the results), or are theoretical statements which can be shown to satisfy these three criteria independently of H. This allows H to contain non-observational terms, but is intended to ensure that there is no incompleteness or circularity in the definition of the meaning of statements containing such terms.[1]

If a statement satisfies these criteria, then it is certainly falsifiable, although if a number of other premisses which are also theoretical statements are involved in condition (3), and if an observation statement entailed by the whole deductive system is false, then it will not always be possible to say which of the premisses has been falsified. This, however, is a correct account of the usual situation in science, where, if an observation expected on the basis of a theory does not correspond with the facts, there is in principle a choice about which part of the theory to discard.

Starting from general requirements of this kind, logicians have endeavoured to provide a formal account of the necessary and sufficient conditions that a system of theoretical statements shall constitute an empirically meaningful explanation of the relevant observations. The task has, however, proved surprisingly difficult.[2] First, purely logical difficulties arose when it was shown by Church that the criteria as stated by Ayer do in fact admit any statement whatever or its negation to be empirically meaningful. Subsequent

[1] cf. A. J. Ayer, *Language, Truth and Logic*, 2nd ed., London, 1946, p. 13. Other criteria are found in, for example, C. G. Hempel and P. Oppenheim, 'The logic of explanation', *Philosophy of Science*, xv, 1948, p. 135; R. Carnap, 'The methodological character of theoretical concepts', *Minnesota Studies in the Philosophy of Science*, I, ed. H. Feigl and M. Scriven, Minneapolis, 1956, pp. 49ff. Carnap's criteria here are somewhat more liberal than the others in referring to theoretical *terms*, not theoretical *statements*. This means that he allows empirical meaning to some sentences which have no possible relation to observation such as is stipulated by (1) to (3) above, as long as they contain only terms which belong to *some* sentences which do satisfy (1) to (3). Thus he allows 'The value of the magnitude M [a significant theoretical *term*] at a certain space-time point is a rational number' although no observational evidence could be relevant to this (ibid., p. 61). But this criterion is too broad, for it would allow any sentence containing English words to be meaningful, for example, 'The cosmic mind dislikes cheese'.

[2] See, for example, C. G. Hempel, 'Problems and changes in the empiricist criterion of meaning', *Revue Internationale de Philosophie*, XI, 1950, p. 41. In 'Between analytic and empirical' (*Philosophy*, XXXII, 1957, p. 112), J. W. N. Watkins has argued that Ayer's criteria are satisfactory only if the observation statements mentioned in them are understood, following Popper, as unsuccessful attempts to *refute* the hypothesis, and not as simple confirmations. I shall assume this amendment made, and speak of theories as being testable (by tests designed to refute them if they are false) or as falsifiable, rather than as verifiable or confirmable.

attempts to modify the criteria to deal with Church's argument have led to a dilemma: either the criteria have to be made so weak that they admit any statement whatever to be empirical, or they have to be so strong as to exclude some theoretical statements which everyone wishes to accept as meaningful parts of science. There is a second difficulty about the formulation of any set of criteria, namely, how are we to distinguish and isolate individual statements from within a closely connected deductive theory? Rigorous formulation of the criteria seems to require that such elementary statements should be defined, and yet it is highly artificial to analyse actual theories into elements in this way. It seems more realistic to apply the criteria to theories as a whole, or at least to detachable and more or less self-contained parts of theories, and not to elementary statements of dubious status. This means, however, that the criteria again become too weak, since *any* theory, however fantastic, which can be made to yield some observable consequences will then have empirical meaning as a whole according to the criteria. Even theories usually called metaphysical may yield some observable consequences, and pseudo-sciences such as palmistry and astrology certainly do so. It must be concluded that the programme of demarcating scientific theories by means of formal criteria specifying only their deductive relation to observation statements has failed.

Any set of formal criteria will be at best necessary and not sufficient for scientific acceptability, for it will admit indefinitely many theories which would not be regarded by scientists as having empirical meaning. This is recognised by some of those who write on the hypothetico-deductive method: for example, Carnap admits that fulfilment of his formal conditions does not guarantee the *fruitfulness* of a theory, but he doubts whether this can in any case be subjected to precise definition.[1] There is something more to be considered here, however, than a distinction between meaning and fruitfulness, and what this is can best be indicated in terms of an account and criticism of one form of the hypothetico-deductive theory which has been suggested. I shall state this as baldly as possible in order to show up its inadequacy, although I do not suggest that anyone would in fact wish to maintain it in exactly this form. But it is an example of a type of theoretical structure which might be made to satisfy all the suggested falsifiability criteria, yet which would still be an inadequate account of the

[1] *Minnesota Studies*, I, p. 62; cf. W. Sellars, 'Some reflections on language games', *Philosophy of Science*, XXI, 1954, p. 204

structure of actual theories, and in trying to show this, I shall also be showing that such criteria do not provide a sufficient account of empirical meaning.

The Dictionary Theory

One difficulty about the hypothetico-deductive scheme as outlined above is that the most recent theories in physics are expressed largely in mathematical terms, or at least in terms of the properties of ' unobservable entities ' such as atomic particles and quantum fields. Now if the postulates of a theory are expressed partly in these terms and partly in terms of operational properties such as length, weight and temperature, it is possible that the deductive system will produce some relations from which the ' unobservables ' have been eliminated, and which can therefore be tested experimentally. But in much of present-day physics the hypotheses themselves contain no reference whatever to operational concepts, moreover even models of ' unobservable entities ' describable in words often have to be abandoned, leaving only an abstract mathematical formulation. How is it possible that such hypotheses can be experimentally tested ?

It is hypotheses of this kind which make plausible an extreme form of hypothetico-deductive system which I shall call the ' dictionary theory '. According to this account, the statements of the hypothesis need have no properties other than their place in the deductive calculus and the relation of this to observation statements.[1] That is, theoretical statements have no meaning in themselves apart from their connection with observation, and this connection has to be made by means of a ' dictionary ' which defines certain signs of the formal hypothesis in terms of observable concepts which enter into observation statements.[2] The observation statements themselves have meaning which is quite independent of the theory, and their truth or falsity can be directly tested. It is implicit in this view that the dictionary must be quite arbitrary, that is, it is not like a dictionary for translating French into English, where both languages have prior meaning which determines the

[1] In the terminology of American philosophy of science, they have *in themselves* syntactic but not semantic relations.

[2] This view is suggested, although not explicitly or in detail, by Ayer, *Language, Truth and Logic*, 2nd ed., p. 13 ; Braithwaite, *Scientific Explanation*, p. 51 ; Ramsey, *Foundations of Mathematics*, pp. 215, 234 ; and J. H. Woodger, *Biology and Language*, Cambridge, 1952, p. 12. N. R. Campbell, who seems to have first introduced the notion of the ' dictionary ' (*Physics, the Elements*, p. 122), does not claim that the meaning of theoretical statements is exhausted by their formal properties, but does assert that the meaning of observation statements is independent of theoretical concepts.

dictionary equivalences, but like a key to a number-code where the code symbols are initially arbitrary and acquire meaning only *via* the key.

According to this view, for example, quantum theory might be developed in purely mathematical terms without any interpretation being given to the signs used, until certain formulae are reached in the deductive scheme which could be translated into directly testable statements. Consider the equation

$$\left(1 + \frac{h\nu}{mc^2}\right) \tan \phi/2 = \cot \theta.$$

This is reached by deductive processes from the formal axioms of quantum theory, and may be translated by the 'dictionary' as follows :

ν is the frequency of X-radiation in a certain Wilson cloud-chamber experiment (the apparatus of which can be exhaustively described in non-technical terms) ;
ϕ and θ are certain angles between two finite non-intersecting white lines on the cloud-chamber photograph, and a third line drawn from one end of one to one end of the other ;
h, m and c are physical constants.

The equation with this interpretation can now be tested experimentally.

It is recognised, of course, in this logical account, that physicists do not generally speak about the theoretical part of the deductive system as if it were purely formal and uninterpreted, for during the process of deduction of the above equation, they will use language about entities such as quanta of radiation, scattering, tracks of photo-electrons, and so on. In other words they still persist in talking in terms of models, even though these are no longer simply mechanical and even though the same entity may sometimes be described by a wave-model and sometimes by a particle-model. It is held by some positivists however that such model-talk is dispensable ; that it is a concession to the desire of the physicist for pictorial representation of his theory, and that as such it may be a useful heuristic or didactic device, but no more, and that it is not necessary to the understanding of the logical structure of theories. When the theory has been found and expressed in formal terms, the ladder of models which was used to attain it can be thrown away.

This account has the merit that it removes at one stroke all puzzles about the meaning and existence of unobservable entities such as electrons, for they now become aspects of collections of equations in the mathematical deductive system, and have no meaning apart from their place in that system and its connections *via* the dictionary with observation. Then so long as there are no inconsistencies in the mathematical scheme, inconsistencies in its partial pictorial representation are unimportant. But N. R. Campbell himself, who invented the notion of the 'dictionary', also pointed out that this account of models will not do, and that theories constructed in conformity with all the formal conditions of meaning already discussed may still be strictly meaningless in a scientific sense.[1] I think, however, that we must go even further than Campbell, and maintain that neither this account of models, which Campbell rejects, nor the distinction between theoretical and observation statements, which Campbell accepts, is justified by scientific practice.

Before developing this argument I shall make a change of terminology in order to be able to refer more easily to my own view and to distinguish it from the usual type of empiricism which I believe to be mistaken. What are generally called 'observation statements' I prefer to call *phenomenal statements*, and I wish to define them somewhat more carefully than is usually done, although I do not think my definition will be objected to by those who use the 'observation' terminology. My reason for disliking this terminology is that it introduces into the discussion of the structure of scientific theories assumptions about what can and cannot be observed which obscure the argument and are not strictly necessary to the logic of the hypothetico-deductive method. The word 'phenomena' has the merit of having been used by Newton and Kant in senses not too far removed from what is here intended, but it should not be taken as carrying any overtones of phenomenalism, if by that is meant a doctrine that all we directly know are sense-data. Here phenomenal statements are those in which observations are described in such a way as to involve no technical terms or knowledge of scientific theories, and so as to be understood by everyone who speaks the natural language in question. All scientific observations can be described in such terms, mentioning apparatus whose specification cannot involve any knowledge of theory, since it has in any case to be understood in the workshop, and describing coincidences of pointers with scales, and clicks,

[1] *Physics, the Elements*, p. 123

flashes, bands of coloured light, and so on. No doubt it is impossible to make precise this notion of being commonly understood in the language, for what is commonly understood among one group of people may not be so among others, but I shall assume, as do most of the exponents of positivism, that it is possible to recognise a statement as being phenomenal in this sense, and that if any alleged phenomenal statement is challenged because it contains some technical phrase such as 'current in the wire', it is always possible to find an acceptable translation into phenomenal terms, such as 'the position of the pointer on that scale'.

The dictionary theory then consists in the assertions that

1. The meaning and truth-conditions of phenomenal statements are independent of scientific theory, and
2. Theoretical statements derive their meaning *only* from their place in a formal hypothetico-deductive system in which some of the conclusions can be translated into phenomenal statements.

The first objection to this thesis is that it will generally be the case that an infinite number of formal deductive systems can be constructed which, with some interpretation, will yield true phenomenal statements. Clearly some other criterion is required to enable the physicist to choose between these systems. This criterion is sometimes said to be formal simplicity,[1] but, while simplicity certainly plays a part in theory construction, it is difficult to apply it as a criterion in detail because any ordering principle defining relative simplicity is bound to be somewhat arbitrary, and it is not in any case an adequate description of what happens in science, because physicists will often use a mathematical scheme derived from a model even if this is comparatively complex. We shall return to this point in later chapters.

The major objection to the dictionary theory, however, arises from the assumption that phenomenal statements can provide tests of hypotheses if the hypotheses have no meaning in themselves, or at least if phenomenal statements are independent of that meaning. Let us consider as a simple example the case of the straight stick apparently bent when placed in water. Suppose that the phenomenal statements in this case are in terms commonly understood in English, containing no concepts which involve any theory of light. Thus, they will contain words like 'surface', 'sun', 'air',

[1] cf. H. Jeffreys, *Scientific Inference*, 2nd ed., Cambridge, 1957, p. 36; and the comments upon the simplicity criterion in K. R. Popper, *Logic of Scientific Discovery*, London, 1959, Chaps. vi and vii.

'water', but not 'light rays' or 'media of such and such refractive index'. Now the purely formal part of the theory of refraction consists of some postulates from which can be deduced the equation

$$\frac{\sin \alpha}{\sin \beta} = \mu.$$

According to the dictionary theory, the symbols α and β are now interpreted as certain angles which can be experimentally measured, and μ as a constant characteristic of air and water, and so this equation can be tested, and if it corresponds with the measurements, the hypothesis which led to it is said to be confirmed. According to the dictionary theory this is in principle all that can be said about the hypothesis which is normally called the wave-theory of light. But if this were really all, how would we know that this particular interpretation is relevant to this theory?

If in the equation $\sin \alpha/\sin \beta = \mu$, α and β were simply undetermined mathematical symbols, they might be interpreted in an indefinite number of different ways, some of which might be shown to be true of phenomena. They might, for example, be the angles between the Pole star and Mars and Venus respectively at midnight on certain given dates; why would not this be a confirmation of the formalism we have mistakenly called the wave theory of light? Conversely, why can we not invent any theory at all which leads deductively to the equation $\sin \theta/\sin \phi = \mu$, and interpret θ and ϕ as the angles of refraction, and take our refraction experiment as confirmation of this theory? The answer is of course that the possible interpretations of α and β are already circumscribed by the theory; the dictionary is not arbitrary, and the symbols of the hypothesis are already interpreted in terms of a model of light waves in such a way that we know what kinds of phenomena will be relevant to it. The connection with phenomena is not in this case made in terms of a dictionary at all, but by means of a model of imaginary wave-fronts and straight lines drawn in physical space and having relations with observable physical objects.

Again, if the hypothesis were merely formal, we should not know that the equation $\sin \alpha/\sin \beta = \mu$ can be interpreted in other ways in order to provide a test of the theory. Suppose we burn a piece of paper with a lens. The assumption that such an experiment is connected with the bent stick is itself a theoretical assumption not contained in bare phenomenal statements, because it involves seeing, for instance, that the same explanation is to be expected for phenomena involving air, water, light, and a stick, and those involving air, glass, heat, and a piece of paper. It may be objected that little theory is required to see the connections in this case, and this may be true, although it clearly depends on the extent to which scientific notions have permeated ordinary language. But there are other examples in which it is certainly not true. To a layman entering an atomic physics laboratory it would not be at all obvious that certain pairs of scintillations over here are connected with the presence of a nuclear reactor over there, and that they confirm a theory which explains why certain other scintillations near the reactor have the character they have. But to a physicist they count as the detection of neutrinos whose presence was suspected because of a loss of energy in certain decay processes in the reactor. The mere report of scintillations cannot proclaim its own relevance to this theory, nor its connection with phenomena on the other side of the laboratory ; the observation becomes a confirmation of the theory only when it is interpreted into language about fundamental particles, that is, into theoretical language. The connection between theory and observation, again, is not *via* a dictionary, but in terms of the particle model, because it is known what happens when macroscopic charged particles (e.g. charged pith-balls) touch a fluorescent screen, and it is assumed that the same happens when the charged particles of the model (electrons) touch it, although the electrons are not in other respects observable as pith balls are.

We must conclude that if statements which describe experiments are to be tests of a theory they must speak the language and use the concepts of the theory, that is, their meaning cannot be independent of that of the theoretical language as the dictionary account suggests.[1] If this is the case, is there any justification for the two-tiered structure of statements with dictionary interpretation described by the formal hypothetico-deductive scheme ? If certain theoretical ideas must be introduced into statements which are to act as tests of theories, must we not characterise the situation quite differently and say that scientific language is *richer* than ordinary language and not even in principle translateable into it ? To interpret an experiment directly in theoretical terms so that it can be a test of the theory is always to say more than the corresponding ' common-sense ' description would say, because such interpretation carries

[1] For a more detailed discussion of these points, see my ' Theories, Dictionaries and Observation ', *B.J.P.S.*, IX, 1958, p. 12, and ' A Note on " Theories, Dictionaries and Observation " ', ibid., p. 128. The dictionary theory is defended by P. Alexander, ' Theory-construction and Theory-testing ', ibid., p. 29.

with it natural expectations about possible but so far unobserved behaviour which the scientist has to *learn*, just as the child learns the contextual overtones of ordinary language. This is why the translation of the equation

$$\left(1+\frac{h\nu}{mc^2}\right) \tan \phi/2 = \cot \theta$$

which was suggested above is not adequate. The so-called observation statement consisting of a description of the cloud-chamber and an apparatus producing X-rays, together with a statement about two white lines and two angles could test nothing because it does not state its relevance to anything. ϕ and θ as measured on the photograph may satisfy the equation, but they cannot be known to be the ϕ and θ mentioned in the theory from which the equation was derived unless the experiment is interpreted in terms of the scattering of a quantum of X-rays which produces a photo-electron in the chamber. This interpretation enables the experiment to become a test of the theory because it states its relevance to the theory, and the relevance is itself theoretical.

When one language is richer in associations than another its significance cannot be fully represented in the other by means of a dictionary translation. The correct analogy for the relation of theoretical and phenomenal languages is not the relation between a number-code and English, or even between simple sentences in English and in French, but the translation of poetry into prose, or, to look at it the other way round, which is the order of discovery, description of an experiment in ordinary language has a relation to the scientist's theoretical description which is similar to that between Holingshed and Shakespeare. Description of observations can never be absolute, and there is every degree of interpretation of an observation from those involving simple common-sense assumptions implicit in the use of the natural language itself, to those involving the highest-level theoretical concepts.

There is however a place where talk about dictionaries is in order, and this can be explained very briefly. It frequently happens that uninterpreted mathematical calculi are used to perform the deductions which are required by a theory.[1] Where uninterpreted mathematical expressions are used, as for example, $x'' + n^2x = 0$, in order to facilitate deductive arguments by calling upon the apparatus of pure mathematics, it is necessary that somewhere a connection should be made between the signs appearing in these

[1] For a detailed examination, see Braithwaite, *Scientific Explanation*, Chaps. ii and iii.

expressions and the scientific language. Here the process of translation is properly said to be carried out according to a dictionary. But the point which has been obscured by exponents of the formal hypothetico-deductive system is that *interpretation need not be into ordinary language*. The above equation is adequately interpreted for use in part of atomic physics by the following entries in the dictionary :

x is the amplitude and $2\pi/n$ the wave-length of stationary waves representing a single electron moving in a finite straight line in a field of constant potential.

It goes without saying that the concepts in this definition are not phenomenal in the sense of ordinary language.

We may therefore distinguish two different classifications of scientific statements which have sometimes been confused in positivist writings :

1. The distinction between theoretical and phenomenal statements, which does not call for *translation*, but for *interpretation* of observations at different interpretative levels involving more or less reference to theoretical concepts.
2. The distinction between formal and interpreted statements, in connection with which it is appropriate to speak of translation by means of a dictionary, but where the interpretation may be into theoretical rather than phenomenal terms.

Some special circumstances must now be noted, however, in which the argument that uninterpreted formal theories cannot be tested breaks down. Suppose the dictionary consists of entries translating expressions of the theory into concepts $\alpha_1, \ldots \alpha_n$, which may be regarded as being operationally defined in the phenomenal language. Suppose a number of independent functional relations between the α's is deducible from the formal theory. It may happen that the theory is constructed after only some of these have been found to be experimentally confirmed ; it is then possible to say that the deduction and translation of the remainder enables the theory-plus-dictionary to be tested.[1] Such tests are distinguished from others by requiring no new entries in the dictionary, and I shall call them *formal tests*.

If the phenomenal language is defined as above, that is, in

[1] This is implicit in Braithwaite's account of his 'four-factor' theory (*Scientific Explanation*, pp. 68ff.).

such a way that it does not contain any concepts which imply correlations or connections not assumed in non-scientific discourse, then formal tests are usually, but not always, trivial. Most *applications* of a theory will be formal tests, for example if operational definitions are given of the time and space coordinates of astronomical objects, then the prediction of the orbit of an earth satellite is a formal test of Newtonian gravitation, and this is more usually regarded as an application than a test. But Newton's comparison of the force required to keep the moon in its orbit with the gravitational force at the surface of the earth was not a formal test, because his *phenomenal* language did not include a concept of force referring indifferently to celestial and terrestrial dynamics. On the other hand, Hertz's experiments on the transmission of electromagnetic waves were formal tests of Maxwell's equations, since they required no new operational definitions in the phenomenal language, and as a matter of historical fact these tests were not trivial. And if the language into which a theory is interpreted for test purposes is not the phenomenal language, but, for example, the entire language of classical physics, as is the case for quantum theory, then formal tests may be more interesting and important. But to admit that classical physics may be used as a language for direct descriptions of observation is already to abandon the empiricist observation language which is usually adopted by those who hold the 'dictionary theory'.[1]

Models

The foregoing discussion should make it clear why physicists persist in speaking in terms of 'models' when engaged in theoretical work. Because a model is drawn from a familiar and well-understood process, such as particle mechanics, it provides the context of natural expectations in terms of which a theory can be tested.[2] If we now consider some of the mechanical models of nineteenth-century physics, not necessarily as literal descriptions of nature as in the naïve realist view, but as devices which were essential for rendering a theory intelligible and testable, it will be possible to describe their logical function more clearly than we have so far done. Such an account will then be found to throw light on the

[1] On the view that 'observation languages' are relative to theoretical interpretation, see P. K. Feyerabend, 'An attempt at a realistic interpretation of experience', *Proc. Aris. Soc.*, LVIII, 1957–8, p. 143.

[2] For more detailed discussions of this question, see Campbell, *Physics, the Elements*, Chap. VI; E. H. Hutten, 'On Semantics and Physics', *Proc. Aris. Soc.*, XLIX, 1948–9, p. 115, and 'The Rôle of Models in Physics', *B.J.P.S.*, IV, 1953, p. 284; M. B. Hesse, 'Models in Physics', *B.J.P.S.*, IV, 1953, p. 198.

more complex problems connected with the use of models in modern physics.

The most obvious property of a satisfactory model is that it exhibits an analogy with the phenomena to be explained, that is, that there is some identity of structure between the model and the phenomena. Now one may say in a straightforward sense that there is an analogy between two branches of physics if the same mathematical structure appears in the theory of both, for example, the theories of heat and of electrostatics can be formulated in the same equations if one reads ' temperature ' for ' potential ', ' source of heat ' for ' positive electric charge ', and so on. When there is an analogy of this kind, one theory may be used as a model for the other, as Kelvin used the idea of heat flow, whose theory was already established, as a model for the field theory of electrostatics, which he was developing for the first time. In an extended sense, the word ' analogy ' may then be applied to the relation between the model itself, for example billiard-ball-like particles, and the entities which are postulated to account for the phenomena, for example gas molecules. To say that there is an analogy here is to assert correspondences between a variety of experimental measurements and certain numbers deduced from the theory of the model. For example if the appropriate calculations based on the theory of mechanics are made about the energy of colliding billiard-balls, we can obtain a series of numerical values which is the same as that given by a thermometer placed in a vessel containing the gas.

The reason why a model such as that implied in the dynamical theory of gases is not just a dispensable way of picturing the appropriate equations, is that the model can be generalised, extended and tested, and if necessary modified, as a purely formal deductive system cannot. The model can be tested, because it is a system of entities and processes whose behaviour is already known apart from the new experimental facts which it is being used to explain. The behaviour of a collection of particles moving at random in a closed vessel is described in the theory of dynamics independently of the experimental results about gases with which it is compared, and this means that further ramifications of the theory of colliding particles can be used to extend and test the theory of gases. Further questions can be asked, such as ' Are gas molecules like rigid balls or like elastic ones ? ', ' What is their diameter ? ', and so on, and the theory is tested and developed by devising experiments to answer questions like these suggested by the model.

In order to function in the way just described, models need

not of course be mechanical. Mechanical models were on the whole preferred during the nineteenth century, but even in classical physics the model of gravitating particles was used for electricity and magnetism, electrical models for the theory of chemical combination, and the model of heat flow for field theory. What is required is not that the model be mechanical but that its properties be already known and described in terms of some, preferably mathematical, theory, and that it should have the 'open texture' which allows modifications and extensions to be made as may be appropriate for the explanation and prediction of new phenomena.

The difficulty that seems to arise in modern physics from this account of the indispensability of models is that no single model of classical type, using charged particles or waves, is adequate to explain the phenomena of the atomic domain, and it is sometimes said that in consequence we must not ask for picturable models, but must be content with formal mathematical hypotheses in which the paradoxes associated with particle and wave models do not arise. Now there are two things to be noticed at this point which indicate that it is misleading to draw such a conclusion. In the first place, physicists do in fact continue to use both particle and wave models, each in appropriate situations, even though they are at first sight mutually contradictory, and this is not only in condescension to readers of popular science, nor merely to assist in the teaching of students. It is an essential part of research in these fields, as a brief glance at original papers will show, and as the above arguments have indicated. But it is true that at a deeper level of theoretical investigation, where particle-like and wave-like behaviour both have to be taken into account, models of the classical type can be dropped, and the theory developed in apparently formal mathematical terms. What then becomes of our insistence that uninterpreted formal systems are not sufficient to provide theoretical explanation?

The difficulty is resolved when it is realised that mathematical theories are not necessarily (or perhaps ever) uninterpreted formalisms, if by that is meant mere collections of signs combined in arbitrary axioms and permitting inference according to arbitrary rules. It is difficult to show this in general terms, but perhaps it may be illustrated by some examples. When the physical model of wave-motion in a material medium had to be abandoned in physics, it left its traces in the kind of mathematics which was used, for this was still a mathematical language derived from the wave equations of fluid motion, and so, for the mathematician, it carried some of the imaginative associations of the original physical picture. Again, when Riemannian geometry is used in general relativity theory, it is not an uninterpreted formalism, but a natural extension of the two-dimensional geometry of the surface of a sphere, which is picturable, to the geometry of a three-dimensional space curved in a fourth dimension, which is not, but in which certain of the interpretations of the symbols such as ' geodesic ' or ' radius of curvature ' remain valid. Just as there may be many levels of interpretation of a set of dynamical equations from sentences involving colliding elastic balls to statements about pressure and volume of a gas, so there may be various interpretations of a theory of pure mathematics on different levels of abstractness, and involving more or less reference to the comparatively concrete statements of Euclidean geometry or of arithmetic. And these interpretations of an apparently formal mathematical scheme provide the open texture which enables the theory to be tested, generalised and modified, just as is the case with more concrete mechanical and electrical models. It therefore becomes appropriate to speak of ' mathematical models ' alongside these more traditional types.[1] It might be thought that the word ' model ' is misleading here, since there is no concrete thing to be built or pictured, but the word is now sanctioned by widespread use in connection with sciences as diverse as cosmology and nuclear physics, brain physiology and Freudian psychology. In the case of fundamental physics at least what are called ' models ' are now always partly or wholly mathematical in type, for example in cosmology, where ' world-models ' are certainly not models in the picturable sense.

The question now arises as to how seriously these various kinds of model are to be taken. In showing that they are, after all, essential to theories, and not dispensable embellishments, have we fallen back into the paradoxes of realism? Not necessarily, for it is not now asserted that there is a *perfect* analogy between the model and the world, only that there is an analogy in certain respects (which we may call its *positive analogy*), and that this may extend further than has hitherto been investigated. There may seem to be little point in talking about ' models ' at all unless there is some respect in which the analogy they exhibit breaks down. Atoms are thought of as being *like* billiard-balls, and not as *being* billiard-balls precisely because it is known that there are some respects (the *negative analogy*) in which they are *not* like billiard-balls. The

[1] Further examples of mathematical models are given in my ' Models in Physics ', *B.J.P.S.*, IV, 1953, p. 198, and others will be encountered in the following chapters.

whole strength of the formalist view of theories lies in its assertion that it is possible to abstract from the model the positive analogy which represents the extent of assured knowledge about the phenomena, and to throw away the negative analogy which might render the model misleading. We have seen that a theory cannot in general be tested or extended if reduced to a bare formalism, but what of a theory which (like Maxwell's) has been tested to breakdown and whose range of applicability and limits are hence known? Hertz declared that Maxwell's theory *is* the formal structure of Maxwell's equations, and it does indeed seem that when we know the exact extent of the analogy which Maxwell's aether-model bears to phenomena, what is true or useful in it can be expressed formally without any 'as if' clause to introduce irrelevances. Clearly the formalists are right to this extent: the purpose of using models is to make them unnecessary by so familiarising ourselves with the new field of discovery that it can be described by means of its own language, without comparison with something more familiar. The metaphorical language derived from the model may then become *dead metaphor* ('attraction', 'tubes of force'), in other words, it acquires a technical meaning from the context of new discoveries and loses its original associations. Or its meaning may retain some of the original associations, only becoming modified gradually as the extent of negative analogy becomes clear, thus 'particle' in physics may come to mean not 'hard, coloured, spherical object which . . .', but 'singularity in the electromagnetic field which . . .', or 'wave-packet which . . .', the dots indicating that an indefinite number of things may be said about any of these entities, just as an indefinite number may be said about ordinary physical objects, and that at any given stage of physics we do not know even implicitly what all these things are.

In fact, no field of inquiry is ever closed in such a way that its formal description exhausts all that physics ever wants to say about it. Even when the formal structure of a limited area is known, physics always strives to find a more fundamental and more general theory to embrace it. Theories which are isolated and, as it were, confined within formal fences are no longer *scientifically* interesting, however useful the *applications* of their formal descriptions may be, and when a new fundamental theory is discovered the description of even such a theory is in principle changed, as billiard-ball mechanics is changed by relativity theory, although in practical applications no formal change may be required.

The question of whether models are intended as real descriptions is however a different one. Because all models ultimately prove to exhibit only a limited analogy with things, and because discoveries of hidden relationships between things can ultimately be expressed in formal terms, or in a modified and technical use of words first used in connection with the model, it does not follow that these relationships are not factual. It was long ago accepted as fact that the world is round, although for Aristotle this was a precarious theory suggested by the model of the sun and moon and justified by phenomenal arguments ; it was accepted that the earth goes round the sun ; that chemical compounds consist of elements ; that magnetism is electricity in motion ; that wireless waves pervade space ; and so on and so on. The frontier of fact is continually shifting, and this is precisely the progressive character of science. But in many cases this progress shows that the various models in terms of which new facts came to be understood and accepted, were themselves literally *false*, because the new facts were not exactly like the old facts with which they were compared. And if they were in fact false, then they could, logically, have been true, and this is sufficient to place all such theory-models in the category of factual statements, and to enable us to make finer distinctions between those which were better or worse approximations to the truth.

It should, however, be noticed that not all models introduced into physics are intended to be true descriptions in this way. Four different types of non-realistic use may be distinguished. Firstly, *archaic models* which are deliberately used for practical purposes, although they are known to be false. The extent of their usefulness depends on the extent of their positive analogy, and the extent to which their negative analogy can be neglected in particular circumstances. Thus a model of heat flow may be used in contexts where it is a sufficient approximation to the kinetic theory, and Newtonian mechanics may be used where the accuracy of relativity mechanics is not required. Secondly, *analogue machines* (of a steel-and-copper or a paper-and-pencil kind) may be deliberately constructed to simulate certain aspects of natural processes, usually in order to act as computers where the mathematical theory of the phenomena is either not known, or is known but intractable. Examples of this use of models are the electronic tortoises, where there is an obvious negative analogy in certain biological and chemical respects between the model and the animal, but a positive analogy of unknown amount in some aspects of behaviour ; or wind tunnels, where the fundamental mathematical theory is known, but is intractable in complicated special cases. These models are used in place of mathematical deductive theories whose details are not yet known, and they are not themselves intended as true descriptions, but only as aids to the discovery of such descriptions. Thirdly, *post hoc models* may be invented to embody an existing mathematical theory largely or solely in order to make the mathematical theory easier to apply, or to demonstrate its consistency. Examples would be the nineteenth-century mechanical aether-models, whose positive analogy was wholly contained in the corresponding equations, and which did not therefore contribute directly to the extension or testability of the theory, and were not intended realistically. Fourthly, there are *complementary models* such as the wave and particle models in quantum physics, which exclude each other in certain respects and which therefore limit each other's positive analogy, but whose potential positive analogy is unexhausted in other respects so that each can still function as a useful model in particular circumstances.

Other examples of all these types will be encountered in the subsequent historical accounts, and no doubt other kinds of model could be distinguished in the practice of physics and the other sciences, but this brief classification serves to suggest a definition of the reality-status of a model. *A model is intended as a factual description if it exhibits a positive analogy and no negative analogy in all respects hitherto tested, and if it has surplus content which is in principle capable of test*, where ' in principle capable of test ' is to be understood in a wide sense which will be explored in relation to some historical examples. Models which satisfy this criterion may be called *descriptive models.* It may seem that continued use of the word ' model ' in relation to this definition is paradoxical, since what is here envisaged is potentially a literal, not a metaphorical, description, and that a model which satisfied these criteria would not require an ' as if ' clause. But in view of the potential but as yet unexplored positive analogy, retention of the ' as if ' is a useful reminder that the model may turn out to be a *false* description, and in any case use of ' model ' may be understood to underline its *intelligible*, not its metaphorical, character. It is a model in the sense of a blueprint which copies phenomena as accurately as possible, not in the sense of an impression or caricature which deliberately distorts. The property of theories of embodying models so that they are rendered meaningful, and can be tested and extended, may be called their *intelligibility*, and this will be a necessary condition for theories in addition to the confirmation and falsifiability criteria already mentioned. Intelligibility is clearly also related to the intuitive idea of explanation according to which we wish not only to correlate phenomena and to be able to make predictions, but also to *understand* their connections, and this desire in large part accounts for the long persistence of models drawn from familiar mechanisms.

A particularly important class of descriptive models or theories in the science of any given period are those which may be called *fundamental*, in that they are more general than others and presupposed by them. A model will be fundamental only in relation to a particular historical situation, for example Democritan atoms, Newtonian attractive and repulsive particles, classical electrodynamics, and quantum electrodynamics, are fundamental relative to their historical context. These models do not fall naturally into the hypothetico-deductive hierarchy in terms of which theories are generally described, because in these terms they seem to be functioning at once as low-level generalisations, as high-level hypotheses, and as rules of inference. Take, for example, Newton's laws of motion in classical physics. In one sense these are low-level generalisations from the experimental facts about moving bodies. In another sense they are high-level hypotheses from which, in conjunction with other observations and generalisations, prediction and explanation are given of diverse physical phenomena. And in yet another sense, they are rules in accordance with which deductions from hypotheses are carried out. Such fundamental models have had little attention in current writing in the philosophy of science, but we shall here be driven to consider them in some detail, since the mode of action of bodies upon each other is one of the general properties which such models exhibit, and indeed the sense of 'action at a distance' or 'contact action' cannot be determined except in terms of the fundamental model and the concepts which it implies. Thus in the following chapters the fundamental models, and transitions from one to another of them, will be described as they make their appearance in the history of physics. At the same time the question arises as to their status, that is to say, do they satisfy the criteria for scientific theories that we have discussed, and in particular, what is the status of the modes of interaction which they imply or allow? In the last resort the logical account of theories must be governed by the actual character of theories which are agreed by everyone to be scientific.

Chapter II

THE PRIMITIVE ANALOGIES

Analogies in Primitive Scientific Explanation

THE focal point of controversy about action at a distance lies in the seventeenth century, in Newton's statement of the law of gravitation. This controversy took place at a time when conceptions of science and scientific method were changing very rapidly, but for the leaders of the revolution the traditional scientific textbooks were Greek in inspiration, and it is to ancient science that we must look for the source of the ideas which formed the climate of seventeenth-century thought.

In considering early science it is easy to be misled by superficial similarities between their conceptions and ours, and if we allow ourselves to interpret their science in terms of its modern equivalents we have the illusion that its progress was tediously slow, and so hampered by confusion and prejudice as to be unworthy of detailed attention. But the Greeks were generally not groping after ideas which we have now clarified, they were more often on a different road altogether; and what we pick out as a forerunner of some modern scientific concept may have had a quite different significance for the Greeks who first wrote of it. Our hindsight is often misled. The Greek natural philosophers were not doing modern science badly, but were concerned rather with seeing how things form part of a connected, rational and aesthetically satisfying system, than with detailed explanation and prediction, and when they discuss atoms or action at a distance or any other concept which looks familiar, it is usually as part of a metaphysical, or even mythological, system, having its own assumptions and methods which are not those of modern science. The arguments which lead them from one theory to another are usually not empirical or theoretical arguments such as we would recognise as scientific, they are arguments arising out of the logical requirements of a metaphysic or out of the pattern of primitive myth.

Though there are these differences between the methods and interests of Greek science and our own, there is one feature which it has in common with modern science, and indeed with science at any period. This is that the attempt to describe natural phenomena is in terms of analogies drawn from processes which are familiar and felt to be better understood. I have argued in the previous chapter that this is not merely the most obvious and convenient way to proceed, but that it is an essential ingredient of intelligible scientific explanation. Ancient science may have lacked other ingredients essential to science in the modern sense, but it had this one, and to describe its theories from this point of view will enable us to bring some sort of order into their bewildering variety, and at the same time to show what kind of continuity exists between Greek science and that of the seventeenth century. Another reason for treating the matter in this way is that vague and general questions, such as whether action can take place at a distance or not, can be made precise only in the context of the kind of explanations that are being sought, and this means in the context of the analogues or models which are presupposed.

The most striking difference between ancient science and that of more recent times lies in the great variety of analogues used in mythology and in Greek natural philosophy, in comparison with the attempt in science between the seventeenth and the nineteenth centuries to make do with only one—the analogy of mechanism. This analogy is concerned with particles of matter stripped of all powers and qualities except the 'primary qualities' which enter mechanics, namely, size, shape, position, and inertia; and in terms of this basic model the question as to whether the particles act upon one another at a distance or only when in contact is capable of precise formulation. In terms of the analogies which enter ancient science however, the question is far from precise, and it does not seem to have become a problem at all until a certain amount of mechanisation of physics had taken place, that is, until the time of the atomists and Aristotle. Part of the history of the problem of action at a distance is therefore that of the growth of a mechanical conception of matter, and the use of mechanical analogues in explaining natural processes.

In the mythopoeic explanations of the ancient world a great variety of analogies are seen between the processes of non-human nature, the functioning of the human body, and human society. Things may act upon one another to produce change and movement in many different ways: the rising and setting sun, the seasons, weather, plant and animal growth and generation, the artefacts of a human workman, the king's command, human strife and friendship; all these familiar agents of change are used analogically to provide more general accounts of the origin and history of the world. Among some of these are to be found the precursors of modern scientific explanations.

Perhaps the most primitive and most widespread imagery used in mythical cosmologies is that of organic growth and reproduction. In ancient Egyptian texts the god Atum generates the god of air and his consort the goddess of moisture, who become the parents of a family of gods and goddesses and ultimately of lower beings and men ; in the Polynesian myths there is a genealogical account of how Darkness and the Cleaving-together of earth and heaven mated, and their child was the Land, then the Land lay with the Sky and their child was the Void, and so on through the generations of gods and men.[1] It is against such a background as this that Cornford understands even the early Greek cosmologies. In Hesiod the genealogies become self-conscious allegories, and in the Ionian philosophers the old cosmologies are ' demythologised ' and expressed in terms of natural events concerning the seasons and the weather, but the analogy with human reproduction is still there : the heaven fertilises the earth with rain, and the earth brings forth living things. For the Ionians the ability to generate life and the power of self-movement, which are characteristics of living beings, also characterise the whole of nature, so that if a natural process can be described in these terms, no further explanation is thought to be necessary. Nature is alive, reproductive and self-moving. This conception as used in scientific explanations is what I shall call the ' analogy of organism '.

A second type of explanation may be called the ' analogy of attraction '. Men experience sympathy and antipathy, attraction and repulsion, love and strife, between themselves and other men, and between themselves and nature, and these are therefore seen as forces which can produce effects in nature. Cornford has pointed out the importance of this for Greek ideas about natural change and motion.[2] The Greeks had no systematic theory of motion, but were content to rely on popular maxims based on the ideas of attraction and repulsion : like attracts like, like nourishes like, like affects like, like perceives like. Magic is the attempt to utilise the same principles, as for instance in the practice mentioned by Plutarch

[1] This and other examples are given in F. M. Cornford's *Principium Sapientiae*, Cambridge, 1952, pp. 202ff.

[2] *Laws of Motion in Ancient Thought*, Cambridge, 1931. Elsewhere (*From Religion to Philosophy*, London, 1912, p. 140) he traces the analogy of attraction to primitive feelings of kinship within a family or tribal group, and of separation from other groups, and remarks that the notion of causality implied is spatial and static rather than temporal and dynamic as ours is : only things classified in groups of kindred can act upon one another, and they do so through a medium of sympathy.

of treating jaundice with the yellow eye of the stone curlew, which was supposed to attract the yellowness out of the patient, and there are numerous examples of this kind to be found in the classical writers as well as in the folk-beliefs of all cultures. Associated with these ideas, and also exploited in magic and witchcraft, is the belief that an image or any sort of representation of a thing or person is not only symbolic, but can actually be made to stand for the thing represented, so that to burn the effigy of a man, or even any property of his, is actually to do him harm. In particular, a man's name can be taken to represent him in this way. This is why, for example, the Israelites were forbidden to utter the name of God, for the name itself is invested with his holiness and power. To use or act in any way upon a symbol is to act by sympathy upon the thing symbolised.

Thirdly, a particularly potent way in which men influence one another is by speech and command, and it is not surprising to find that in primitive belief speech is universally endowed with effective force which can act at a distance with no apparent material intermediary. In some of the ancient Egyptian myths a name uttered by a god is itself an act of creation, in the Old Testament God says of his word : ' so shall my word be that goeth forth out of my mouth : it shall not return unto me void, but it shall accomplish that which I please, and it shall prosper in the thing whereto I sent it '.[1] The analogy is with the word of command uttered by a ruler, an analogy to which the centurion explicitly appeals in his encounter with Christ : ' speak the word only, and my servant shall be healed. For I am a man under authority, having soldiers under me : and I say to this man, Go, and he goeth ; and to another, Come, and he cometh ; and to my servant, Do this, and he doeth it '.[2] Hence also arises belief in the power of words of blessing and cursing, and also in the power of prayer to move a god, by analogy with petitions addressed to a king.

Fourthly, there are various analogies from the techniques and artefacts of the human workman. All societies have their labour-saving gadgets and their products of creative art, and it is natural to find the workings of nature likened to the products of technique, especially as those techniques themselves are only efficient means of utilising the workings of nature. As technique develops and becomes more complex there are increasing possibilities of comparison between natural and technical processes, as for instance the use of the analogy of irrigation systems to explain the circulation

[1] Isaiah 55:11 [2] Matthew 8:8, 9

of bodily fluids, which seems to have been widespread in medical texts at the time of Socrates. But there are more primitive examples. The artificer who makes something intelligently and purposefully is an obvious analogue for the Creator of the world, and is used in the creation myths of many cultures. In one of the Babylonian legends, the god Ea creates man from the blood of a slain god which he mixes with clay ; in Genesis 2 God ' formed man of the dust of the ground, and breathed into his nostrils the breath of life ' ; in the self-conscious mythology of Plato's *Timaeus*, the divine artificer creates the world by imposing form on pre-existing matter, as a sculptor creates a statue. One characteristic of the work of an artificer is that it is purposeful, having a definite end in view. So if nature is created in this way, one can go on to ask for what purpose it was created, what are the final causes of things. This is exactly what we find in the developed artificer analogy of Plato, and following him, of Aristotle. For them, things in nature are for a purpose, ' to be the best possible '.

It is not necessary, however, to bring out the creative and purposeful aspects of human technique. There are also analogies from the sheer labouring activities of men, which have on the whole a humble place in myth and in Greek philosophy, but which come nearer than any of the others to what we understand by mechanical explanation. It is the influence of these analogies which has persisted throughout the history of physics. Aristotle, for instance, bases an important part of his theory of motion on an analogy with a gang of workmen hauling a ship,[1] and asserts almost without argument that a body can move another only if they are in contact. Analogies of this kind help to explain the persistence in physical science of the idea that bodies can act upon one another only by contact.

In primitive mythology and in the writings of the philosophers there is no appeal to one of these analogies to the exclusion of the rest. Most of the peoples whose myths have been collected have several accounts of the creation of the world and of men, and the accounts combine some or all of the analogies we have mentioned, and range from the crudely concrete to the comparatively abstract. In the book of Genesis we have an early account of creation in the second chapter which uses the artificer analogy, and a later, more sophisticated, account in the first chapter in which it is the word of God which creates : ' God said, " Let there be light ", and there was light '. In Egypt, myths of sexual repro-

[1] *Physics*, 250*a*

duction occur alongside those in which the utterance of a name is an act of creation. On this multiplicity of images the Frankforts remark :

> ' Natural phenomena, whether or not they were personified and became gods, confronted ancient man with a living presence, a significant " Thou ", which, again, exceeded the scope of conceptual definition. In such cases our flexible thought and language qualify and modify certain concepts so thoroughly as to make them suitable to carry our burden of expression and significance. The mythopoeic mind, tending towards the concrete, expressed the irrational, not in our manner, but by admitting the validity of several avenues of approach at one and the same time.' [1]

No incompatibility was felt between the different accounts—each added to the descriptive richness of the imagery.

Nature as ' Thou ' : Indefiniteness of the Problem of Interaction

How far did these various types of explanation by analogy imply acceptance or rejection of action at a distance? In the absence of systematic philosophy the question can hardly have been made explicit, so that if one were to attempt to answer it one would have to look for suggestions of spatial continuity of causes in cases where at first sight there is no such continuity. To take an example from modern anthropology, Professor Evans-Pritchard reports [2] that the Azandi believe witches to operate on their victims by sending out the soul of their witchcraft to feed on the soul of the victim's organs, while the witch himself is asleep on his bed. Witchcraft itself they believe to be a substance housed in a particular part of the witch's body, and although this substance as such does not pass to the victim, something (its ' soul ') passes, and they allege that flashes of light can be seen at night to mark its passage. The witch and the adulterer are compared : one uses his soul to work corruption, the other his body, and since the second is not an action at a distance, neither presumably is the first. One cannot of course assume that the partial rationalisations of modern primitive peoples

[1] H. and H. A. Frankfort, *The Intellectual Adventure of Ancient Man*, Chicago, 1946, p. 19 (English ed., *Before Philosophy*, London, 1949, p. 29). The linguistic situation in modern physics is interestingly similar : there two accounts of the fundamental particles (the wave and particle descriptions) are necessary as long as physical thought and language are insufficiently developed to carry the burden of new experiences and new meanings.

[2] *Witchcraft, Oracles and Magic among the Azandi*, Oxford, 1937, pp. 33 and 269

are representative of ancient mythopoeic beliefs, but the example is suggestive in indicating that where there is apparent acceptance of action at a distance, further investigation may reveal a desire to fill in the spatial gaps in the causal chain by intermediate events of some kind, even if they cannot in the nature of the case be material events. Professor Onians gives examples[1] from the Homeric writings to show that thoughts are regarded as words housed in the lungs and breathed forth in speech. 'Breaths' thus pass between mind and mind when communication takes place, but this, as we shall see later, does not imply that there is a mechanical carrier of action, like sound waves, between persons who speak to one another, for 'breath' is not lifeless mechanical air. Nevertheless, when persons influence each other by speech, some intermediate entities are felt to be required, and they are conceived as carrying power from the speaker, and may not only act upon persons but also upon things, as in magic.

The important question to ask, then, is, What kind of intermediate events are contemplated? If something has to travel through the space between agent and recipient, what is the nature of that something? This question becomes significant only when matter is explicitly distinguished from something non-material, or at least when different kinds or grades of matter are recognised. It is precisely this distinction which is not made in mythopoeic thought. In their Introduction to the book mentioned above, the Frankforts suggest that for primitive man nature is a 'Thou', that is, his relationship with it is personal, like his relationship with other men. Nature can be addressed, commanded, implored, and will respond; it has purposes of its own which may be sometimes reasonable and sometimes capricious, just as those of a man may be. It is not that nature is deliberately personified in primitive belief, or that human and animate categories are imposed upon experience of the non-human and inanimate, it is rather that nature is directly experienced as dynamic relationship, and no ultimate distinction between personal and impersonal, animate and inanimate is conceived at all. Primitive man seems to people the world with spirits and to be dominated by religious hopes and dreads, but on the other hand, he conceives the soul as the material breath in his body, he leaves food in tombs to satisfy the hunger of the dead, and he believes in the real efficacy of material things—bread, water, wine, oil, blood—in inducing mental and spiritual states. From our point of view the spirituality and the materialism

[1] *The Origins of European Thought*, Cambridge, 1951, p. 67

are contradictory, but not for him, for he finds a unity of all the potentialities of nature, human and non-human, spiritual and material, animate and inanimate, in every concrete experience. To take an example suggested by the myth of Noah, the experience of seeing a rainbow, which for us may be only an exercise in tracing the paths of light rays, is for the primitive mind an episode in a whole drama of dependence on rain and sun, of destruction and death or new life and hope, in which his emotional as well as intellectual response is called forth. The categories of material, mental, and spiritual are meaningless in the context of this unity of experience, for it is not a deliberate synthesis, but an unself-conscious response prior to distinction and abstraction.

Some important illustrations of this lack of distinction between mental and physical are to be found in the etymology of words denoting mental functions. In most languages words denoting the higher human faculties, such as ' mind ', ' soul ', ' spirit ', have a multiplicity of meanings and very complex etymologies, which no doubt reflect, if we can learn to read the signs correctly, the development of man's ideas about himself and whatever suprahuman powers there may be, and also about matter and its natural forces. Consider, for example, the Greek ' thymos ', ' psyche ', and ' pneuma '.

Thymos and psyche occur in Homer, the earliest surviving Greek literature, and already show considerable complexities of meaning. Thymos is usually interpreted as ' blood-soul ', since its root seems to be θύω, and this can mean ' to offer sacrifice ', or ' to rage, seethe ', like the rushing up of blood in a sacrifice. On the other hand, in Homer it is also clearly associated with ' breath ', for when consciousness is lost as in fainting or death, the thymos is said to be ' breathed forth ', presumably as material breath. It also stands for life and consciousness, especially in the aspects of strong emotion, desire and courage, as with the English ' heart '. Now it is meaningless to ask whether the primary connotation of the word is material or mental or spiritual. It is clearly all three. As Cornford puts it :

> ' It is an easy fallacy (encouraged by dictionaries) to suppose that a word has at first a single sense—the sense that happens to be uppermost at its first occurrence in written records—and later accumulates other meanings. It is nearer the truth to say that the original meaning is a complex in which nearly all the later senses are inextricably confused.' [1]

[1] *The Unwritten Philosophy*, Cambridge, 1950, p. 42

This view evidently lies behind the suggestion of Professor Onians regarding the meaning of thymos.[1] Its various contexts are all associated with the functions of the human chest: the material breath and the process of breathing, the blood and the beating heart. The motions of breathing and of the heart are those most affected under the stress of strong emotion, and are therefore naturally associated with the more robust aspects of life and consciousness. Thymos is not mere breath, nor mere blood, nor even the 'blood-soul' as opposed to the 'breath-soul', but a complex of all these, located in the region of the chest and manifesting itself in the various physical events which seem to take place there.

Psyche, again, is not to be thought of as a single material substance. Onians describes its associations, in Greek literature and custom, with the head and with generation, with bodily fluids such as the cerebro-spinal fluid, seed, tears, and sweat, and with the sap of plants or the juice of their fruit: what seed is to man, wine is to the vine.[2] It could also be a material 'breath', and is identified with air by the philosophers Anaximenes and Diogenes of Apollonia. Because of these material associations, psyche carries also the connotation, later abstracted from them, of life-bearing principle or soul.

Pneuma provides a similar example. In its early usage, it has clearer material connotations, meaning primarily any kind of 'wind' or 'breath'. Some philosophers identify it with psyche,[3] and in Stoic philosophy all qualities, whether physical or mental, are thought to be gaseous substances permeating bodies, and are called 'pneumata'. With Philo and Plutarch pneuma becomes superior to psyche and takes on a definitely 'immaterial' connotation; Philo uses it to translate the Hebrew 'ruah', which is the prophetic breath of divine inspiration, and Pneuma is the word used for the Holy Spirit in the Greek of the New Testament.

All this may seem to be mere equivocation, but it does indicate that one should be wary of identifying the meaning of such words as these in any particular context either with definite physical substances or with immaterial functions or states. The material substances and the life and consciousness for which they are essential

[1] *Origins of European Thought*, pp. 44ff. [2] ibid., Chap. i, pp. 217ff.

[3] e.g. Anaximenes, who speaks of air as the life-breath (pneuma) of the cosmos, 'embracing it, as our own soul (psyche), which is air, holds us together' (Fragment 2. The wording ascribed to Anaximenes in this fragment is almost certainly not genuine, but the parallelism implied probably is).

are interchangeable, by a sort of unself-conscious metaphor which is characteristic of mythopoeic thought. The multiplicity of metaphors, not all consistent with one another, is the mythopoeic way of expressing abstractions, as the Frankforts point out in the passage already quoted. There are many other examples : soul-substance can be air or blood, but in Heraclitus and Leucippus it is also equated with fire ; again, when Thales declares that all things are water, this primal water must not be thought of merely as the chemical substance H_2O, but as retaining associations with the life-giving and generative power which is attributed to the gods or spirits of rivers and springs in mythopoeic thought, and with the various bodily fluids involved in generation. The notion that matter has intrinsic powers, and conversely that active influences and forces are in some sense substantial, was never entirely absent from Greek thought, and although apparently banished in seventeenth-century science, it may be said to have returned, heavily disguised, in contemporary physics.

The problem of how things act upon one another is, then, closely related to a gradual emergence of distinctions between material and immaterial, animate and inanimate; and to the establishment of a view of nature according to which things happen only according to certain patterns of interaction, and in which analogies cannot be drawn at random between different orders of natural events. Such a view of nature was elaborated by the Greek natural philosophers from the time of the Ionians of the sixth century B.C., and was taken for granted by the time of Plato and Aristotle. The process can be seen to begin already in the mythopoeic period with the distinction of the gods as superhuman personalities from the natural forces which were first felt as personal in themselves. The storm, for example, from being itself a malevolent personal power, became, in Babylonian mythology, the revelation, or instrument, of a being behind the storm—the god Enlil. This allowed a certain amount of abstraction and generalisation to take place, for Enlil, being distinguished from the storm as such, became the executive of the other gods, their leader in war, the subduer of chaos and the creator of the world.[1]

[1] cf. T. Jacobsen in H. and H. A. Frankfort, *Before Philosophy*, pp. 153–7. Or perhaps, as Cornford suggests (*Principium Sapientiae*, p. 237), the process of abstraction is to be understood in terms of primitive ritual. In ritual it is the divine king who is responsible for the proper functioning of nature. The gods would then be, not personifications of natural processes, but personifications of the various functions of the king, which, since the king is mortal, need to be detached from any particular holder of the office when he dies.

The Pre-Socratics : Distinction between Matter and Force

Ionian philosophy seems to have been a reaction against this type of crudely anthropomorphic mythology as represented in Greece by the writings of Hesiod. In a sense the Ionians return to the primitive view in which supernatural powers are not distinguished from their revelation in nature itself. For the Ionians the whole of nature is alive and conscious, so that movement and change are inherent in things, and no external explanations of them are sought. But now there is a difference, for the philosophies, unlike the myths, are seeking unity and generality. Aristotle describes the Ionians as ' monists ', for they sought to trace the development of the world back to a single immanent principle (which Aristotle calls ' arché '), and to describe how the qualitatively differentiated world as we know it arose from that principle. Thales suggested that the first principle was water ; Anaximander called it ' the boundless ' or ' unlimited ' ; Anaximenes called it air, and suggested a physical process by which it was differentiated to form the world, namely by condensation and rarefaction.

This monism of the Ionians cannot properly be called materialism, since materialism cannot arise until non-material forces or powers have been distinguished from matter. It is rather that, as in mythopoeic thought, the whole complex situation experienced by sense perception carries the overtones which later ages come to associate with the ' spiritual ', but which are not yet distinguished from physical matter. The process of distinguishing ' body ' and ' not-body ' out of the primitive unity seems to begin when the functions of conscious life : thinking, perceiving, feeling, willing, and moving, are assigned to one group of substances rather than to the whole. A distinction then begins to be felt between the ' dead matter ' of flesh and bones, and the substance, be it air or blood or fire, which has the appropriate qualities to make it represent the moving and guiding soul. Then the alternative ways of materialism and immaterialism begin to part : one may either say, with Anaxagoras and the atomists, that the soul is just a special kind of matter, or one may think of it as the pattern or form of bodies, somehow independent of their matter, as in the Pythagorean and Platonic traditions.

Progress towards such a distinction was accelerated by the radical criticism of Ionian speculations initiated by the philosopher Parmenides of Elea. Following the logic of monism to what he took to be its inescapable conclusion, he asserted that the first principle, or One, must be absolutely without qualities or differentiation, and that it can initiate no change nor movement. The world that we perceive as moving and changing must therefore be illusory. Parmenides was prepared, with the other natural philosophers, to tell a ' likely story ' about it, in what he called the Way of Opinion, but for him the Way of Truth left no hope and even no subject-matter for natural philosophy. His One Being, having been stripped of motion, life, and consciousness as well as all other qualities except imperishability, spelt the end of monism as a basis for natural philosophy. If reality changes, then it cannot ultimately be one ; the truth must lie in a plurality of ultimate elements ; and it was upon this assumption that subsequent philosophy proceeded.

One way out of the impasse was exploited by theories of a plurality of elements [1] together with separate moving causes ; another was that of the atomists, who admitted only atoms and the void and no moving causes. The chief exponents of theories of the first type were Anaxagoras and Empedocles.

Anaxagoras held that matter is infinitely divisible, and that each infinitesimal part contains the ' seeds ' of all things. In common with Empedocles and the atomists, he replaced the primitive organic conceptions of ' generation ' and ' corruption ', which had been shown to be paradoxical by the arguments of Parmenides, by a mechanical conception of ' mixture ' and ' separation ' which seems to have been first suggested by contemporary medical writers. The material seeds, according to Anaxagoras, have not in themselves the power of initiating change, but this power resides in a separate moving force which he calls Nous (Mind or Intelligence). The word, like ' psyche ' and ' pneuma ', denotes a complex idea neither strictly material nor immaterial. Anaxagoras says ' it is the thinnest of all things and the purest ',[2] but it is not mixed with the other elements in every infinitesimal part of matter ; it is found only in some things, and then it is pure and unmixed. Mind has all knowledge and all power, and it is Mind which orders all things, which started the cosmic revolution whose force separated out the elements, and which causes the motion of the heavenly bodies. Anaxagoras is criticised by the Platonic Socrates and by Aristotle [3] because, while in postulating Mind he seems to be giving an account of

[1] The word ' elements ' must not, of course, be interpreted in the sense of the modern chemical elements. It is used here to mean, in the case of each philosopher, ' that with which the world is ultimately constructed ', that is, the *arché*, as Aristotle calls it, whatever that is conceived to be.

[2] Fragment 12

[3] *Phaedo*, 97*a* ; *Metaphysics*, 985*a*

intelligent design in nature, in fact he makes use of Mind only as an *ad hoc* mechanical agent when he is at a loss for ordinary mechanical explanations.

Empedocles takes as his bodily elements the four which became traditional: earth, water, air, and fire; and his moving causes, which have as it were the same status as the elements in the cosmic economy, he calls Love (Philia), personified as Aphrodite, and Strife or Hate (Neikos), recalling Hesiod's goddess of strife, Eris. The bodily elements also are called by the names of pairs of gods: Zeus and Hera, Aidonius and Nestis; but there is no question of a genealogy of gods here as in Hesiod. The six gods of Empedocles's Olympiad are a democratic society like the society for which he is reputed to have fought in his native Akragas, and he says of them that they are 'equal and of the same age', that is, they are each deserving of the same honour. In other respects, however, Empedocles is clearly following Hesiod, who also gives prominent parts in the cosmic drama to the gods of Love and Hate. For Hesiod it is Eros, one of the oldest of the gods, who joins by his power the first couple, Earth and Heaven, and these then generate the succession of the gods. In Empedocles it is Philia ('affectionate friendship') who takes over the functions of Eros, and is the cause of all combination and pairing in nature, whether organic or inorganic.

There is a further parallelism between Empedocles's four bodily elements and his two moving causes: just as (in accordance with the analogy of attraction) the elements are traditionally believed to be perceived by like elements in the sense organs, so he says, we perceive Love by love 'whom even mortals recognise as implanted in their members, whereby they have thoughts of love, and accomplish the works of union',[1] and we perceive Hate similarly by hate within ourselves. In other words Empedocles is describing the nature of the forces which move the cosmos: they are the powerful forces which move men, Love uniting and harmonising, Hate separating and destroying.[2] There is evidence too of an important

[1] Fragment 17

[2] I can find nothing in the fragments to justify J. Burnet's assertion that Love and Hate are 'corporeal' since 'nothing incorporeal had yet been dreamt of' (*Early Greek Philosophy*, London, 4th ed., 1948, p. 232). The Greeks did not need a theory of immaterial substance in order to recognise human passions, and it is clearly human passions that Empedocles is describing. Love and Hate are certainly said to be equal in weight and length and breadth to the elements, but this may imply only that they are all-pervasive, not that they are corporeal, and this remark is immediately followed by the exhortation to contemplate Love with the mind, not with the 'dazed eyes' (Fragment 17). Evidently there were literalists in Akragas too.

As Cornford points out, and as is evident if one reads through the fragments, there is more of the seer and the mystic in Empedocles than the rationalistic natural philosopher of nineteenth-century tradition (*Principium Sapientiae*, pp. 121ff.).

organic analogy in his description of the periodic dispersal and destruction of the universe by the victory of Hate over Love, and the reversal of the process with the return of Love causing union and contraction. M. Jammer points out [1] that this is ' an adaptation of the ancient idea of the world as a breathing organism ', and goes on, ' Empedocles, being a naturalist, did not feel the need for a mechanical explanation of the cosmic systole and diastole. As for all early Greek science, the animal organism is, in principle, simpler than any artificial man-made mechanism.'

Atomism

The pluralists, then, continued the process of distinguishing force and spirit from matter, and thereby of defining a conception of matter as in itself powerless. But their break with the primitive analogies was not as radical as that of Parmenides, or of the atomists. The atomists in a sense remained closer than the others to the spirit of Parmenides by admitting only two ultimates : ' non-being ', which they identified with void space, as well as ' being ', which now became a multiplicity of material atoms continually in motion in the void. The atoms are as near as possible to the ' real ' in the Parmenidean sense of having no qualities except imperishability, although they have also now to be endowed with shape, and the existence of void enables them to move and produce change. The idea of change as generation is replaced by a purely mechanical and geometrical theory of shapes and arrangements of the atoms, and these are not thought of as identical, or spherical, in the fashion of later atomic theories, but vary in shape, some having hooks so that action can be communicated between them by pulling as well as by pushing, and solid bodies can cohere. But the action is purely by contact ; the atoms are moving irregularly in the void, colliding with one another and sometimes becoming entangled in compound bodies. Aristotle summarises the theory thus : ' The atoms act and suffer action whenever they chance to be in contact . . . and they generate by being put together and becoming intertwined '.[2]

There has been some doubt as to whether Democritus, who is the earliest writer on atomism about whom we have detailed information, assigned any external cause for the atomic motions. Aristotle criticises him for not doing so, but also states that he associated weight with the size of the atoms, and Theophrastus

[1] *Concepts of Force*, Cambridge, Mass., 1957, p. 27
[2] *On Generation and Corruption*, 325*a*

suggests that he did in fact make weight the cause of motion. The evidence seems to show, however, that he spoke of weight only in connection with the tendency of ' heavier ' bodies to go to a centre of rotation, just as bulky pieces of twig and leaves tend to collect in the centre of an eddy of water. It seems to be correct to say that there is no beginning or cause of motion in Democritus, but that ' vibration ' of the atoms in all directions is natural to them, like the random motion of motes in a sunbeam, with which he compares it, or like the motion of the living stuff of the Ionians which required no cause. It is not until Epicurus that a natural tendency to fall in one particular direction ' downwards ' is attributed to the atoms. The innate tendency to move which Democritus ascribes to atoms is not the only concept arising from the analogy of organism in his otherwise mechanical philosophy, for he also speaks of the growth and destruction of cosmic whirls or universes in terms drawn from the life and death of the human body, and the idea is elaborated by Lucretius.[1]

In spite of archaic touches such as this, atomism clearly illustrates a new tendency towards mechanical explanations in natural philosophy. I say ' illustrates ' rather than ' initiates ', for, as we shall see presently, mechanical explanations were not confined to the atomists, but were also attempted by the medical writers, and by Aristotle who was far from being a disciple of Democritus. The atomists did however introduce mechanism into the stream of philosophical argument, and the break that that involved with the more primitive kinds of explanation calls for further consideration.

If the major problem of early Greek philosophy was that of the qualitative differentiation and continual flux of the natural world, then what is remarkable about the atomists' solution of the problem is not their attempted construction of the world out of elementary bodies, for this notion was shared by the other cosmologists, but the particular qualities with which their elements are endowed. Looking back through two thousand years of development of mechanistic physics it seems that to take as basic the geometrical and mechanical qualities of bodies—size, shape, arrangement, motion, impact—was an obvious and reasonable step, bringing clarity and simplicity into the confusion of early cosmologies. But without foreknowledge of the subsequent progress of physics, what made it appear a reasonable step to the atomists?

Democritus at least seems to have been influenced by the kind

[1] *De Rerum Natura*, Book v, 235ff.

of arguments which, in the seventeenth century, led Galileo and Locke to distinguish between the primary and secondary qualities of bodies. In one of the fragments he says 'Sweet exists by convention, bitter by convention, colour by convention; atoms and void (alone) exist in reality',[1] by which he appears to mean that sensation of the 'secondary' qualities, sweet, bitter, colour, is produced by the impact of atoms on our sense organs, and is not caused by these qualities being in the bodies themselves. The atoms do however have geometrical qualities: size, shape, arrangement, and motion. Now it is not very clear even in the seventeenth-century arguments how one can distinguish, purely on the grounds of how we perceive them, between the primary and secondary qualities, for as Berkeley saw (and perhaps also Parmenides, since he regarded all qualities alike as unreal), anything which tends to show that the secondary qualities are dependent upon the conditions of perception, and are therefore subjective, can be made to apply equally to the primary. Thus the colour that we see in an object certainly depends on the light in which we see it, but the size we perceive it to be depends equally on its distance, on the intervention of lenses, and so on. And the notion that we can 'think away' the secondary, but not the primary, qualities of an object while still conceiving the object is unsatisfactory, for how can we think of an object without, for example, its colour, or at least *some* colour? In the seventeenth century there is no doubt that the distinction was made plausible by the existence of a science of primary qualities, namely mechanics, which claimed to embrace, in principle, all other qualities, but this means that it is circular to explain the emergence of a mechanical science among the Greeks by appealing to the distinction between primary and secondary qualities. Some independent reason for the distinction must be found, and in the case of Greek atomism, and also in the doctrine of Galileo, this is to be traced to the influence of the Pythagoreans.

This school, which was probably founded by Pythagoras in the sixth century B.C., built their cosmology on the notion of harmony pervading the universe. Impressed particularly by the simple numerical ratios involved in the production of musical harmony, they asserted that 'All things are (or, are like) number'. Now 'number' in sixth- and fifth-century mathematics was hardly distinguished from its concrete embodiment in groups of things and in geometrical measures, so to say that 'all things are number'

[1] Fragment 9

was, in a sense, platitudinous, since all things can, in some respects, be counted or measured. But the Pythagoreans were probably saying more than this : they were asserting that the properties of things are reducible to, or explicable by, the properties of numbers. Numbers, furthermore, were at first understood as integers, or ratios of integers, so the discovery that some phenomena were produced by simple ratios of integers led naturally to the principle of ' number atomism ', namely, that things are made up of equal, unit atoms, or ' numbers '. Thus they thought of the musical string, and indeed every physical line, as being composed of an integral number of these basic units. Already we have the emergence of the ' primary ' qualities as alone real, and of geometry, by which number can be manipulated, as the basic science.

Some modifications in the scheme of number atomism were made necessary by further developments in mathematics and philosophy. First, the discovery, by an application of the so-called theorem of Pythagoras, of the irrationality of $\sqrt{2}$, meant that geometrical lines could not be regarded as composed of integral numbers of the same unit atom, for in that case the side and diagonal of a square would be expressible as a ratio of integers, and this had been proved not to be the case. Second, the paradoxes of Zeno, though their exact significance is obscure, seemed to show that there is something unsatisfactory about the assumption that bodies and space are composed of indivisible units. (He showed as well that the alternative assumption of a continuum led to paradoxes, thus creating a dilemma for subsequent philosophy which was never satisfactorily resolved by the Greeks.) But whatever Zeno may have been attacking, it is clear that some of the mathematical and logical difficulties arising in Pythagorean number atomism are evaded in the atomism of Leucippus and his disciple Democritus, by the admission of a void and of atoms of different shapes and sizes.

For an explanation of the mechanical, as opposed to the geometrical, qualities with which the atoms are endowed, namely impenetrability and the transmission of action by, and only by, impact, we must return to the problem of Parmenides and recall that atoms are ' the real ' and void ' the non-real '. The atoms are imperishable and impenetrable, for the Parmenidean real cannot be annihilated or changed in any way, and so the only mode of change possible for the ' atomised real ' is motion. Change of motion will be the product of the action of atoms on each other, and it is evident that no action can be transmitted *through* the non-real except by passage of the real, so if there is no action at a distance, interaction can take place only by collisions of atoms and not *via* any other kind of medium. As for the possibility that there is action at a distance, it does not seem logically impossible to conceive of action 'jumping' across the void from one atom to another, but the conception would clearly be difficult, for the only quality of an atom that can be changed is its motion, and we have very little direct experience of one body changing the motion of another at a distance. The action of a magnet would in fact be the only apparent example known to the atomists, and it would have been implausible to make this singular instance as fundamental as the familiar experience of collisions. It was much more plausible to ascribe the apparent action at a distance of the magnet to invisible collisions.

Burnet well summarises the atomic doctrine as follows: 'Atoms are Pythagorean monads endowed with the properties of Parmenidean reality'.[1] This places atomism in its philosophical context, and draws attention to the fact that its interest for the Greeks was chiefly as an intelligible and self-consistent account of the nature of reality, and not as providing detailed explanations in a modern scientific sense. This is perhaps why no mechanical science of physics developed from atomism until the seventeenth century, and although it did then certainly provide some of the conceptions of the new physics, it did so within a quite different context of interests and aims, which were practical rather than purely rational.

Immaterial Causes

For atomism even souls and gods are atoms and act by atomic impacts. So although the motive of the doctrine was not, as we have seen, primarily atheistic, it soon became associated with all those trends in fourth-century Greek philosophy which seemed to deny the existence of divinity and purpose and will, and to reduce all things to materialism. Now that matter had been so clearly defined by the atomists and others, it was possible to speak of two orders of being: material and spiritual, where the spiritual could now be defined in terms of all or some of the qualities which were denied to matter. Curiously enough, it is the Pythagorean tradition which gave the clue, not only to the primary qualities of matter in a mechanical philosophy, but also, by a different twist, to the possibility of conceiving the non-material. The line of thought was somewhat as follows: the natural world may seem to embody primarily numerical and geometrical qualities, but it is

[1] *Greek Philosophy, Thales to Plato*, London, 1914, p. 98

also clear that it does not embody them perfectly. There is no geometrically accurate straight line or circle in nature. So a distinction began to be made between perfect mathematical form which is apprehended rationally, and the copies of this form which exist in nature and are known by the senses, necessarily imperfectly, because of the recalcitrance of the material in which they are embodied. In Plato's theory of Forms this notion is developed into an account of the relation between all kinds of qualities or forms as they are in themselves—Goodness and Beauty and Colour and the rest, as well as Number and Circularity—and the matter of the sensible world which they inform.

The world of immaterial form is not however, in Plato, sufficient to account for natural change and movement, since his Forms are static, eternally unchanging. Plato recognises as well mechanical causation of change, but claims that mechanical motions are never self-produced, and are ultimately parasitic upon what he calls 'the motion which is self-moved'.[1] Just as the fact of Socrates awaiting death in his cell is not ultimately explained by the mechanical motions of his limbs which brought him and keep him there, but by the conscious choice of his rational soul, so all physical motions are said by Plato to be ultimately soul-caused, for the soul is the motion which is self-moved. Soul is variously described in terms of the psychic causes of earlier philosophers, particularly the Nous of Anaxagoras, which is for Plato the rational cause of motion, and Ananké, the 'errant' or erratic, acosmic cause, which is analogous to the Strife of Empedocles. Soul is the oldest of all things and prior to body, not immanent in body like the 'living stuff' of the Ionian naturalists, nor like the immanent tendency to the natural place later substituted by Aristotle.

In Plato's only cosmological work, the *Timaeus*, the universe is said to be 'a living creature', endowed with soul (psyche) and reason (nous), and this world-soul governs the world as our soul governs our body. The motion appropriate to Nous is the uniform circular motion of the heavens; on earth this is disturbed by erratic rectilinear motions produced by the errant cause, but both are psychic, not mechanical, causes. The souls of earthly things are understood to be intermediate between the divine, immortal, intelligible and non-sensible Forms, and the living, changing, unintelligible perceptual things; they partake of both being and becoming, for they are immortal and intelligible like Forms, but also alive and therefore partaking in the flux of time like perceptual

[1] *Laws*, 893ff.

things. Plato, however, nowhere gives any satisfactory account of how soul acts upon body to produce motion, apart from the bare statement of an analogy between the human soul and body and the souls and bodies of other moving things. He mentions, for example, three possible theories about how the moving soul of the sun acts : it may be in the sun as our soul is in our body ; it may provide itself with a body ' of fire or air, as some affirm ', and propel the body of the sun externally (a suggestion which would seem to involve an infinite regress, for how does the soul move this body of fire or air ?) ; or ' being incorporeal, she [the soul] has some extraordinary and wonderful guiding power '.[1] But Plato comes to no conclusion about which, if any, of these theories is to be preferred.

Neither is it any use looking to Plato for detailed development of the tradition of mechanical explanation which had become well established by his time and which is implicit in much of his work. He was very far from rejecting this kind of explanation, in fact in some respects he was more of a mechanist than many of his contemporaries, as I shall try to show in the next chapter, but he believed that interest in these things was not the primary function of the philosopher. He puts the point clearly in a passage of the *Timaeus* where he is speaking of the mechanism of vision. Of the visual currents which he supposes to be projected from the eye he says :

> ' Now all these things are among the accessory causes which the god uses as subservient in achieving the best result that is possible. But the great mass of mankind regard them, not as accessories, but as the sole causes of all things producing effects by cooling and heating, compacting or rarefying, and all such processes. But such things are incapable of any plan or intelligence for any purpose. For we must declare that the only existing thing which properly possesses intelligence is soul, and this is an invisible thing, whereas fire, water, earth and air are all visible bodies ; and a lover of intelligence and knowledge must necessarily seek first for the causation that belongs to the intelligent nature, and only in the second place for that which belongs to things that are moved by others and of necessity set yet others in motion. We too, then, must proceed on this principle : we must speak of both kinds of cause, but distinguish causes that work with intelligence to produce what is good and desirable, from those which, being

[1] *Laws*, 898

destitute of reason, produce their sundry effects at random and without order.'[1]

It is unimaginative to complain about this system of priorities as some historians of science have done. Plato's place in the history of thought does not depend upon judgments such as these, and it is in any case most misleading to suggest that he has no place in the history of science, for the metaphysical framework within which scientific explanations are formulated is often as important as the detail of the explanations themselves, and the subsequent influence of Platonism was decisive in encouraging the use of mathematics in science, and in keeping alive the notion of non-mechanical causes in physics.[2] Mathematics has been generally beneficial to the progress of science in the modern sense, and the notion of non-mechanical causes, although sometimes reactionary, has had its triumphs at certain crucial periods, as we shall see in the course of this study.

If the prohibition against action at a distance is expressed in its medieval form : 'matter cannot act where it is not', clearly the problem as to whether this is true or not becomes definite only when we know what is meant by 'matter'. I have tried to show how the notions of matter and natural force emerged in early speculative philosophy and how they still depended to a great extent upon the analogies in terms of which primitive myth seeks to understand and explain the natural world. Types of scientific explanation in later centuries have been more dependent upon the 'model' of nature expressed by these analogies than upon the details of the metaphysical logic which sought to rationalise it, and so it has happened that the imaginative ideas, for example of atomism or of Platonism, have persisted, although their metaphysical grounds have been forgotten. The ideas associated with them come to have a deceptive obviousness which can be misleading in two ways. First, we must not assume that they appeared obvious to the Greeks, and in order to understand their approach to the problem of matter and force, it is necessary to try to reconstruct

[1] *Timaeus*, 46. cf. *Laws*, 891, where Plato goes further, and maintains that natural explanations of the soul undermine the gods, and therefore the law of the state.

[2] As A. N. Whitehead continually insisted : 'In a sense, Plato and Pythagoras stand nearer to modern physical science than does Aristotle. The two former were mathematicians . . .' (*Science and the Modern World*, Cambridge, 1926, p. 41) ; '[Plato] evokes interest in topics as yet remote from our crude understanding of the interplay of natural forces. The science of the future depends for its ready progress upon the antecedent elucidation of hypothetical complexities of connection, as yet unobserved in nature' (*Adventures of Ideas*, Cambridge, 1933, p. 195).

their metaphysical arguments—this is the purely historical motive. But also, we must not assume that what appears obvious to us is necessarily true or helpful as a principle of science. It is good to be reminded that the obviousness of world-models is historically relative, for we are then more ready to criticise our own.

Chapter III

MECHANISM IN GREEK SCIENCE

Mechanical Analogies among the Cosmologists and Medical Writers

THE first examples of mechanical explanation in Greek physics are to be found in the use, among natural philosophers and medical writers, of analogies from techniques in their accounts of natural processes. The employment of material things for technical purposes no doubt encourages the idea of a distinction between inanimate, passive matter, and living, active beings, and makes it ridiculous to endow matter with active powers or to fear or worship sticks and stones.[1] Familiarity with techniques therefore leads to explanations in terms of the purely mechanical properties of matter : the forces of expansion and contraction, suction, pressure, centrifugal force, and so on, and explanations of this kind may be said to constitute the first mechanical theories of science, although ' mechanical ' is taken here in a rather more general sense than is customary in more recent physics.

It is difficult to be sure quite how influential mechanism was in Greek science, because this was not one of the philosophical questions that interested the compilers and commentators, and very little original material has survived. In the philosophy which has come down to us, mechanical explanations were rarely thought to exhaust the possibilities of physical explanation, and sometimes descriptions of mechanical processes are hardly more than illustrations of general principles arrived at on other grounds. There are, however, certain hints, in writings which are mainly devoted to something else, to indicate that from the time of the earliest Hippocratic writers there was a continuous tradition, mainly associated with the practice of medicine, in which attempts were made to account for various processes in purely mechanical terms. Certain unifying concepts and hypotheses emerged from these attempts which were still taken seriously in the seventeenth century, when the rest of Greek natural philosophy had been discarded.

[1] cf. *Isaiah* 44:9–20 : ' I have burned part of it in the fire ; yea also I have baked bread upon the coals thereof ; I have roasted flesh and eaten it : and shall I make the residue thereof an abomination ? shall I fall down to the stock of a tree ? ' (v. 19).

Natural science is usually said to begin with the explanations given by the Ionian philosophers in terms of condensation, rarefaction, evaporation, freezing, melting, burning, and so on, and it may well be the case that these were drawn from contemporary techniques as well as from more obvious meteorological phenomena. But it is in the Hippocratics that the influence of techniques becomes clearer. There we find many descriptions of what are really processes of evaporation, capillarity, and osmosis, which were themselves imperfectly understood, but which were used as analogues for movements of fluids within the body. Examples are comparisons of the pressure of air which causes yawning before a fever with the pressure of the steam of a boiling pot, and of the sweat of a fever to steam condensed on a cool surface.[1] In another, probably post-Platonic, writer, the process of bodily ' toning-up ' which an athlete undergoes is likened to the techniques of iron smelters, fullers, cobblers, builders, curriers, basketmakers, goldsmiths, corn grinders and potters.[2] During the Socratic period and later there are many examples of analogies between parts of the body and mechanical tools and inventions[3]: eyelashes are sieves, ears are troughs, the nose is a wall, the passages of the body which conduct fluids are irrigation channels. Xenophon's Socrates uses illustrations of this kind to show that nature is not the work of blind chance, but provides evidence of purposeful design or art (techné), and Jaeger has compared this method of explanation with that of the Enlightenment ' with its efforts to understand the corporeal side of man's nature as a system of purposeful mechanical arrangements comparable to true machinery: the heart as a pump, the lungs as bellows, the arms as levers, and so forth '.[4] Another example is Empedocles' famous description of a girl playing with a water-lifter,[5] which he uses to illustrate the flow of air and blood in the body; while Aristotle prefers the bellows as a mechanical analogue for respiration.[6]

Natural processes involving circular motion such as eddies of water and techniques of centrifuging were widely used as explanatory illustrations. Various mechanical principles are involved here, and these are not always clearly distinguished, either in the originals or in the commentaries. When solid matter is caught up in an eddy it may move towards the centre or towards the circumference,

[1] Hippocrates, *On Breaths*, VIII [2] Hippocrates, *Regimen*, I, XIII ff.
[3] In Aristophanes, Euripides, Xenophon, Aristotle, probably based on the teaching of Diogenes of Apollonia. See W. Jaeger, *Theology of the Early Greek Philosophers*, p. 168.
[4] ibid., p. 168 [5] Fragment 100 [6] *Parva Naturalia, On Respiration*, 474*a*

depending on its density and also on the extent to which the motion of the surrounding fluid is imparted to it.[1] If a body is less dense than the rotating water it will always go to the centre of the eddy because, by an application of Archimedes' principle to the field of the centrifugal force, the inward pressure of the water is greater than the centrifugal force acting upon the body. This is seen in the behaviour of twigs and leaves floating on the surface of an eddying stream. If, however, the body is denser than water, one of two things may happen. It may acquire the rotational velocity of the eddy completely, and then it will move outwards because the centrifugal force upon it is greater than the inward pressure; but if for some reason, for example friction with the bed of the stream, it does not acquire a velocity as great as that of the eddy, its own centrifugal force may not be great enough to take it to the circumference, and it may either move to the centre, or remain in rotational equilibrium at a fixed distance from the centre. The various possibilities explain how Anaximander, Anaxagoras and Leucippus[2] could apparently use the analogy of heavy bodies going to the centre to explain how the earth is at the centre of the cosmic vortex, surrounded by the lighter bodies, water, air, and fire, whereas Anaxagoras also uses centrifugal force to explain how the planets were thrown out from the earth to their present orbits.[3] According to Aristotle, Empedocles uses the same analogy to explain why the earth does not fall 'downwards'.[4] The technique of centrifuging with a rotating sieve is used by Democritus to support the traditional belief that like is attracted to like:

> '. . . for among animals too, like herds with like, doves with doves, and cranes with cranes, and so on with other beasts: and similarly in the case of inanimate things, as we can see with seeds in a sieve and pebbles on the shore: for in the whirling of the sieve beans range themselves apart with beans and barley with barley and wheat with wheat, and by the motion of the waves large stones are driven to the same place with other large

[1] The problem was elucidated by Huygens in a tract 'On the cause of gravity'. See *infra*, p. 108

[2] J. Burnet, *Early Greek Philosophy*, 4th ed., London, 1948, pp. 61, 269, 338

[3] Fragments 9, 12, 13, 16

[4] *On the Heavens*, 284*a*, 295*a*. T. Gomperz, following Aristotle's suggestion (*Greek Thinkers*, I, London, 1901, p. 242) thinks that Empedocles is appealing to the analogy of a trick seen in fairs, where cups filled with water are fixed on a hoop with their mouths facing inwards and rotated rapidly so that the water does not spill. If this is correct, the application to the stability of the earth is far from clear, and Heidel (*The Heroic Age of Greek Science*, Washington, 1933, p. 187f.) thinks the example was meant to explain the position of the heavenly bodies as in the theory of Anaxagoras, and that Aristotle's interpretation was mistaken.

stones and round and round, just as if similarity among them had an attractive force.'[1]

The Hippocratic writer of 'The Nature of Children' explains the growth of parts of the body by differential deposition from a moving fluid: like elements tend to go to like as they are carried round on the stream of breath, just as if one were to introduce a mixture of earth, sand, and water into a bladder, and then blow through a tube attached to the bladder so that the water circulates:

> 'At first all the substances mix in the water; then after a certain time, thanks to the blowing, lead goes to lead, sand to sand, earth to earth; and if the bladder is allowed to dry and is torn open, one finds that like has gone to like.'[2]

In the *Timaeus*, Plato explains the attraction of like for like by analogy with the winnowing basket, which separates by being merely shaken. This accounts for the separation of the four elements into distinct regions of the universe, for according to *Timaeus* the 'receptacle', that is, physical space, is continually shaking under the influence of the irrational cause Ananké, so that the elements

> 'were perpetually being separated and carried in different directions; just as when things are shaken and winnowed by means of winnowing-baskets and other instruments for cleaning corn, the dense and heavy things go one way, while the rare and light are carried to another place and settle there.'[3]

The Horror Vacui and Antiperistasis

There is some evidence of a controversy, in which Plato took part, between various views of the nature of 'attraction'. The Hippocratic writers use the concept to explain a variety of processes which we ascribe to evaporation, capillarity, expansion and contraction, oxidation, and so on. For instance, it is noted that hollow vessels tapering towards the mouth, such as medical cupping instruments, can be made to 'attract' fluid if they are warmed and placed mouth downwards in the fluid. This process is described in terms of the power of heat to 'draw fluid' to itself, and it is sometimes explained that the heat causes expansion of the air in the

[1] Fragment 164 [2] *On the Nature of Children*, XVII [3] 52*e*

vessel, which then contracts on cooling, drawing up liquid in order to avoid a vacuum, it being generally assumed that there can be no void. The bladder, the head, and the womb are of the required shape, and are supposed to attract liquid in this way, and there is similar action in spongy parts like the spleen, lungs, and breast.[1] Other illustrations of the attractive power of heat are provided by the action of the sun in drawing up water from the earth[2] and the attraction of a current of air towards a flame.[3] Thus it is said that 'fire can always move all things', and that the foetus is formed by fire which draws in nourishment from the food and breath,[4] and the heart, which is the hottest part of the body, can draw in air.[5]

Some writers, however, deny that the *horror vacui* can exercise any attractive force, and account for the apparent suction in medical cupping instruments by means of a theory of 'circular thrust', called *antiperistasis* by Aristotle and his successors.[6] This theory seems to go back to Empedocles, who adopted it in the attempt to account for motion in the Parmenidean plenum. If there is no void, motion can take place only if it is circular, each body moving into the place previously occupied by a neighbouring body, and the whole cyclic motion taking place instantaneously. Thus a vacuum is avoided not by suction, but by impulsion communicated round the circle. The theory clearly commends itself to those who wish to maintain that there are no 'pulls' but only 'pushes' in nature. Its classic statement is in the *Timaeus*, where Plato is describing the circulation of air and blood in the body by an 'irrigation-channel' hypothesis similar to that of his medical contemporaries. He uses the occasion to attack those physicians who maintain that the movement of fluids in the body is caused by the 'attraction' exerted by various of its parts,[7] and digresses to show that not only respiration, but many inanimate physical processes, can be explained instead by antiperistasis :

> 'To this principle, moreover, we may look for the explanation of what happens in the case of medical instruments for cupping, of the process of swallowing, and of projectiles, which keep moving after their discharge either through the air or along

[1] *Ancient Medicine*, XXII ; *Weeks*, XIV, XIX ; also Aristotle, *Generation of Animals*, 739*b*

[2] *Airs, Waters, Places*, VIII ; *The Nature of Children*, XXV

[3] *Fleshes*, VI

[4] *Regimen*, I, III, IX

[5] *Fleshes*, VI

[6] In Plato, *Timaeus*, 79*c* etc., it is called περίωσις.

[7] For evidence of the controversy among medical writers over the attractive powers of the head, the womb, and the bladder, see M. Wellmann, *Fragmentsammlung der Griechischen Aerzte*, I, Berlin, 1901, p. 37.

the ground. . . . There are, moreover, the flowing of any stream of water, the falling of thunderbolts,[1] and the 'attraction' of amber and of the lodestone at which men wonder. There is no real attraction in any of these cases. Proper investigation will make it plain that there is no void; that the things in question thrust themselves round, one upon another, that the several kinds of body, as they are disintegrated or put together, all interchange the regions towards which they move; and that the results which seem magical are due to the complication of these effects.'[2]

The reference to bodies 'disintegrated or put together' is an allusion to Plato's theory, previously expounded,[3] that although all things move to their like, this does not eventually lead to a stalemate, because the elements are continually being transformed into one another by the destruction and recombination of atomic arrangements. In this 'inequality', or heterogeneity, and in the tendency of like for like, resides the motive power which maintains the circular thrust.

Three of Plato's examples are of particular interest: the medical cupping-instruments, the flight of projectiles and the attraction of the lodestone.

Cupping instruments were used to extract blood and pus from the flesh by being heated and then placed mouth downwards on the affected part. We have already seen that some of the Hippocratic writers gave a partially correct explanation of the process, namely that the air in the cup is rarefied by the heat, and cools when placed on the flesh, drawing up fluid to prevent the occurrence of a vacuum. They did not see that the effect is really due to differential air-pressures inside and outside the cup, and therefore concluded that the process involved real 'attraction', or pulls rather then pushes. Others, like Plato's Timaeus, agreeing that there is no void, nevertheless denied that there can be any attraction, and were therefore thrown back on the theory of antiperistasis. In his commentary on this passage of the *Timaeus*, Plutarch[4] explains that the heated air evaporates through the pores of the metal cup, displacing neighbouring air, and causing it to transmit pressure under the flesh so that the humours are expelled into the cup, thus completing the cyclic motion.

The flight of projectiles presented a problem because it was

[1] Unnatural, because fire moves naturally upwards. [2] *Timaeus*, 79*e*–80*c*
[3] ibid., 52*e* ff. [4] *Quaestiones Platonicae*, VII, 1004f.

thought that any motion in a direction other than downwards required a moving cause in continuous contact with the moving body. How then does the projectile continue to move after it has left the projector? Clearly, if there is no void, there must be cyclic motion of the air as the projectile passes through it, and from the thrust of this displaced air Timaeus and others tried to derive the force which kept the projectile in motion. Aristotle [1] mentions two theories of this type: one says that the displaced air gathers in behind the projectile and so pushes it on; the other says that the first mover moves the air which itself becomes a mover and imparts motion to the projectile. But, says Aristotle, whichever explanation is adopted, the difficulty arises that the air would cease to move simultaneously with the first mover, and so would the body, so the theories do not provide a solution. Aristotle is forced into the artificial suggestion that the first mover gives to the air the power of being itself a mover, and this is passed on in a series of impulses until it is finally exhausted and the projectile falls to the ground.[2]

The explanation of magnetic attraction in terms of antiperistasis seems to go back to Empedocles:

> 'The effluxes from the lodestone repel the air at the entrance to the pores of the iron, and when this is eliminated the iron follows the violent efflux from its own pores. The effluxes from it enter the pores of the lodestone, being proportioned to them, and fit into them, and the iron is borne along with these effluxes.' [3]

Plutarch [4] explains that only iron is attracted by the lodestone, because it alone has pores of the right size and shape for the air to get entangled in the iron and so draw it along towards the magnet. According to Theophrastus,[5] this 'proportion of the pores' was also the explanation given by Empedocles for perception by each sense organ of its proper object, for the objects of other sense organs are either too large or too small to be held in its pores.

Aristotle evidently found the attraction of lodestone difficult to assimilate to his physics, and he mentions it very seldom. In *On the Soul* [6] he quotes Thales as saying that the 'Magnesian stone'

[1] *Physics*, 215*a*, 266*b*; *On the Heavens*, 301*b*

[2] The medievals pointed out that no such explanation will work for a spinning top.

[3] Diels, *Fragmente der Vorsokratiker*, I, Fragment A. 89; trans. A. E. Taylor, *Commentary on Plato's Timaeus*, p. 579

[4] *Quaestiones Platonicae*, 1005*d*

[5] *On the Senses*, 7

[6] 405*a*

has a soul because it causes movement in iron.[1] In the *Physics* he mentions the magnet in passing in his discussion of projectiles. This is the passage considered above, where Aristotle points out, rightly, that antiperistasis alone will not explain the projectile's motion, because 'all the things moved would have to be in motion simultaneously and also to have ceased simultaneously to be in motion when the original mover ceases to move them', and then he adds 'even if, like the magnet, it makes that which it has moved capable of being a mover'.[2] He then goes on to suggest that parts of the air transmit to each other, not only motion, but also the power of being movers, and that this power is retained after the original mover has come to rest. The suggestion is reminiscent of the power of iron to retain its magnetism after the original magnet is withdrawn, but Aristotle is not referring to this retentive property, for neither the Greeks nor the Romans seem to have known of it. Cornford suggests [3] that what he has in mind is the magnetism transmitted through a series of pieces of iron, all of which lose their attractive power when the first magnet is withdrawn, and quotes Plato on

> 'the power in the stone which Euripides calls the Magnesian stone. . . . This not only attracts rings that are made of iron, but puts into them the power of producing the same effect as the stone and attracting the other rings in turn. Sometimes there is quite a long chain of rings hanging from one another; but all the power depends on the stone.' [4]

In all this there is clearly a desire to eliminate quasi-magical attractive forces from physical and physiological theory, and to replace them by mechanical analogies drawn from familiar observations and contemporary techniques. But these analogies were not strong enough to bear the weight of explanation laid upon them, and five hundred years after Plato, Galen is forced to reject much of the mechanical theory of the medicine of his day, in favour of the

[1] The successive stages of emancipation from animism are well illustrated by Diogenes Laertius's report of Aristotle's report of Thales (quoted by M. Jammer, *Concepts of Force*, p. 24*n*): 'Aristotle and Hippias say that Thales attributed soul also to lifeless things, forming his conjecture from the nature of the magnet, and of amber.' Aristotle imagines that Thales was looking for a psychic explanation for magnetism, Diogenes that he got the idea that all things are ensouled from the behaviour of the magnet. It is likely that both are mistaken, and that Thales, having a fundamental apprehension of all things as ensouled, merely pointed to the magnet and amber as striking illustrations of this fact.

[2] *Physics*, 266*b*

[3] Aristotle, *Physics*, II, trans. Wicksteed and Cornford, London, 1934, p. 416

[4] *Ion*, 533*d*

notion of attraction, and he criticises Plato's physiological theories of circular thrust as unworkable. He remarks on the false theories 'set forth even by physicians who would not await the findings disclosed by dissection but stated conjectures founded on analogy as if they were based upon observation',[1] and ridicules those who, in their passion for mechanical explanation, had tried to describe all the physiological processes of the body in terms of suction by the vacuum. Some processes, he agrees, are certainly explicable in this way, but in other cases he says that every body must possess 'a faculty by which it attracts its proper quality'.[2] He speaks of the 'faculties' by which lodestone attracts iron, those by which parts of the body are replenished by attracting the appropriate parts of food, and those by which poisons and thorns are extracted from the body by appropriate medicaments. Although his protest against too easy use of mechanical analogies was no doubt justified by the state of medicine at the time, it cannot be said that his alternative theory of faculties was any more satisfactory than the maxims that 'like attracts like', 'like nourishes like', and so on, upon which traditional explanations had been based.

There are, then, plenty of examples of the use of analogies from technique in Greek science, but, as Galen at least was aware, this in itself was not sufficient to ensure the establishment of a science based on mechanical principles. For such analogies to constitute explanations, rather than merely striking illustrations, it was necessary that the scope of the analogies should be tested much more rigorously than was possible with the somewhat superficial observation with which the Greek philosophers contented themselves. It never became apparent to them, for example, that the attractive powers exhibited on the one hand by cupping instruments and on the other by magnets are only analogous to a very limited extent, and it did not occur to them to test the analogy by the kind of experiments which would immediately come to the mind of a modern investigator: for example, is there any evidence of the alleged production of a vacuum in the neighbourhood of a magnet? any effect on smoke from the supposed streaming of magnetic particles? and so on.[3] But the description of superficial analogies seemed to satisfy them. The crucial tests for a mechanical theory of matter which seemed to them to justify the analogies were not experimental, but rational, that is, the theory itself had to be

[1] *De Placitis Hippocratis et Platonis*, ed. Müller, I, p. 165

[2] *On the Natural Faculties*, I, XIV (Loeb ed., London, 1947, p. 85)

[3] cf. the experiments devised by Gilbert and Boyle to test the vacuum-suction theory, *infra*, pp. 88, 116

seen to be self-consistent, and if this were so, it needed to have only the most vague and general correspondence with the physical world.

Aristotle : Matter and Form and the Primary Qualities

On grounds of consistency and reasonableness it was not the thorough-going mechanical philosophy of the atomists that had the greatest subsequent influence, but the theory of Aristotle. This itself contained elements of mechanism, at least in its account of local motion, but Aristotle found reasons for rejecting many of the tenets of the atomists, particularly their assertion that there is void, and their denial of all causes and forms of interaction other than mechanical.

The philosophy of Aristotle, like that of his predecessors, takes as its primary problem the question of how change and differentiation of qualities is possible in the world. It was his solution of this primary Greek problem which fixed for succeeding centuries the standard ways of describing matter and change and forms of action between bodies, not perhaps because this solution was intrinsically so much more satisfactory than others, but because of the sheer weight and impressiveness of his logical method. Every aspect of every question is dealt with in meticulous order and detail ; the system is close-knit, and to live in its world and apprehend its interweaving patterns is itself a fascination. Here it is only possible to describe in detail those themes from the *Physics* which are most relevant to our subject, together with some other references which help to clarify these.

In the *Physics* Aristotle begins by dismissing the Parmenidean theory that Being is one and motionless. This theory, he says, is obviously not a contribution to natural science, for physicists must take for granted that there is change and motion.[1] He goes on to criticise the solutions of the Ionian monists and the pluralists, but finds in them a clue to his own solution, in that all of them assume some pairs of contrasting qualities in their account of change, for example the rare and dense of Anaximenes, or the being and non-being of the atomists. These particular pairs of opposites Aristotle replaces by a single, general notion of the contrast between form and deprivation of form, together with the substratum, matter (hylé), in which form inheres. By this manoeuvre he does not, however, avoid a plurality of qualities in his system, for he has

[1] *Physics*, 185*a*. The word κίνησις is usually rendered 'motion' in translations, but its meaning is wider than the English 'motion', and includes qualitative and quantitative change.

to assume that every quality or structural pattern found concretely in the world exemplifies some form, and he asserts that the forms are independent of each other : when change takes place, it is not that one form changes into another, but the same substratum comes to embody a different form. What changes is the degree of *actuality* of the form which is *potentially* present in this matter, as when an acorn grows into an oak, the acorn is said to have potentially the form of a fully grown oak, and this potentiality becomes actual as it grows ; or when brass is made into a statue, brass is said to have a potentiality for the form of the statue, which is actualised when the statue is created. Thus Aristotle concedes to the Eleatic philosophers that nothing can come to be out of what is not, but saves the possibility of change by saying that something that was potential becomes actual by the operation of certain causes.

Unlike Plato, Aristotle does not regard the forms as existing in a supernatural realm of ideas independent of their occurrence in the world. For him they are always embodied in matter (with the exception of unmoved movers to which we come later), and matter always exhibits some form, so that any individual existing substance is matter plus form. Elementary or primary matter is, in fact, unknowable, because it implies existence without qualities, and this can neither be experienced, nor understood. Form however can be thought of apart from matter, just as a species can be thought of apart from any particular member of it. The forms are considered to be real existents in the sense that there are real specific kinds in the world, not just collections of similar things associated for convenience by human classification.

It may be thought that by postulating as many forms as there are different kinds of substance, Aristotle is going back on the whole development of natural philosophy, which had previously sought to reduce the multiplicity of things to a few elementary principles. This is indeed the criticism which Aristotle himself brings against the Platonic version of the theory of forms,[1] and it is difficult to see how his own version is an improvement in this respect. However, Aristotle does not neglect the search for elements, and is prepared to assert that some qualities and substances to be found in the world are simpler and more fundamental than others. Strictly speaking, his elements are matter, form, and deprivation of form, but these are only elementary in the sense that matter is at the base of a hierarchy of increasing complexity of form, while form and deprivation are fundamental because they are qualities

[1] *Metaphysics*, 990*b*

of all possible pairs of qualities. What particular contrary qualities are the elementary or primary ones is discussed in *On Generation and Corruption*,[1] where it is asserted that these must be tangible qualities, presumably because, for Aristotle, touch is the primary sense upon which all the others depend.[2] Thus, seeing, hearing, and smelling, although they seem to take place at a distance, are possible only because there is a medium between the object and the sense organ which is modified by the object and modifies the organ by contact. So tangible qualities are presupposed by visual and other qualities, but Aristotle does not assert that the other qualities are *reducible* to the tangible, as Democritus thought they were reducible to geometrical arrangement. Colour and the rest, although they are perceived by means of a mechanism which involves touching, are in their own right potential qualities in the object, which are actualised when perceived by a subject.[3]

Having decided that the primary qualities must be tangible, Aristotle lists various pairs of such qualities : hot-cold, wet-dry, heavy-light, fine-coarse, viscous-brittle, and hard-soft. He eliminates heavy-light because these are not true contraries which destroy each other when brought together, and then he argues that the last three pairs are all modes of wet-dry, that is, they are all aspects of the distinction between liquids and solids. Hot-cold and wet-dry are not however reducible one to the other, so he concludes that these four are the primary qualities. This is not however to be understood in the sense that these are the only real qualities, as is the case with the primary qualities of the atomists, but merely that they are at the base of a hierarchy of existing things in which the higher levels presuppose the lower, and at each level new forms appear which are not wholly reducible to forms at lower levels. Next in order of complexity after the primary qualities come the traditional four elements, which are formed by the four possible combinations of the primary qualities : earth is cold and dry ; water, cold and wet ; air, hot and wet ; fire, hot and dry. Thus these four elements are not immutable as Empedocles thought, but can be transformed into one another by action of the appropriate qualities.

Comparing the arguments by which Aristotle reaches his primary qualities with those by which the atomists reached theirs, it is remarkable that Aristotle relies on common experience of the actual properties of bodies, however superficially he may interpret this, while the atomists on the other hand were influenced by the

[1] 329*b* [2] *On the Soul*, 415*a*, 435*a* [3] ibid., 426*a*

most sophisticated metaphysical speculations. This example and others like it make it somewhat ironical that in the seventeenth century it is atomism which is regarded as progressive and empirical, while the Aristotelian tradition carries the stigma of non-empirical speculation.[1]

Aristotle : Theory of Motion and Change

Returning to the *Physics*, Aristotle continues with a classification of the four types of cause which produce change. They are all in different ways answers to the question ' why ' such and such an event took place, and usually one event will have causes of all four types, although sometimes one element in an event will be its cause in more than one sense. It is noticeable that the examples which Aristotle gives are almost always from the products of human art, or from living beings, not from the inanimate world. The four causes are :

1. The material cause, or ' that out of which '. ' Matter ' is here used in a relative sense ; it is not the unknowable substratum, although that is certainly the ultimate material cause, but that out of which a thing is made, as, for example, bronze is the material cause of the statue relative to the form of the statue.
2. The formal cause or pattern, for example : the form of the statue ; or, numerical relationships such as the ratio 2 : 1 which is the formal cause of the octave. This might be said to come nearer to the sense of ' cause ' in modern mathematical physics than any of Aristotle's other causes, for here cause is understood in terms of a mathematical pattern to which events conform.
3. The efficient, or moving, cause, which Aristotle calls ' the primary source of the change or coming to rest ', and ' what initiated a motion '. This is often taken to be nearest to our sense of ' mechanical cause ', as of one billiard-ball causing another's motion by hitting it, but the examples Aristotle gives in the *Physics* are all non-mechanical, for instance the father is cause of the child and a raid is the cause of going to war.[2]
4. The final cause, or ' that for the sake of which ', as health may be the final cause of taking exercise, or power the final cause of going to war. Aristotle argues at length that there are final causes in nature, as against those (the atomists and others) who maintain

[1] Galileo knew better : in his dialogues it is Simplicio, the Aristotelian, who represents the common-sense approach, while the spokesman of Galileo himself is aware that the new science needed more than empiricism.

[2] *Physics*, 194*b*, 198*a*. cf. *Posterior Analytics*, 94*b*

that things which appear to be designed for an end, have come about spontaneously by other causes, but it should be noticed that in giving examples like the making of nests and webs, and the production of leaves to provide shade, and roots to provide nourishment, he does not assert that there is here evidence of art or deliberation as in men's purposive activities, but merely that the goal-seeking behaviour is immanent.

The *Physics* goes on to develop a theory of the efficient causes of motion. Here there are four principles which Aristotle either assumed or attempted to prove, which were greatly to influence future discussion on modes of action. These are :

1. The denial of the void.
2. Every motion has a moving cause.
3. The mover must be in contact with the thing moved.
4. For every motion there is an unmoved first mover.

Aristotle begins his discussion of the void, as is his habit, by stating and criticising the views of his predecessors. There are those who try to refute the existence of void by showing by experiments that air is in fact a body, but this, says Aristotle, is not conclusive proof that air or some other body is present everywhere. Then there are those who maintain with the atomists that there must be void because without it there could be no change nor motion. To this Aristotle replies that a plenum can suffer qualitative change, and that locomotion can take place by antiperistasis : 'for bodies may simultaneously make room for one another, although there is no interval separate and apart from the bodies that are in movement. And this is plain even in the rotation of continuous things, and in that of liquids.' [1]

Most of Aristotle's positive arguments against the void turn on his belief that heavy and light bodies have a natural motion towards or away from the centre of the earth. In void there can be no 'up' and 'down', nor can there be any reason for a moving body to stop anywhere, in fact all bodies would move through the void with infinite speed, since he believes that velocity is inversely proportional to resistance. Again, on the principle that all moving bodies require a mover in contact with them, Aristotle remarks that no projectile could move through a void, since all motion other than up and down relative to the surface of the earth is unnatural for inanimate bodies, and requires a moving cause

[1] *Physics*, 214*a*

continuously in contact with the body. There could be no such mover in a void. He apparently fails to notice that this argument is inconclusive, since everyone admits that the atmosphere is not a void, and it is only in the atmosphere that we know the motion of projectiles to be possible.

There is no need, he continues, to postulate a void to account for compression and rarefaction. This may be due not to discrete atoms being more or less loosely packed, but to a continuous matter which is capable of existing in any density. As Joachim puts it in his commentary on *Generation and Corruption* : according to Aristotle ' We must not think of a " dense " body as one in which there are few or small " pores ", and of a " rare " body as one with large or many gaps interspacing its corporeal particles. We must rather conceive of ὕλη [matter] as a material capable of filling space with all possible degrees of intensity.' [1] Neither must we conceive of hyle as a continuous compressible fluid in the manner of modern hydrodynamics, for to say that matter becomes more dense or more rare by the motion of part of it into or out of a given region is to assign intrinsic qualities and motions to matter, and this is precisely excluded by Aristotle's definition of it : ' By matter I mean that which in itself has neither quality nor quantity, nor any of the other attributes by which being is determined.' [2] It is rather that the unknowable, unchangeable substratum he calls matter acquires the *form* ' dense ' or ' rare ' just as it acquires ' heat ' or ' cold ' or ' colour '. There is conservation of matter only in the sense that the substratum, which is neither dense nor rare in itself, cannot be created nor destroyed, but this is not because the quantity of matter remains constant, for matter has no ' quantity ', it is because matter is postulated as the substratum of change, and must therefore be changeless by definition, otherwise there would have to be an infinite regress of ' matters '.

The second of Aristotle's principles is that every motion requires a moving (that is, an efficient) cause. He complains that this has been neglected by the Ionian monists and by the atomists and Plato. It is not enough to say with Democritus that there has always been motion ; we should look for a first principle to explain it, just as we look for more ultimate principles to explain the properties of a triangle, even though these are always true.

In the case of natural motions, the moving cause is to be found in the thing moved. The sources of motion in animals and plants

[1] *Aristotle On Coming to be and Passing away*, Oxford, 1922, p. 124
[2] *Metaphysics*, 1029*a*

are their own souls, but the natural up or down motions of inanimate things must differ in some way from animate motion, because the power of self-movement is precisely what distinguishes living from non-living. Aristotle's solution of this difficulty lies in his theory of 'proper place': the four material elements, earth, water, air, and fire, are fully actual only when they are distributed round the earth in concentric spherical shells with earth at the centre and fire at the circumference. When any element is displaced from its proper sphere, it has a potentiality or natural tendency to return to it, so that when other obstacles are removed, naturally light and heavy objects will begin to move towards their proper places. The potentiality is shown in earth and water by their relative heaviness, and in air and fire by their relative lightness, and it is this potentiality, together with the removal of obstacles, which is the cause of their motion: '. . . in all these cases the thing does not move itself, but it contains within itself the source of motion—not of moving something or of causing motion, but of suffering it.' [1]

There has been some discussion as to whether or not Aristotle is here asserting that 'proper place' exerts on bodies an attraction at a distance, and certainly his statements will bear more than one interpretation. On the one hand he uses language about the 'desire' of things to fulfil their potentiality, which suggests that he is thinking of the attraction of the earth for heavy bodies as being like the attraction of a pool of water for a thirsty animal—in both cases the tendency to move may be compulsive and unconscious, but the moving cause is in the moving body, not in the centre of attraction. On the other hand there are suggestions that the centre of the earth exerts an active force on heavy bodies at a distance, or rather that 'place' itself does so, for Aristotle maintains that if the whole earth were removed to the moon, earthy bodies would move not to it, but to the place it previously occupied.[2] Aristotle does not in any case believe that the attractive force exerted by proper place is mechanically explicable, for, in discussing the secretion of semen in *On the Generation of Animals* in terms of motion to its proper place he says: 'Each of the residues [e.g. semen] is carried to its proper place without the exertion of any

[1] *Physics*, 255*b*

[2] *Physics*, 208*b*, *On the Heavens*, 310. M. Jammer (*Concepts of Space*, Cambridge, Mass., 1954, p. 17) likens Aristotle's 'space' to a field of force. This conception attributed to Aristotle was criticised by the sixth-century commentator Philoponus on the grounds that space cannot possess any such inherent power (ibid., p. 55). The comment of Aquinas on this passage of the *Physics* is that although place has an influence on bodies, 'it is not shown from this that place has an attractive power, unless it is said to attract as an end' (*De Physico Auditu*, Naples, 1953, ed. Angeli and Pirotta, 794).

force from the breath and without compulsion by any other cause of that sort, although some people assert this, alleging that the sexual parts draw the residue like cupping-glasses.'[1] Here he is presumably denying that motion to the proper place can be accounted for by any mechanical theory of suction by the vaccuum or anti-peristasis.

Aristotle: the Principle of Action by Contact

In the case of unnatural motions of lifeless objects, and of living objects moving by compulsion and not by their own will, there must be an external mover, and moreover this mover must be in continuous contact with the moving body. This is stated as a generalisation from empirical facts, as is shown by the kinds of examples Aristotle gives, such as a team of men hauling a ship.[2] There is nowhere any argument to support the generalisation, and Aristotle seems to regard it as self-evident, for he repeatedly asserts that the moving body 'must be either in contact with or continuous with that which moves it . . . as we see to be universally the case',[3] or again, the mover 'is always together with that which is moved by it. . . . This is universally true whenever one thing is moved by another.'[4] Phenomena such as projectile motion, which only fit into this scheme with great difficulty, are not allowed, as we have seen, to overthrow the general principle that there must be a mover which acts by contact as long as the motion continues. Aristotle has, of course, given reasons for rejecting the void, but these are not conclusive for action-by-contact, because it is possible, although perhaps less natural, to conceive actions taking place at a distance even though the intervening space contains matter, for this matter might not affect, or be affected by, the action.

Aristotle continues his discussion by showing how the principle of action by contact applies to the various kinds of change. In the case of pure motion in space, or locomotion, 'the motion of things that are moved by something else must proceed in one of four ways: for there are four kinds of locomotion caused by something other than that which is in motion, viz. pulling, pushing, carrying and twirling. All forms of locomotion are reducible to these.'[5] He goes on to show that throwing, and all the processes of combination and separation such as packing, combing, inhaling, exhaling, spitting, and other motions of the human body, are all forms of pushing and pulling. Also, a thing carried must be pulled or pushed or twirled,

[1] op. cit., 737*b* [2] *Physics*, 250*a* [3] ibid., 242*b*
[4] ibid., 243*a* [5] ibid., 243*a*

and twirling is a combination of pulling and pushing—here he is probably thinking of twirling a stone on the end of a sling, when the sling has to be pulled towards the centre of the circle and at the same time pushed round in the circle. Thus all forms of locomotion are reducible to pulling or pushing. But in pushing and pulling the mover and moved are adjacent: 'So it is obvious that there is no intermediary between the mover and the moved in the case of locomotion.'[1] There is no attempt here to reduce pulling to pushing, but we are presumably to understand that since there is no void, pulling involves a motion of antiperistasis, and therefore the object moved is, in a sense, pushed from behind.

We see Aristotle here arguing from a list of phenomena in common experience, as he did in his discussion of primary qualities, and not from any *a priori* principles. But this is just the procedure by which he believes knowledge of scientific first principles is obtained. The method which he calls induction in the *Posterior Analytics* is that of seeing by intuition in a class of similar particulars the common features which constitute their essential nature, and then rising from this to the features common to various species which define their genus, and so on until we come to the necessary first principles of the science.[2] We have ascended by intuition which, Aristotle asserts, is *not less certain* than scientific demonstration; we can therefore descend again to particulars by syllogistic demonstration, and our conclusions will remain necessary. Thus Aristotle presumably regards the necessity for contact-action as intuitively given in sense-experience, and hence according to his own principles he is able to universalise it.

It would be easy to point out the obvious difficulties of this theory of intuition, but is it the whole explanation of his insistence upon action-by-contact in unnatural motions? There are certainly also remnants of a primitive kind of anthropocentric view in his assimilation of all locomotion to that which we can effect by manipulation, and Jammer suggests[3] that he is here following the view expressed by Plato in his statement that 'the definition of being is simply power',[4] and that therefore anything that has power to affect anything else has substantial existence. With Aristotle this becomes the dogma that power to act is spatially inseparable from substance, although it is not clear that this follows from the Platonic definition. Perhaps, however, we should take Aristotle not so much to be denying that bodies act upon one another at a

[1] ibid., 244*b*
[2] *Post. An.*, 99*b* ff.
[3] *Concepts of Force*, pp. 36, 30
[4] *Sophist*, 247*e*

distance, as asserting that where power is seen to be active, there is substance. His view of substance is more liberal than that of, for example, the atomists; and if we remember that his matter is ubiquitous because there is no void, and in itself qualityless and not presupposing any particular manifestation of qualities under any particular circumstances, we may take him to be saying no more than that matter is informed by power of some kind at places adjacent to a body in forced motion. This by itself does not commit him to any particular mechanism for the transmission of this power from agent to distant patient, and indeed the many kinds of immaterial emanations and substantial forms which were later postulated to account for it could rightly be claimed to be consistent with Aristotelian teaching. There is here an ambiguity which, as we shall see, runs right through the history of concepts of action at a distance, namely that if the medium of transmission is not defined too closely, the assertion that action must be continuous through the medium may be no more than a directive to look for some continuity of qualities no matter what, and the search is almost always successful, although the result may be quite unlike matter with the qualities we know from experience of gross bodies.

The medium which Aristotle resorts to is often air, as in his explanation of projectile motion. When he goes on, in the *Physics*, from locomotion to quantitative and qualitative change, he asserts that these also take place only when the cause and subject of the alteration are adjacent, and that this 'can be shown by going through the possible cases'.[1] He cites sensation, which is an alteration in living bodies: 'colour is continuous with the light and the light with the sight. And the same is true of hearing and smelling: for the primary mover in respect to the moved is air.'[2] Elsewhere[3] he maintains that sight would be impossible without body intermediate between the object and the eye. And in general, bodies only change when there is a spatially continuous set of causes: 'Fire heats not only when in contact, but also from a distance. For the fire heats the air, and the air—being by nature such as both to act and suffer action—heats the body.'[4]

There remains the fourth type of change, namely that of substance. This is the most radical, since it involves change of the essential qualities of a thing, those which make it what it is, as in the generation and death of living things, or the combination of elements in compound bodies. Qualitative change, or alteration,

[1] op. cit., 244*b*
[2] ibid., 245*a*
[3] *On the Soul*, 419; *On the Senses*, 440*a*
[4] *On Generation and Corruption*, 327*a*

on the other hand, involves only accidental qualities ; thus the greying of a man's hair is an alteration which does not change his substance, but death is a *transformation* in which his substance is dissolved into other substances. Aristotle mentions the question of contact in relation to substantial change in *On Generation and Corruption,* where he maintains, again, that the agent of change is always in contact with the patient.[1] But though this is necessary, it is never a sufficient condition. When substances undergo transformation, as in what we would call chemical change, or in the change of one of the four elements into another, mere mechanical contact of the ingredients is not sufficient, because according to Aristotle homogeneous substances are infinitely divisible, therefore they cannot be constituted by the mere juxtaposition of unchanged ' atoms ' of their ingredients, but every infinitesimal part of these ingredients must be transformed. He does not elaborate an alternative theory on the level of efficient causes, probably because he does not believe that the problem is amenable to that kind of solution, but it is discussed further in the *Metaphysics* in terms of formal causes. Here he argues that, for example, a man is not simply ' an animal ' plus ' two-footed ', but requires something more if the combination is to be actually a man.[2] What this something more is, he concludes, is the form or actual substance of man, so that the particular form ' man ' is essential to the combination, although it is not one of its elements, for if man is divided into elements, they will be found ultimately to be matter and the primary contraries, and the form which makes him man will have disappeared. One can appreciate Aristotle's problem by considering that one of his major concerns was with the biological phenomena of reproduction, and here it arises in an acute form : how is it possible that the elements which go to make up the embryo reproduce the specific form of the parents ? It seems to Aristotle that the reply must be that the form is somehow handed on, over and above the material elements, and he extends this conception to all kinds of combination or change of arrangement where new form appears. I shall not pursue his argument here, but mention it only to indicate that his notion of substantial change is by no means a simple mechanical one of mixture and separation like that of the atomists.

Aristotle : the Unmoved First Mover

Finally, we come to the mode of action of Aristotle's unmoved first mover,[3] which is not conceived to act by contact in the same

[1] ibid., 322*b* [2] *Metaphysics,* 1041*b*, 1043*b* [3] *Physics,* Book VIII

way as every other kind of efficient cause. Aristotle arrives at the necessity for an unmoved first mover by the logic of his principle that every motion must have a moving cause : everything, therefore, that moves is moved either by itself or by something else. If it has a source of motion in itself, that source, in animals and plants, is a soul which may be called an unmoved first mover in a limited sense, although a soul is also moved accidentally by physical causes such as respiration and digestion. But souls are not eternal, and Aristotle concludes that their successive generation and destruction require the existence of an unmoved first mover which is eternal, and which must therefore be different in kind from an organic soul. This line of thought reinforces an argument from locomotion, which is as follows : everything that is in forced local motion must be moved by something else, which may or may not be itself capable of motion. If it is incapable of motion it is the unmoved mover of the motion. If it is capable of motion it must be in motion, because it must remain in contact with the moving body. So the mover must be moved by a third thing, to which the same argument applies. Now Aristotle argues that there cannot be an infinite series of movers, since that would imply an infinite quantity of motion, which is impossible. There must therefore be a first mover in the series which is itself unmoved.

Aristotle now sets out to show that locomotion is the primary type of motion.[1] All other types of change presuppose it ; for instance when the qualities of something are altered, the cause of the alteration must approach or recede, and again all change of size involves change of place. Furthermore, locomotion is continuous, and it may be eternal and exhibit the greatest perfection possible for anything that changes, for uniform circular motion is eternal, and describes a perfect figure. An example of such motion exists in the world in the rotation of the heavenly spheres around the earth, and therefore the most fitting candidate for the unmoved first mover of all things is that which moves the spheres. This is also the ultimate efficient cause of change and motion and even generation and destruction on earth, because it is the motion of the sun which causes seasonal variations, as Aristotle explains in *On Generation and Corruption*[2] and the *Meteorology*.[3] In a sense, then, Aristotle is here asserting that even the souls of animate things are not unmoved movers in the chain of efficient causation, but that this chain can always be traced back to the rotation of the heavenly bodies.

[1] *Physics*, 260*a* [2] 336*a* [3] 346*b*

Does this unmoved first mover act by contact? Aristotle is determined to preserve his principle of contiguity at all costs, even for the first mover which he has proved to have no parts and no magnitude, and he does so by appealing to the metaphorical sense of 'touch'. Usually in our experience movers are in contact with the bodies moved, and move with them. There is reciprocal touching of mover and moved, and reciprocal action between them. But in the case of an unmoved mover 'it may touch the "moved" and yet itself be touched by nothing—for we say sometimes that the man who grieves us "touches" us, but not that we "touch" him.'[1]

This language prepares us for the use of human analogies in accounting for the mode of action of the first mover. It acts as mover by being an object of desire: 'by being loved, while all other things that act as movers do so by being moved'.[2] The heavenly spheres, then, move by Eros, by desire to achieve as far as possible the perfect good of the unmoved mover, and this they do by executing the most perfect motion, namely, uniform circular motion. Love, in fact, makes the world go round. This is not the only occasion in Aristotle's philosophy on which he introduces love or desire as an explanatory cause; it is one of his fundamental analogies for final causation. He speaks, as he says 'metaphorically', of the potentiality of matter for acquiring form as desire: matter desires form 'as the female desires the male, and the ugly the beautiful',[3] and all motion to the natural place is actuated by the desire of the potential to become actual, and therefore involves, on the level of final causation, attraction at a distance.

So although organic action on earth is subordinated to locomotion and action by contact as regards its efficient causation, organic and human analogies come into their own in the discussion of final causes, and of the ultimate efficient and final cause of all things, where the two kinds of causes are identified in the unmoved mover, Aristotle's God. And it has to be remembered that for Aristotle, as for Plato, explanation in terms of efficient causation alone is less than half the story, and that their main interest was not in extending mechanical explanations, of which there seem to have been no lack in fourth-century Athens, but in showing how Nature strives after the good, and how rationality is exhibited in all things.

In the discussions of mechanical causation which Aristotle does give, he is content to assume that all things are what they seem, and to generalise from what some things seem to what all things are.

[1] *On Generation and Corruption*, 323*a* [2] *Metaphysics*, 1072*b* [3] *Physics*, 192*a*

If bodies appear to our observation to be homogeneous and indefinitely divisible, then homogeneous and divisible they are, and these are the properties which have to be explained. If it appears from common experience that all kinds of change involve locomotion, then locomotion must be a universally indispensable aspect of change, and if all the most obvious processes that we experience seem to involve a causal agent in contact with the patient, then this too must always be the case, even if, as in the cases of vision and the flight of arrows, the mechanism is not immediately obvious.

Aristotle's principle that agent must be in contact with patient, does not, then, spring from any carefully thought-out philosophical argument, and does not generate any particularly interesting philosophical problem. Neither does he contribute much to the kind of detailed mechanical explanation attempted by the Hippocratics. What is important and influential in his *Physics* is his classification of motion and change in which locomotion has a prominent place, indeed the supreme place, and in which at least some of the ingredients of mechanical explanation are present. He ensured that no serious student of his works could overlook the simple motions of bodies, and when, in the twelfth century, the full Aristotelian corpus was recovered in the West, it was the attempt to apply Aristotelian principles to particular motions like those of falling bodies and spinning tops that led in the fourteenth century to the first steps towards modern theoretical dynamics. But it is a long way from theoretical dynamics to a fully mechanical theory of matter, and here Aristotle's influence was indecisive. By trying to do full justice to organic as well as mechanical change he could not provide inspiration for those in the seventeenth century who, mistrusting any but ' clear and distinct ideas ', wished to carry mechanical explanation as far as it would go, even at the expense of blurring the distinction between organic and mechanical. It was necessary to retreat into the narrow world of pure mechanism before its limitations could be fully appreciated, and this retreat involved the abandonment of Aristotle's theory of matter and motion as a conceptual framework for science, even apart from the fact that philosophical criticism soon began to expose its inadequacies.

Chapter IV

THE GREEK INHERITANCE

The Primitive Analogies in Medieval Belief

It has become something of a platitude to speak of the seventeenth century as a watershed in the history of thought and ideas, more important even than the Renaissance and Reformation; but when one studies the literature of the period in science and philosophy one still cannot fail to be astonished by the facts. In 1600 natural philosophers are discussing the same problems that had exercised the scientific minds of Greece, using the same arguments and the same categories of thought, quoting the same authorities, describing the same simple experiments. By 1700, most of these problems had either been solved, often as trivial special cases of newly formulated physical laws, or they had been shelved as lacking in interest, or, what often appeared to be the same thing, as being too complex for discussion in the existing state of science. Not one, but two revolutions had taken place, both comparable in importance with the rise of Greek philosophy itself from primitive myths. Not only had every type of explanation known to the Greeks, with the exception of atomism, been entirely discredited, but the new orthodoxy, an alliance of Cartesianism and atomism, was already in decline. Newton had invented a new analogue in his mathematical theory of central forces, and a new analogue is rarer in the history of science than new theories or new methods. By the end of the century a particular type of subject-matter and a particular method of investigation were taking charge of science, and critics, though vocal, were ineffective.

In spite of much historical discussion in recent years, the century continues to wear an air of paradox. One can read most of the work of Galileo and Huygens, Descartes's *Principles of Philosophy*, and Newton's *Principia*, and feel that their concern to clarify, to mechanise or mathematise, the physical world, is the same as our concern, and that their methods, though clumsy, are in the modern manner. But when one turns to Descartes's *Passions of the Soul* or to the works of such fathers of modern science as Francis Bacon or Pierre Gassend or Robert Boyle, it is apparent that this is still an age in which it is exceptional not to believe in alchemy and astrology, magic and witchcraft, and it seems less paradoxical that an experimental scientist such as Boyle can write long tracts with titles like 'The Excellency of Theology' and 'The Christian Virtuoso', or that Newton can devote years to interpretation of the Apocrypha.

The natural philosophy which the seventeenth century inherited from classical, medieval, and Renaissance thought was the orthodoxy against which the new science revolted, and its characteristics have some importance for an understanding of the results of the revolution. To speak very broadly, its chief concern was to understand man, and nature in relation to man, rather than nature as an objective and independent existent in which man is subordinate and dependent. Physical processes, when they were considered at all, were used as incidental illustrations of general metaphysical concepts. It was more important to understand how form inhered in matter and how the soul was related to the body than to have accurate theories of the flight of projectiles or the action of the vacuum, and because metaphysical explanations tend to be comprehensive rather than specialised and detailed, anthropomorphic and organic analogies flourished. On less sophisticated levels of thought, too, the inevitable tendency was for the primitive analogies to remain the framework within which nature was interpreted, and during the long period between the decline of Greek culture and the rise of seventeenth-century science, all but a very small minority were dominated by views of nature in which pre-Socratic modes of thought were only slightly modified by the Platonic and Aristotelian traditions.

For example, the ancient conception of air and fire as bearers of vital and rational activity was universal. There was no clear-cut distinction between the 'spiritual' and 'material', but rather the universe was seen as a hierarchy of substances, ascending from gross matter, through vaporous and gaseous matter, to the pure, immaterial form or intelligence which was ascribed to God and the angels alone. This hierarchic conception stemming from Plato and Aristotle had been modified in the Stoic tradition by rejection of the notion of pure form : for the Stoics everything was corporeal and spatial, and all spiritual being and activity were referred to *pneuma*, or gaseous matter. God, the soul, the mind, were described indifferently as fire or breath, and attributes such as virtue, knowledge, emotions, skills, and actions were ascribed to the properties of subtle air-currents pervading gross bodies. The idea that qualities are substantial coverings of the bodies in which they inhere seems to be very old : Professor Onians quotes examples from Homeric literature, which seem to have been more than metaphorical, of the use of ' cloud ' covering men stricken with grief, the ' cloak ' of sleep or death, and the substantial ' mist ' of darkness.[1] This way of conceiving the inherence of forms in matter is more primitive and perhaps more easily understood than the sophistications of Aristotelian and Neo-Platonic philosophy, and helps to explain the tendency of the Aristotelian ' forms ', which for Aristotle were inseparable from matter, to become the independent ' substantial forms ' or qualities of medieval speculation.

Some of the Stoic writers, however, began to describe the action of pneuma in mechanical terms. They discuss propagation in a continuum, using the analogy of water waves to suggest the spherical propagation of sound in air, and Chrysippus is reported as giving an account of sight in terms of a tension in the air between eye and object, so that the object is felt as with a stick, an analogy later used by Descartes. Professor Sambursky[2] sees in this theory the continuum counterpart of atomism : instead of ascribing all qualities to the shape and arrangement of atoms, the Stoics ascribe them to tension, pressure, and wave-motion in the continuous and pervasive pneuma. Thus the first suggestions towards both of the fundamental contact theories of modern physics were made within Greek science.

Pneuma or *spiritus* did not, however, retain in general the mechanical connotations introduced by these Stoic writers. It was still used in succeeding centuries with a multiplicity of meanings, to denote the Holy Spirit of the New Testament, or the soul or souls which preside over the human body, or any type of volatile substance such as ' spirits of wine ', or ' spirits of mercury ' produced by alchemical distillations. The ambiguity was not seen as a source of confusion, but merely as an expression of continuity in the hierarchy of natural substances. Man in particular was seen as the microcosm of the whole ; he is a body informed by spirits which are at the same time the air he breathes and the divine reason which he shares with God. Medical practice and theory were dominated by the concept of spirits or subtle fluids which were responsible for all vital and rational activities of men, and which had to be kept in harmonious balance if health was to be maintained.

The primitive lack of distinction between animate and inanimate was still apparent in such universal beliefs as that precious stones

[1] *Origins of European Thought*, p. 422

[2] *The Physical World of the Greeks*, London, 1956, pp. 134–40

'grow' in the earth, taking nourishment from the rocks and soil, and that organisms are spontaneously generated from putrefying material. The attempts of the alchemists to transmute base metals into gold were always described as 'growing' gold from its own seed, a process which needed warmth and the 'breath of life' (pneuma), just as did the growth of vegetable matter in the earth. Stoic philosophy too, in spite of its denial of immaterial substance, was by no means a materialism in the modern sense, for its 'corporeal being', which is all that is, is not inanimate matter: the 'primary being' was likened to seed; and pneuma and fire, as efficient causes, were living and even divine. Animistic ideas such as these retained their hold throughout the Middle Ages and Renaissance, and were the presuppositions of the almost universal belief in alchemy, astrology, and magic.

The Emanation Theories

While it is true that physical processes were chiefly of interest as illustrations of metaphysical principles, and as analogues for spiritual activity, there is at least one example which shows that this very interest eventually led to a study of the details of the physical process itself. During the Hellenistic and medieval periods an analogy between the propagation of light and the activity of soul, or reason, was developed with profit to both physics and metaphysics, and the notion of 'emanations', which began as a concept of Stoic and Neo-Platonic metaphysics, became by the seventeenth century a theory of the propagation of light, heat, sound, and magnetic and electric attractions. First, metaphysics borrowed her language from physical processes: in Stoicism the diffusion of reason, law, providence, and destiny throughout the universe was spoken of in terms of radiation from fire, or as pervasive air-currents flowing from the ruling part of the world and reaching every part of it. The same image appeared in Neo-Platonism, although the general standpoint of that philosophy was very different. There, 'emanation' was a technical term for the way in which lower levels of spiritual being proceed from higher; the Nous or Divine Mind from the One, the World Soul from the Nous, and Nature from the World Soul; and the process was described principally in terms of light-symbolism, as in much religious imagery of all ages. Emanation was said to resemble the radiation of light or heat from the sun, coolness from snow, or perfume from its source; the light of Nous was said to illumine the Soul, and when Nous contemplates the One, it is as if the eye looks at the light itself, not at objects illuminated by it; souls were said to 'go out in rays', carrying the power to initiate action where they are present; Nature, the principle of life and growth, was said to mingle with bodies as light with air, or fire with the warmed body.

It was a reading of the Platonists that helped Augustine to conceive an immaterial idea of God, and like them he expresses the idea by analogies with light and mental activities:

> 'For that as the body of this air, which is above the earth, hindereth not the light of the sun from passing through it, penetrating it, not by bursting or by cutting, but by filling it wholly: so I thought the body not of heaven, air, and sea only, but of the earth too, pervious to Thee, so that in all its parts, the greatest as the smallest, it should admit Thy presence, by a secret inspiration, within and without, directing all things which Thou hast created.' [1]

Since in Greek thought it is soul which initiates change and movement, and in Augustinian theology it is the immanence of God, and since soul or spirit is here conceived like light, no question of action at a distance arises. Soul acts where it is present, and its causal action is propagated in rays as the sun propagates light and heat. From this notion it is a short step to the statement of Robert Grosseteste in the thirteenth century that the cause of movement and change and the basis of extension in three dimensions *is* light, which carries active power and 'unfolds' itself in space. And Giordano Bruno, in the Stoic tradition, compares the diffusion of airy souls and spirits throughout the universe to the diffusion of light and sounds, for both can exert influence at a distance without transmission of a body and without mutual interference between different sources.

So far, the properties of light are used to express the properties of soul, but the primary importance of light in Grosseteste's metaphysics led in the thirteenth and succeeding centuries to an examination of the laws of the propagation itself, and hence to the foundation of geometrical optics. But Francis Bacon five hundred years later still has cause to complain that while the laws of radiation and perspective have been studied, the 'forms of light', by which he means its physical causes, have been neglected:

> '. . . the manner in which light and its causes are handled in Physics is somewhat superstitious, as if it were a thing half way

[1] *Confessions*, Book VII, 2

> between things divine and things natural ; insomuch that some of the Platonists have made it older than matter itself',

which, he says, is contrary to Holy Scripture.[1] And Gilbert complains that ' insensible rays ' are used to explain everything which is not understood, by ' men who . . . in very many cases often wretchedly misuse rays, which were first introduced in the natural sciences by the mathematicians ',[2] that is, in geometrical optics.

Multiplication of Species

One of the culprits in the eyes of Bacon and Gilbert was no doubt Francis's namesake Roger Bacon, who in the thirteenth century had attempted to discuss the physical causes of radiation. His theory was a development of that of Aristotle, and is an attempt to wrestle with the relation between the radiation, which was not regarded as material, and the medium, which was, and which was necessary to preserve the spatial continuity of causes required by Aristotle. In most of the Greek theories there was no difficulty about preserving action by contact in radiation, for light and heat were thought of as themselves material substances, sometimes equated with elementary fire. For the atomists, they were corporeal emissions, and in the theory of perception of Lucretius [3] physical images or replicas of an object were supposed to peel off its surface and pass to the eye or other sense organ. Aristotle, on the other hand, denied that light is material,[4] arguing that since space is a plenum, this would involve two bodies existing in the same place at the same time. He therefore made light a *potentiality* of the transparent plenum, which is actualised by fire or by the sun, so that the plenum thus modified can act upon the eye. Objects at a distance are perceived by the senses through the medium of air, upon which their forms are stamped in a series of impressions, just as a series of objects transmit an impulse, or as water is disturbed at a great distance by dipping an object into it. The senses therefore receive the forms of objects without the matter, as wax receives the impression of a signet-ring, but without the metal of the ring.[5] The number of different metaphors which Aristotle uses here to explain the transmission, indicates that he is not interested in its exact nature—whether it is a series of impulses (like longitudinal waves), or like ripples in water (transverse waves), or like the communication

[1] *Advancement of Learning*, IV, iii ; *Works*, ed. Ellis and Spedding, IV, p. 403
[2] *De Magnete*, London, 1600, p. 64 (English trans. S. P. Thompson, London, 1900)
[3] *De Rerum Natura*, Book IV [4] *On the Soul*, 418*b* [5] ibid., 424*a*, 434*b*, 435*a*

of pressure through wax—and the whole conception comes nearer to the modern wave theory than does the alternative explanation in terms of corpuscles only in the sense that the 'form' of light is transmitted without the passage of matter.

This theory is elaborated by Roger Bacon in his doctrine of 'multiplication of species', which he uses to explain not only radiation of light and heat, but also transmission of sound, magnetic attraction and astral influence. He ascribes these transmissions to the power of a radiating body to multiply its qualities, or 'species', in the surrounding space, and convey them by radial propagation to distant bodies. He argues that the species cannot be material parts of the radiator, because if they were, the sun, for example, would gradually wear away. On the other hand the species must be 'of the nature of substance', otherwise they would not be able to produce effects in substance. So, since the space between radiator and receiver is a plenum, and two bodies cannot occupy the same space, the substance of the species must be that of the medium itself, and the transmission of the species consists in the momentary modification of successive parts of the medium : 'it does not move from place to place, but is continually generated anew, just as the shadow does not move, but is renewed in various places'.[1] The transmission is more like the pulsation of sound than the flow of water : 'it is not made by a flux from the luminous body, but by calling forth the potentiality of the material air'.[2] The propagation takes time, for no finite change takes place in an instant, as Aristotle himself remarks in the *Physics*.[3]

So far Bacon's theory follows Aristotle's very closely, and he is not attempting any more than Aristotle to describe a mechanism by which the medium transmits the form. It is therefore misleading to interpret Bacon's doctrine in terms of modern theories of transmission in a medium, and this becomes even more obvious when he goes on to generalise Aristotle's account and give it wider application. The medium, he says, is passive ; it is the power of the agent which is able to assimilate a weaker substance to its own nature, or to impose its form on another substance. Bacon thinks that all substance has this power in some degree, and so explains magnetic attraction as due to the power of the magnet to endow iron with its own motive virtue by the passage of a 'species magnetica', which causes it to move towards the magnet. There is here no suggestion

[1] *De Multiplicatione Specierum*, appended to the *Opus Maius*, ed. Bridges, Oxford, 1897, II, p. 504

[2] *Opus Maius*, V, I, IX ; ibid., p. 72

[3] In the course of his discussion of continuity and Zeno's paradoxes in Book VI.

that the magnetic species acts mechanically,[1] it simply transmits a form to the iron, and since the form is that of a magnet which has the power of self-movement, this power appears in the iron. How far the whole conception is from any modern theory is indicated by the fact that Bacon elaborates it as a theory of *generation*, in which the ability of organic beings to ' propagate their species ' is the major example of the universal tendency of substances to assimilate others to their own nature. Thus a magnet might be said to generate its offspring in iron, or a perceived object to generate its likeness in the mind, as the male generates its kind in the female.[2]

The terminology of the doctrine of multiplication of species was used by several later medieval writers, by the sixteenth-century writer Fracastoro, and by Francis Bacon, to whom we shall return later. But unlike Roger Bacon, they did not always distinguish between the transmission of species and corporeal emanations, and the the theory became confused with atomism, particularly the atomism of Lucretius, and by doctrines of the attraction of like for like. For example Fracastoro [3] postulated sympathies and antipathies between bodies, which act by corpuscles emitted by some and entering through the pores of others. He says that perception is caused by ' species ' or images entering the percipient, and puts forward a germ theory of disease according to which ' seminaria ' are given off during the putrefaction caused by the disease, and propagate their like through the air, causing putrefaction in that one of the bodily humours to which they are most similar.

Before the sixteenth century, atomism itself, and the idea of mechanical explanation in general, found little support. Although action at a distance was almost universally rejected, the doctrine of multiplication of species shows that action by contact can be described in terms of analogues other than mechanism. Aristotle's influence also told explicitly against atomism ; and there were other reasons for its neglect, for, as expounded by Epicurus and Lucretius, it had become associated with atheism, and furthermore it had not shown itself capable of providing satisfactory scientific explanations. Its

[1] Bacon explicitly distinguishes the attraction of magnet for iron from that of the cupping-glass : see P. Duhem, ' Roger Bacon et l'horreur du vide ' in *Roger Bacon Essays*, ed. A. G. Little, Oxford, 1914, p. 257.

[2] Compare the account of the thirteenth-century writer John of St Amand : ' But then there is the more difficult question how the magnet attracts iron, and also the magnet the magnet, since nothing can evaporate from it. . . . I say that it does it by multiplying its like and, without any evaporation, exciting the active power which exists incomplete in iron, which is born to be completed by the form of the magnet ' (L. Thorndike, ' John of St Amand on the Magnet ', *Isis*, xxxvi, 1945–6, p. 156).

[3] *De Sympathia et Antipathia Rerum, De Contagionibus et Contagiosis Morbis*, Opera Omnia, 2nd ed., Venice, 1574

scientific weaknesses had long been recognised : for example, in the generation following Aristotle, Strato had found it impossible to conceive how solid bodies cohere if they consist of discrete atoms related only by impact and pressure, and had rejected the hypothesis of atomic hooks as undemonstrable. But arguments such as this did not prevent the revival of atomism in the sixteenth and seventeenth centuries, and it was probably the general lack of interest in mechanical theories before that time which was mainly responsible for its eclipse. In this connection one must remember the influence of Galen as well as that of Aristotle. In Greece it had been the medical writers who had first explored the possibility of mechanical explanations. But Galen had found good reasons for rejecting most of the mechanical theories of his contemporaries in medicine, and it was Galen, together with Aristotle, who was most studied in the medieval medical schools. So although there was considerable discussion of scientific method in these schools, it is not surprising that there was no stimulus towards the development of mechanical explanations.

The Theory of Vacuum-suction

Atomism was in fact the only theory of Greek physics to be retained in the new science, and we shall explore its revival in the next chapter. But it was the discussion and explicit refutation of other kinds of explanatory analogies that provided the starting point of seventeenth-century physics, and among those inherited from the Greeks one of the most important for the explanation of attractions and repulsions was the doctrine of suction by the vacuum.

The *horror vacui* was widely discussed in the Middle Ages in relation to the *Pneumatica* of Hero of Alexandria and a work of Philo of Byzantium. These were known in the twelfth century, and the *Pneumatica* was read in many Latin and vernacular translations in the sixteenth century. Both works describe experiments with pumps and siphons, in which artificial vacua are produced, and the *Pneumatica* contains an introduction in which Hero draws some conclusions regarding the nature of air and the vacuum. He argues that the material nature of air, and the fact that force is needed to produce a vacuum, show that continuous vacuum does not exist naturally, but, on the other hand, air cannot be a plenum because it is compressible. This argument shows that he adopts the atomist view of 'conservation' of matter, rather than the Aristotelian. He concludes that there must be small void spaces between the particles of air, like the spaces between grains of sand, into which the particles are pressed under compression, and which are enlarged when the air expands. When the force of compression is withdrawn the air returns to its original volume by ' the elasticity of its particles as is the case with horn shavings and sponge ',[1] and when the force of rarefaction is withdrawn the particles come together again by the action of the intervening vacua, ' for bodies will have a rapid motion through a vacuum, where there is nothing to obstruct or repel them, until they are in contact '. Hero seems here to be seeking some kind of mechanical explanation for the ' suction ' of the vacuum in terms of the possibility that particles will come into contact again (and then stick together by means of their hooks ?) as a result of their random motion through the interspersed void. He mentions that this is also the explanation of the action of cupping-glasses, but later he has an inconsistent description of the air escaping ' through the pores of the sides of the glass ',[2] which suggests an explanation in terms of circular thrust. His analogies of shavings and sponge for the elasticity which restores shape after compression seems to make some form of elasticity an ultimate property of the particles, and are not therefore consistent with a Democritan atomism in which the atoms always retain their shape as well as their size.

Hero adds further arguments for the existence of interspersed vacua: for instance, if water did not contain small void holes, light and heat would not be able to penetrate, in fact the rays of the sun would cause a full vessel of water to overflow ; and it is by means of the interspersed vacua that wine is able to spread uniformly through water.[3]

Throughout the Middle Ages observations and arguments such as these were taken as proof that ' nature abhors the void ' ; that a continuous vacuum can only be produced by force ; and that the facts of cohesion and elasticity can be explained by interspersed vacua. Galileo discusses the adequacy of this theory in his dialogue *Two New Sciences*, first published in 1638.[4] It was known to his contemporaries (he says he learnt it from a workman) that no pump will raise water above a height of about thirty feet, and this, Galileo thinks, must be a measure of the greatest force a vacuum can exert. But it is clearly not sufficient to account by itself for the

[1] op. cit., trans. Woodcroft, London, 1851, p. 3. Hero's dates are uncertain, but he is thought to have lived in the first century A.D.

[2] ibid., p. 6 ; cf. the account of how it is possible to suck wine upwards, ibid., p. 13.

[3] Many of Hero's arguments are found in Strato, a successor of Aristotle at the Lyceum. cf. G. Rodier, *La Physique de Straton*, Paris, 1890.

[4] op. cit., trans. Crew and Salvio, New York, 1914, pp. 16ff.

cohesive force of many solids. On the other hand, it is not plausible to explain cohesion by the presence of some sort of material glue between the particles, because such a glue would surely be destroyed by extreme heat, and yet when gold or glass are melted and then cooled again they retain as great cohesion as before. And if glue be the explanation, what holds together the particles of glue? Galileo can only suggest that the force of the vacuum does provide the explanation, and that the combined force of a great, perhaps infinite, number of small vacua somehow exceeds the limiting force of the artificial continuous vacuum, and is then sufficient to explain cohesion.

All these arguments were, however, overthrown by Galileo's pupil, Torricelli, the inventor of the barometer. By discovering that the vacuum in the closed arm of a U-tube can support in this arm a column of water 30 feet high, or a column of mercury about 26 inches high, and can maintain itself without the exertion of any force, he showed that the cause of the ' force of the vacuum ' is not an internal attractive power, but the external weight of the atmosphere. This is the only natural resistance to the production of a vacuum. In a letter of 1644 Torricelli declares ' I have endeavoured to explain by this principle [the weight of the atmosphere] all sorts of repugnances which are felt in the various effects attributed to vacuum, and I have not yet found any with which I cannot deal successfully.' [1]

The motions previously ascribed to ' suction by the vacuum are then, results of external impulsion, not attraction, and this result lent support to the mechanical theories of Descartes and the atomists, who would allow action to be communicated only by impulse or pressure. With regard to cohesion and similar effects such as those of capillarity and surface tension, the barometric experiments suggested that the explanation might be found in terms of external pressure by an aether of subtle particles. This was the theory put forward in various forms by Descartes, Malebranche, James Bernoulli and Huygens, while the pure atomists had to be content with the traditional explanation of atomic hooks.

The Greek mechanists had sought to eliminate the attraction of the vacuum by reducing it to a circular thrust or antiperistasis. The word ' antiperistasis ' was also used in a rather different sense in Aristotle's *Meteorology* [2] and in the pseudo-Aristotelian *Problems*,[3] to describe certain processes in which it seems as if two contraries, such as heat and cold, when placed together, tend to intensify one another, or flee from one another. For instance it is observed by

[1] *Opere*, III, Faenza, 1919, p. 188 [2] 347*b*, 348*b*, 382*b* [3] 867*b* et al.

several Greek writers that caves and the water in wells are colder in summer than in winter, that hail is commoner in spring and autumn than in winter, that cold can produce the sensation of burning, and so on. These things are explained by the mutual repulsion of heat and cold, so that, for example, in warmer weather the surrounding heat of the earth causes cold to shrink into the lower parts of the earth away from the heat, and hence to be intensified. The theory assumes that heat and cold are both substances which may be intensified by becoming denser. These explanations are repeated by many sixteenth- and seventeenth-century scientific writers, and heat and cold are not the only qualities said to repel by mutual antiperistasis : Boyle mentions a theory that the tendency of water to form spherical drops is due to the flight of wetness from dryness, which causes water to concentrate within the smallest possible surface area.[1]

It is not clear whether there is any real connection between the two theories of Aristotle which go by the name ' antiperistasis ',[2] but in any case the seventeenth-century writers had no need of a theory of antiperistasis to explain projectile motion, since a theory of internal impetus or of inertia was by then almost universally accepted,[3] but the theory of repulsion of contrary qualities remains, and 'antiperistasis' is used by them exclusively in this sense. Francis Bacon still appears to believe that in physics the energy with which a principle acts is often increased by the antiperistasis of its opposite, and a metaphorical sense of the word passed into the English language to denote the well-known phenomenon that the more opposition one encounters, the more obstinate one becomes.[4] In physics, the doctrine is attacked by Boyle, who argues that contrary qualities do not in fact intensify each other, they destroy each other,

[1] 'An Examen of Antiperistasis' (1665), *Works*, ed. Birch, II, p. 663

[2] H. D. P. Lee, in a note to his translation of the *Meteorology* (Loeb. ed., p. 82) detects certain ambiguities in some places, where the meaning hovers between (1) ' mutual repulsion or compression ' (the theory of the flight of contrary qualities), and (2) ' mutual replacement ' (the theory of projectiles and magnetic effluvia). cf. Aristotle, *On Sleep*, 456*b*–458*a*, where the meaning is sometimes (1) that cold *concentrates* heat within the body, as a more powerful force surrounds and concentrates a weaker, and hence intensifies the heat ; and is sometimes (2) associated with the convection effect when hot vapour rises in the body, is cooled by the brain, and descends again, as moisture is evaporated by the sun, cooled, and descends. The latter process suggests the circular-thrust theory, but the two accounts are inconsistent, for in (1) the effect would be to make the lower parts of the body *hotter* and intensify the temperature difference, and in (2) they would become *colder* and the temperature would be equalised.

[3] Although Hobbes is still to be found in 1655 describing projectile motion in terms of circular thrust of air (*De Corpore*, III, xxii, 12 ; *Works*, ed. Molesworth, I, p. 343).

[4] The word is described as archaic in the O.E.D. and the examples given there are from literary writings where it is used metaphorically. Its literal use in seventeenth-century scientific works is much more common.

as in the case of heat and cold, and in any case the theory is an anthropomorphic one, transforming 'physical agents into moral ones'.[1] All the phenomena in which repulsion by antiperistasis is alleged to take place can, he thinks, be understood more easily in terms of the motions of corpuscles.

Gilbert's 'De Magnete'

The strong and persisting influence in the early seventeenth century of classical and Renaissance analogues for attraction, and particularly that of the tendency of like bodies to unite, is illustrated by what William Gilbert has to say in the *De Magnete*, published in 1600, about various kinds of physical action, and particularly attraction and repulsion at a distance. Gilbert is often thought of solely as an experimenter, and it is true that he was one of the first to design experiments specifically to test theories, but, as has often happened in the history of science, his theoretical ideas have been forgotten, because of a mistaken view that a theory which is later superseded is not interesting, whereas carefully performed experiments are part of the permanent structure of science.[2] But in order to appreciate the genius of the seventeenth century fully, we must know what the new theories replaced, and why the old ones required replacing, and, as we shall see, experimentation alone was not sufficient in Gilbert's case to wean him from traditional theoretical ideas, and is not sufficient in general to explain why radical changes of theoretical framework take place.

Gilbert begins by summarising the opinions of the ancients and of his immediate predecessors, the Renaissance writers, regarding the attractive powers of electric bodies and of the lodestone. He mentions various similes for attraction : in the Orphic hymns it is stated that 'iron is attracted by lodestone as the bride to the arms of her espoused';[3] Aristotle thought the magnet is ensouled; among Renaissance writers, Ficinus, Paracelsus and others ascribe magnetism to the power of the stars, and the compass property in particular to the stars near the pole of the heavens 'even as plants follow the sun, as Heliotrope does'[4]; Cornelius Gemma holds that the lodestone attracts by insensible rays 'to which opinion he conjoins a story of a sucking fish and another about an antelope'; Baptista Porta speaks of a combat between iron and stone in the lodestone : 'the iron, that it may not be subdued by the stone,

[1] *Works*, II, p. 664
[2] cf. my 'Gilbert and the Historians', *B.J.P.S.*, 1960, XI, pp. 1, 130
[3] *De Magnete*, p. 61 [4] ibid., p. 6

desires the force and company of iron, that being not able to resist alone, it may be able by more help to defend itself ',[1] thus lodestone draws iron or another lodestone for the sake of its iron, but does not draw stone alone. Cardan explains the attraction of amber as a dry substance imbibing a fatty humour, but Gilbert objects to this that there is no growth of the amber, nor diminution in the attracted body. Other writers speak merely of attraction deriving from the 'essential nature' of the lodestone, that it moves iron towards its 'perfection' as bodies move towards the earth, that 'the iron is borne by a wonderful yearning . . . in uniting itself with its own principle.' Gilbert has some respect for the opinion of Aquinas, who suggests that the lodestone gives a certain quality to the iron.[2] 'With his divine and clear intellect,' says Gilbert, he 'would have published much more, had he been conversant with magnetic experiments.' [3]

To the traditional theory of the attraction of similars Gilbert makes the obvious objections. Similars do not attract each other 'as stone stone, flesh flesh, nor aught else outside the class of magnetics and electrics'.[4] On the other hand, all types of bodies are drawn by amber. It is true that the lodestone draws lodestone, and excited iron draws iron, but it is wrong to use this as many physicians do as a proof of Galen's theory that purgative drugs and medicaments for drawing out poison act by similarity of substance. The action of drugs is quite different from that of magnetic bodies.

It is, however, to the mechanical explanations of the Greeks that Gilbert devotes most attention. He disagrees with them, but thinks them worthy of explicit refutation, sometimes by means of well-designed experiments. He mentions the theory of the Hippocratics that attractions take place by heat, and proves by experiments that the theory is groundless. If amber is merely warmed by fire, or even forced into flame, it does not attract straws, but only when it is rubbed. Many other bodies do not attract at all, whether warmed by fire or by friction, although a flaming torch or a glowing iron or coal do draw and consume air. The attraction of the cupping-glass is not due to heat, but to rarefaction of the air inside it. Neither does the sun attract humours from the earth, but only rarefies them and turns them into air, which rises above the denser air around. Again, various theories of the atomists imply that electric and magnetic attractions take place by the motion of air

[1] ibid., p. 63. The marvellous story of the adhesive powers of the echeneis, or sucking fish, is to be found in Pliny (*Naturalis Historia*, XXXII, i).
[2] *De Magnete*, pp. 63, 64 [3] ibid., p. 3 [4] ibid., p. 50

and resulting suction. In the case of magnetism Gilbert refutes this by showing that nothing corporeal passes between the lodestone and the iron ; and in the case of electric attraction, he refutes it by an experiment with a candle flame which is not disturbed by the presence of a strongly attracting piece of amber, as it would be if the amber produced a current of air. Plutarch and others had suggested that the amber emits a thin effluvium which rarefies the atmosphere in its neighbourhood so that other bodies are propelled towards it by denser air. But Gilbert objects that in this case one would expect a little time to elapse before the motion begins, that the motion would slow down as the distance between the bodies decreased, that a body would not be drawn upwards by this force, and that the power to attract would disappear soon after the rubbing. None of these effects are observed to occur : motion begins at once, is quicker the nearer the bodies are to each other, and an electric held over a light body a considerable time after it has been rubbed retains its power to attract. Finally, he says, the opinions of the Aristotelians who ascribe magnetism to ' the four elements and the prime qualities, we relinquish to the moths and the worms '.[1]

Gilbert's summaries of the theories of his predecessors indicate the diversity of speculation which was current at the beginning of the century, and his own comments upon them are good examples of the method of hypothesis and experiment which was soon to carry all before it. The theories which he himself suggests show the same careful attention to experiment, and show equally that the time had not yet come for radical changes in types of hypotheses, for Gilbert goes on to explain electric attraction in terms of effluvia and a natural tendency of all things to unite, and magnetism in animistic categories more ancient still.

Having demonstrated that electric attraction is not due to the motion of air, Gilbert asserts that, nevertheless, something must pass between the bodies, and calls this a ' breath ' from the attracting body, which reaches a body within the radius of the effluvia and unites the two. The natural tendency of all things to unite is a principle he ascribes to Pythagoras, and he seems to think that this is sufficient explanation of the cohesion of bodies and of motion towards an attracting body, once the influence of its effluvia is felt. But there must be actual contact with the effluvia : ' For since no action can take place by means of matter unless by contact, these electrics are not seen to touch, but, as was

[1] ibid., p. 64

necessary, something is sent from the one to the other, something which may touch closely and be the beginning of that incitement.' [1] Gilbert compares the emission of effluvia to that of scent, and remarks that though the effluvia of electric bodies are different from air, yet air itself is the effluvium of the earth, and causes bodies to fall towards the earth :

> 'And this also takes place in all primary bodies, the Sun, the Moon, the Earth, the parts betaking themselves to their first origins and sources, with which they connect themselves with the same appetence as terrene things, which we call heavy, with the Earth. So lunar things tend to the Moon, solar things to the Sun, within the orbes of their own effluvia.' [2]

Gilbert regards these effluvia as 'humid', and compares their action to that of drops of water, which tend to flow together and to unite two wet bodies which are free to move towards each other. He describes experiments with floating sticks which show this effect.

Thus the phenomena of electric attraction, cohesion, gravitation, and surface tension, all of which seem to require some kind of atrractive force between parts of matter, are ascribed by Gilbert to the same cause : a tendency towards unity among like substances, which, however, only manifests itself when effluvia from the attracting body actually touch that which is attracted. Non-electric bodies do not emit the appropriate kind of effluvium and therefore do not attract.

There is no suggestion as yet that the attraction of heavy bodies to the earth, and the motions of the heavenly bodies themselves, are due to the same physical cause. Gilbert follows Copernicus in believing in a daily rotation of the earth, and devotes a large part of his Book VI to arguments in its favour, but he nowhere states positively that the earth also has an annual revolution round the sun. But in any case it is clear that he wishes to ascribe all heavenly motions, whether rotation or revolution or both, to the magnetic powers of the heavenly bodies. His major discovery is that the earth is a 'great magnet', and he believes that it, together with all spherical lodestones and all other heavenly bodies, possesses a primary magnetic form which manifests itself in five kinds of magnetic motion : the coition of unlike poles, the directive or compass property, variation of magnetic pole from the pole of rotation, dip,

[1] ibid., p. 56 [2] ibid., p. 229

and spontaneous rotation about an axis. In the case of the earth, he believes that the axis of rotation is maintained in a constant direction in space by a directional property similar to the effect exerted by the earth on compass needles. Gilbert also suggests that the earth's rotation is prompted by the magnetic and luminous power of the sun and other stars, turning the earth for its own good, so that alternate faces are presented to the sun and moon in order to receive their virtues. The whole moving system of planets is likewise ordered by mutual magnetic influences which ensure the harmonious working of the whole. But how does this omnipresent magnetic power act? Gilbert discusses this question in relation to experiments with lodestones and iron.

If there are magnetic effluvia similar to the electric effluvia, then, he says, they must be capable of penetrating iron as 'quicksilver has entrance into gold',[1] because the magnetic power can be transferred from a lodestone to iron. But iron is affected even through thick screens of dense matter, through which electric action cannot penetrate. Therefore Gilbert concludes that magnetism cannot be conveyed by material particles as he believes electric attraction to be. Iron certainly loses its magnetisation when it is made red hot, but Gilbert regards this, not as the burning up of a material magnetic substance in the iron, but as a 'deformation' of the iron, just as the human body loses the faculties of its soul when it is burnt, although the soul itself is not material and therefore cannot be burnt. In fact the most appropriate analogy for magnetic action seems to him to be that of self-moving and ordering soul. The lodestone is ensouled, as Thales and others have said; iron and lodestone come together by that concord 'of the perfect and homogeneous parts of the spheres of the universe to the whole . . . tending to soundness, continuity, position, direction, and to unity'.[2] Later he says that the lodestone is 'like a living creature'.[3] The ancients believed that the heavenly globes have souls, and for Aristotle it is only the earth which has no soul, but Gilbert thinks that the whole universe including the earth must be ensouled. As an illustration of the organic nature of the magnet, he cites the tendency of a divided magnet to join in one particular way, with the north pole of one half against the south pole of the other, and compares this with the art of grafting twigs together the right way round, so that they grow into unity.[4]

[1] ibid., p. 66 [2] ibid., pp. 67–8 [3] ibid., p. 208
[4] ibid., p. 130. cf. J. Agassi, 'Koyré on the history of cosmology', *B.J.P.S.*, IX, 1958, p. 240

The lodestone also has the power of restoring the confused magnetic form of iron placed within its sphere of influence, and this Gilbert regards as a true action at a distance. He likens it to the transmission of light, but distinguishes his theory of both from a contact theory such as that of multiplication of species :

> 'And yet we do not mean that the magnetic forms and orbs exist in air or water or in any medium that is not magnetical, . . . for the forms are only effused and really subsist when magnetic substances are there.' [1]

Hence it would be misleading to ascribe to Gilbert any kind of continuum or field theory.

Gilbert goes on to describe what various writers have to say about examples of attraction other than those of electric bodies and the lodestone and iron. For instance, diamond was said to attract iron ; magnets to attract gold, silver, and other metals, flesh and water ; naphtha to attract fire ; and quicksilver to attract metals. He regards many of these reports as false, and in other cases remarks that there is no attraction, but only action by contact, as with some gluey substances which are said to attract flesh. Naphtha appears to attract flame only because it gives off an inflammable vapour, and quicksilver is imbibed by metals only when in contact with them. Again, those who believe in sympathies and antipathies say that there are repulsive forces, for instance between oil and water and between water and mud. But Gilbert maintains that these are only separations between unlike bodies without any antipathetic force, and will only admit one case of apparent repulsion : that between like poles of two lodestones. Electric bodies, he says, never exert repulsive force.[2]

Francis Bacon's Classification of Actions at a Distance

As far as his theoretical ideas are concerned Gilbert clearly belongs to the old rather than the new. For an example of the transition period between old and new views of what constitute acceptable analogies for the transmission of physical action, we turn to Francis Bacon's discussions of radiation and magnetism. These show him struggling with the question of whether what is transmitted is

[1] *De Magnete*, p. 205

[2] ibid., pp. 16, 55, 68, 113. Electric repulsion was first described as such by Sir Thomas Browne (*Pseudodoxica Epidemica*, London, 1646, II, p. 82). He says that if electric wax is held over fine powder, the particles of powder will ascend 'and be as it were discharged from the Electrick to the distance of two or three inches, which motion is performed by the breath of the effluvium issuing with agility'.

corporeal or not, but he shows a distinct reluctance, which is not found in Gilbert, to admit immaterial causes, or to accept accounts of phenomena which seem to demand them. In general Bacon complains that the physical causes of phenomena have not been sufficiently studied, and will not admit that Aristotelian qualities and such logical subtleties are sufficient explanations of things, although in some places he does allow that the mutual desire of bodies for contact, the *horror vacui*, the desire to congregate in kindred masses, and so on, ' are truly physical kinds of motion '.[1] Elsewhere, however, he seems to imply that ultimately the forms of things should be described in terms of particle motion, and that most so-called specific virtues, perceptions, sympathies, and so on are the effects of more fundamental motions. This corpuscularian theory was undoubtedly influenced by the fact that a beginning was being made in the discrimination of various kinds of material gases from each other and from air. Bacon declares that

> ' the cavities of tangible things do not admit of a vacuum, but are filled either with air or the proper spirit of the thing. But this spirit, whereof I am speaking, is not a virtue, nor an energy, nor an actuality, nor any such idle matter, but a body thin and invisible, and yet having place and dimension, and real. Neither again is this spirit air (no more than wine is water), but a rarefied body, akin to air, though greatly differing from it.' [2]

These spirits, he says elsewhere,[3] are always in motion, and are responsible for most of the chemical processes of nature. The motions of these ' first congregations of matter ' are like a general assembly of estates, giving laws to all bodies.

Bacon cannot satisfy himself, however, that all processes of nature are to be explained by corpuscular motions. Some bodies seem to work rather ' by the communication of their natures ',[4] and here we have an echo in his work of the doctrine of multiplication of species. While most powers act by contact, as ointments, plasters, and objects of taste and touch, there are some which act at a distance, and in his posthumously published *Sylva Sylvarum* he gives a catalogue of phenomena showing apparent action at a distance,[5] which is of great interest, since it shows a stage of transition between the types of fundamental action recognised by the Greeks

[1] *Novum Organum* (1620), I, lxvi, *Works*, ed. cit., IV, p. 68
[2] *History of Life and Death* (1623), *Works*, ed. cit., V, p. 321
[3] *Sylva Sylvarum* (1627), *Works*, II, p. 380
[4] ibid., 268, p. 430 [5] ibid., 904ff., pp. 643ff.

and the corpuscularian concepts of the later seventeenth century. The discussion of each type of action in this catalogue is brief, but it can be supplemented from elsewhere in Bacon's works.

He begins by remarking that the philosophy of Pythagoras first planted the 'monstrous imagination' that the world is a living creature with soul and spirit, which led to the belief that events at one part of the body are felt throughout, so that 'no distance of place, nor want or indisposition of matter, could hinder magical operations'. His purpose is, he says, to

> 'inquire with all sobriety and severity, whether there be to be found in the footsteps of nature any such transmission and influx of immateriate virtues; and what the force of imagination is',

and to separate the natural from the superstitious and magical. There follows this classification of actions at a distance:

1. Actions which are clearly corporeal, such as the transmission of odours and infectious diseases, which come from the 'thinner and more airy parts of bodies'. Bacon concludes that unlike light and sound, odours are corporeal, because they do not spread far, they adhere to bodies, and they last a long time. Infections, he says, may pass from man to man in the air or by touch. There is no problem about action at a distance in such transmissions.

2. The transmission of what Bacon calls spiritual species, that is, visual images, sounds, and radiant heat.

He distinguishes two kinds of heat, corresponding to the two ways in which he thinks bodies act upon one another: 'by the communication of their natures, or by the impressions and signatures of their motions'.[1] Thus, heat is produced by motion or friction, and transmitted by radiation from one body to another. The latter he describes in the language of the theory of multiplication of species as being an 'assimilation or self-multiplication'—a communication of quality without substance. In this it is similar to light.[2]

Light, radiant heat, and sound do not seem to require the emission of any corporeal substance, and, unlike scents, they spread to great distances, but it is probable that they do in some way affect the medium through which they pass, because transmission takes place only in certain kinds of media. This is what distinguishes them from magnetism. Like magnetism, however, many separate images, sounds, and rays of heat and cold can be transmitted in the same

[1] *Sylva Sylvarum*, 268, p. 430 [2] *Novum Organum*, II, XX, XXVI

medium without mutual interference, and they can pass through very small crannies ; they must therefore ' have the whole species in every small portion of the air, or medium '.[1] Sounds seem more dependent on air motion than does light, and Bacon suggests that their ' sudden generation and perishing ' are either produced like circular ripples on water, or are due to sounds being received by the air as a combustible body receives flame and is soon quenched. The latter suggestion however cannot be intended as more than a metaphor, for flame was universally regarded as a body, whereas the conclusion on the nature of sound which immediately follows is that it ' is a virtue which may be called incorporeal and immateriate ; whereof there be in nature but few '.[2]

3. ' Emissions which cause attraction of certain bodies at distance ', such as electric attraction, the attraction of gold for quicksilver, the attraction of heat and of fire for naphtha. Also the attraction of herbs for water, of purgative drugs for the bodily humours, and the mutual attraction of two bubbles.

Here Bacon is giving the same examples as those discussed by Gilbert, and he follows Gilbert in distinguishing them from magnetic attraction, for they, unlike magnetism, are influenced by the medium through which they pass. He ascribes some of these attractions to ' sympathy ', but this does not mean that he despairs of corporeal explanations for them, because in general he uses the words ' sympathy ' and ' antipathy ' to describe the merely apparent characteristics of phenomena whose detailed explanation would be in terms of corpuscular motions.[3]

4. ' Emission of spirits, and immateriate powers and virtues, in those things that work by the universal configuration and sympathy of the world.' Examples are magnetism, gravity of bodies towards the earth, the disposition of bodies to rotate from east to west, and the tides. All these influences have the property of passing through any medium.

Magnetic attractions are not affected by any medium, do not interfere with each other, and (if they are the cause of the tides as Kepler believed) they must act at great distances. Bacon therefore thinks that magnetism is the best attested example of action being transmitted at a distance without the intervention of any material

[1] ' Consent of Visibles and Audibles ', *Sylva Sylvarum*, 256

[2] *Sylva Sylvarun*, 290, p. 436

[3] See for example ' Experiments . . . touching the sympathy and antipathy of plants ', ibid., 479ff., where attempts are made to account for the behaviour of plants by the motion of ' juices '.

body : '. . . there is a natural virtue or action subsisting for a certain time and in a certain space without a body'.[1] This, he suggests, provides a proof of the existence of incorporeal substance, for if action can exist without a body during transmission 'you are not far from allowing that it can also emanate originally from an incorporeal substance'. He has no consistent theory about how magnetism acts, sometimes ascribing it to the attraction of like for like, as when a magnet shielded by iron attracts iron more strongly than when it is not shielded, and sometimes likening it to the generation of heat in the iron, or the working of leaven, yeast, poisons, and so on. Neither is there any theory about gravity, although he seems to follow Kepler rather than Gilbert in suggesting that this is similar to magnetic attraction, and not due to material effluvia.

There follow four types of action at a distance which seem to be essentially immaterial. Bacon professes himself sceptical about the existence of some of them, but admits others as scientifically demonstrable. The phenomena which he is most ready to ascribe to action at a distance without any material medium are those which savour most of witchcraft, magic, astrology, and telepathy, and since these were the beliefs most discredited by the subsequent advance of physical science, the fact that action at a distance was discredited with them is not surprising. Magnetism and gravity are the only *physical* phenomena for which Bacon can conceive no corpuscular explanation.

5. The operation of the mind of man upon other minds. Bacon gives as examples : parents knowing children they have never seen ; telepathic communication between great friends or great enemies ; the power of masterful personalities ; the influence of melancholy, jovial, lucky or unlucky men upon the company they keep ; the rejuvenation of the old by consorting with the young ; the effects of audacity and confidence in business ; love or envy which works through the eyes ; the contagious effects of fear and shame, and of gaping, yawning, stretching, and laughing ; strong belief affecting another's actions or thoughts, for example in card guessing tricks ; and so on and so on.

Bacon's early suggestion about phenomena like these, in the *Advancement of Learning* (1605) makes a direct analogy between communications from one mind to another and 'the passage of contagion

[1] *Novum Organum*, II, xxxvii

from body to body, the conveyance of magnetic virtues'.[1] A mechanical explanation is suggested by Joseph Glanvill sixty years later,[2] when he compares communication between minds to the communication of a sound between resonant strings, and concludes that motions in the brain may be transmitted through the aether as sounds through air.

6. Influences of the heavenly bodies other than heat and light, and the possible influence of the moon on the tides. Bacon discounts most but not all of these alleged influences, and in any case, with an eye to avoiding the violation of man's free will by the stars, he insists that they are only inclinations and not compulsions. He mentions four such influences of the moon : it draws forth heat, induces putrefaction, increases moisture in plants and in the brain, and excites the motion of the spirits in lunacy.[3]

7. Operations of sympathy in natural magic, such as the alleged virtues of precious stones, the power of parts of a living creature to endow men with that creature's specific qualities, and so on. Bacon here betrays a certain Puritanism in deploring the use of religious ceremonies and images to influence the imaginations of worshippers, for, he says, 'magic of this kind proposes to attain those noble fruits which God ordained to be bought at the price of labour by a few easy and slothful observances.'[4]

In general Bacon is not disposed to repudiate the claims of magic out of hand, but believes that with more knowledge they might be shown to have some foundation in fact.

8. Transmission of virtue between the severed parts of a whole, for example, if a part is consumed or putrefied, so are the other parts, and if one person's skin is grafted on to another, the second person is sensitive to stimuli affecting the first.

Under this heading Bacon includes the very remarkable belief in the sympathetic cure of wounds. This seems to have originated with the Paracelsian doctors, and is mentioned again and again in the sixteenth and seventeenth centuries as a well-authenticated phenomenon, enjoying greater patronage among normally critical writers than most of the other fantasies of sympathetic medicine. 'It is constantly received and avouched', says Bacon, 'that the anointing of the weapon that maketh the wound, will heal the wound itself.'[5] He goes on to describe the preparation of the

[1] op. cit., IV, iii, *Works*, ed. cit., IV, p. 400
[2] *The Vanity of Dogmatizing*, London, 1661, p. 199
[3] *Sylva Sylvarum*, 890ff., p. 636
[4] *De Augmentis Scientiarum*, IV, iii, *Works*, IV, p. 401
[5] *Sylva Sylvarum*, 998, p. 670

ointment, and the various tests to which the practice has been subjected, all of which seem to show that the cure is obtained only when the ointment is applied to the weapon and not to the wound, that the weapon may be at a great distance, and that the cure does not depend on the patient knowing of the anointing. Bacon is impressed by this evidence; all the same he is not yet 'fully inclined to believe it'. The same serious discussion and semi-sceptical attitude is found in Boyle,[1] who quotes this among other apparently well-authenticated examples of the action of sympathetic medicines. Sir Kenelm Digby has a version of the cure in which a cloth covered with blood from the wound is soaked in a solution of the 'powder of sympathy' (iron sulphate), and he provides an atomic explanation based on the principles that like attracts like, and that atoms are attracted by magnets, by siphons and by heat. Blood atoms are drawn out of the cloth by sun and light, and travel back to the wound, carrying some atoms of the powder with them, thus effecting the cure, as Digby says, naturally and without magic or recourse to Demons or Spirits.[2] Glanvill thinks that the cure has been 'put out of doubt by the Noble Sir K. Digby and the proof he gives in his ingenious discourse on the subject, is unexceptionable',[3] although Glanvill is not so sure about the atomic explanation. Mention of the cure even occurs in Sprat's *History of the Royal Society*,[4] in a reply to certain questions sent by the Royal Society to a correspondent in Batavia. The writer remarks that a reported effect on wounds of a certain poison from Macassar which touches blood from the wound 'is not strange to those who study sympathy; and set belief in that much renowned Sympathetical Powder of Sir Kenelm Digby'.

It is impossible to know what to make of these stories, and of the evidence so carefully collected for them by the fathers of the scientific revolution. One can only accept them, with whatever implications one finds in them, as illuminating sidelights on the environment in which modern science was born. In any case it is clear that the experimental method plus the hypothesis of atomism did not yet add up to the corpuscular philosophy. A more thorough purgation of the imagination was required, and that was effected almost wholly by one man—Descartes.

[1] *The Usefulness of Natural Philosophy* (1663), *Works*, London, 1772, II, p. 165

[2] *La Poudre de Sympathie* (1658), English trans., London, 1669 (bound with *Of Bodies*), pp. 147–205

[3] *Vanity of Dogmatizing*, p. 207

[4] London, 1667, p. 165

Chapter V

THE CORPUSCULAR PHILOSOPHY

Falsifiability as a Seventeenth-century Criterion for Theories

THE seventeenth-century attack upon ancient science took place along two fronts. One was the experimental testing of many things hitherto alleged to be facts, which failed to pass the empirical test. Examples of this have already been noted, and could be multiplied from the writings of almost any adherent of the 'new philosophy'. The importance of this factor should not, however, be exaggerated. Few philosophers before the seventeenth century allowed their theories to be entirely uncontrolled by fact; what they lacked were adequate theories and adequate methods of expressing the facts precisely, and it is clear from a study of seventeenth-century science that the revolution owed more to a change of theoretical framework than to the discovery of new facts or the exposure of old fictions. There is another point about the factual basis of the new science that is not so often noticed. There were phenomena which were described by ancient and Renaissance writers and were undoubtedly founded upon fact, but which came to be ignored by orthodox science because they did not, and still do not, fit into the generally accepted physical framework. Examples are to be found among the list of physical actions discussed by Francis Bacon, including such phenomena as hypnosis, telepathy, and the influence of the moon in lunacy,[1] about which there is still a large volume of testimony, but as yet very little theoretical explanation.

The other ground of attack upon traditional science was methodological, and perhaps more important. Here again the point has frequently been misunderstood. It is often expressed by saying that the *a priori* speculations of science before the seventeenth century gave place to *a posteriori* descriptions of the Newtonian type, which 'made no hypotheses'. But this is not an adequate account of the method of the new science, for there is a sense, as I have

[1] The last of these may appear doubtful, but it is interesting that Georges Sarton, who was not initially disposed to believe in it, remarks after investigation in mental institutions that 'It may be, nay, it is highly probable, that moonstruckness in all its varieties is a purely subjective phenomenon; yet it is not impossible that there may be deep in it a kernel of objectivity. It may be worthwhile to dig for that kernel' ('Lunar influences on living things', *Isis*, xxx, 1939, p. 507).

tried to show, in which basic categories and analogues in science are always *a priori*; and the new principles were understood partly as the replacement of the old metaphysics by new, and were argued on metaphysical grounds. But the really important change that took place was that the first systematic attempts were made to formulate theories in such a way that they ran the risk of refutation by empirical tests. It began to be held that theories which were so vague and adaptable that they could be made to cover any conceivable facts, were not explanations of the facts at all, but at best redescriptions of the facts in different words, or at worst groundless and untestable speculations. It is this requirement of falsifiability, to use the modern term, that is missing from pre-seventeenth-century science, and that lies behind the accusations of barrenness and anthropomorphism which the seventeenth-century writers brought against their predecessors. A science consisting of plausible *ad hoc* descriptions, usually couched in vague and obscure language, was barren because it could not be brought into detailed relationship with the facts, and hence neither falsified nor made to yield correct predictions. Into this category came the explanations I have called analogies of organism and attraction. These were said to be anthropomorphic and condemned, not really because of their anthropomorphism, but because comparisons with human and animal life could not be given clear and distinct expression and could not be dealt with in mathematical terms, and therefore could not be compared in detail with the facts.

The requirement of mathematical precision also tells against organic analogies, since even if the analogues were reducible to precise mathematical expression, they would be intrinsically more complex than the phenomena they are intended to explain. The seventeenth-century notion of 'simplicity' was conditioned by a revival of interest in Pythagoreanism and atomism, that is in the expression of things in terms of simple mathematical concepts (the notion that Nature is a book written in the language of mathematics [1]), and that inanimate matter has in essence none but the simple geometrical and mechanical qualities.[2] Such an understanding of simplicity [3] in terms of the minimum number of

[1] cf. Galileo, *Il Saggiatore* (1623), *Opere*, IV (1844), p. 171

[2] cf. R. Boyle, 'The Excellency and Grounds of the Mechanical Hypothesis' (1674), *Works*, IV, p. 68

[3] For an account of the relation between falsifiability and simplicity in this sense, see K. R. Popper, *Logic of Scientific Discovery*, pp. 140ff. This is not a necessary definition of simplicity in science. Nigel Walker suggests ('How does Psycho-Analysis Work?', *Listener*, 6 Oct. 1955, p. 544) that if intelligibility by the patient is the aim, Freudian or post-Freudian theories which mention concepts like repression, catharsis of the memory,

mathematical variables consistent with the subject-matter, at once rules out organic analogues which generally contain more variables than the inorganic situation with which they are compared.

It is now clear why a writer like Gilbert, in spite of his careful experimentation, seems to belong to the old era rather than the new. He rejects, for instance, the old theory of sympathies and antipathies, not because of its anthropomorphism, but because it does not describe the facts; and his own theories, especially his theory of magnetic attraction, are typically anthropomorphic ones, which do not provide explanations in the modern sense because they cannot easily be made detailed and hence falsifiable. Gilbert is operating with a different logic of explanation, namely one of elimination. For him there is a limited number of ways in which distant bodies can be conceived to act upon one another, and after eliminating some of these by empirical test, he is left with two possibilities: action by contact of material effluvia, 'since no action can take place *by means of matter* unless by contact',[1] and action by means of immaterial soul. In the case of magnetism the former explanation can be eliminated experimentally, and so the attractive power must be ascribed to self-moving soul. The fact that this cannot be further tested does not, for Gilbert, detract from its value as an explanation, for it does something to make the phenomenon of magnetism at least intuitively intelligible. Gilbert is perfectly aware of the extent of the *negative* analogy between the magnet and a living creature,[2] and it should be noticed that it is in the production of order and harmony that he sees the positive analogy (more obvious in magnets than in man!), not in non-rational aspects of organisms.

But the Aristotelians had worn this kind of intuitive explanation too thin, and later seventeenth-century writers are not satisfied with it. They hardly stop to inquire whether sympathies and antipathies and the rest of the occult qualities correspond with the facts, for obviously in some cases they do and in others they do not; they dismiss such explanations at once as *ad hoc*, obscure, and anthropomorphic. The requirement of falsifiability, together with the

[1] *De Magnete*, p. 56 (my italics—M. B. H.) [2] ibid., pp. 208ff.

---

self-knowledge, mental habits, etc. may well be simpler than the physico-chemical accounts of the brain physiologist: 'My point is that psycho-therapy is a field in which a good metaphore is worth more to the technician than the literal truth. If this is so, then what we are looking for is not so much the literal explanation of the changes wrought by psycho-analysis as the best possible metaphor, the one that fits in most facts and leads to the greatest improvements in technique' (ibid., p. 545).

combined Pythagorean and atomist demands for simple mathematical and mechanical explanations, led to severe restrictions on the kinds of explanatory analogy which were to be admitted in physics. It became universally accepted that only those actions which produce, or tend to produce, locomotion were to be accepted as fundamental, and furthermore that change of motion in bodies was to be explained only by communication from outside, and not by any innate power or striving within the bodies themselves, that is, the ascription of self-moving soul to apparently inorganic bodies was to be avoided. It was Kepler who, in Collingwood's phrase, 'took the momentous step of proposing that in treating of physics the word *anima* should be replaced by the word *vis*: in other words, that the conception of vital energy producing qualitative changes should be replaced by that of a mechanical energy . . . producing quantitative changes'.[1] Galileo was the first to exploit the principle of quantitative correlations systematically in the study of the motion of inert matter; the rejection of animism was explicit in Descartes's physics; and from that time there was a gradual elimination of analogies from living beings and the teleological categories which pervaded Aristotelian and Renaissance physics. The whole development of physical theory in the seventeenth century can in fact be written in terms of the replacement of the basic Greek analogues of self-moving soul, sympathy and antipathy, the *horror vacui*, antiperistasis and multiplication of species[2] by the three models which have been fundamental in physics ever since: particle motion, wave motion, and gravitational attraction.

The clearest break with ancient thought came with the distinction in the philosophy of Descartes between mental and bodily substance. In this Descartes certainly succeeded in his declared intention of laying completely afresh the foundations of science, relying only on the simplest and clearest propositions which seemed

[1] R. G. Collingwood, *The Idea of Nature*, Oxford, 1945, p. 101. Kepler substituted the word *vis* for *anima motrix* in his notes to the 1621 edition of his *Mysterium Cosmographicum* (*Opera*, ed. Frisch, I, p. 176). The immediate occasion for the change was his discovery that the velocity of the planets, supposedly caused by the sun, diminishes with distance from the sun, therefore, he concluded, the force exerted must be corporeal 'at least in a certain sense'. In other words, the notion of a moving soul which is affected in exact proportions by distance has become too implausible, and Kepler is not prepared to modify his conception of soul indefinitely. In this sense, his conception is a falsifiable one. It is clear, however, that it is quite a different one from that of Gilbert, who sees the analogy with soul precisely in terms of order and proportion.

[2] Hobbes calls the roll: 'For as for those that say anything may be moved or produced by *itself*, by *species*, by *its own power*, by *substantial forms*, by *incorporeal substance*, by *instinct*, by *antiperistasis*, by *antipathy*, *sympathy*, *occult quality*, and other empty words of schoolmen, their saying so is to no purpose' (*De Corpore*, *Works*, I, p. 531). But he believes in the *horror vacui* as an explanation of cohesion and of the retention of water in a perforated vessel (ibid., pp. 414ff.).

to him to admit of no doubt, and he thereby eliminated from scientific explanation all the variety of organic and human analogues. The result was a radically new way of looking at the operation of the spiritual and mental in the natural world, and it had, as Descartes intended, implications far beyond the mind-body paradoxes of the philosophical textbooks, for it carried with it the corollary that scientific theories about matter and action need have no relation to metaphysics or theology. This declaration of independence by science was not entirely new: Aquinas had paved the way for it by his separation of natural from revealed knowledge, and it had been more explicit in the nominalist school of the fourteenth century. William of Ockham had then held a view of science which comes near to modern positivism, namely that the chief aim of natural science is the correlation of observed data, and that therefore there is, for example, no *a priori* reason why bodies should not be described as acting upon one another at a distance, or why species should be postulated to transmit action, or why moving bodies should require an external force accompanying their movement. According to his dictum that entities should not be multiplied without necessity, he held that science is concerned with observable motions only, and that hidden causes should not be invoked indiscriminately in order to explain these motions.

Descartes's Mechanical Continuum

Aquinas and Ockham were, however, more concerned to vindicate faith by thus separating it from science, than to develop science itself, and the time was not then ripe for their ideas to penetrate into scientific thinking. Even the revival of atomism was not sufficient to effect the change, for Gassend, who is commonly credited with this revival and who was a contemporary of Descartes, has no clear distinction between matter and spirit such as Descartes initiated. Gassend thinks of soul in animals and plants as corporeal, although it is not to be equated with dead matter but is rather an organising principle which in some sense produces in the animal body a unity which makes the whole to be more than the sum of its parts. In man there is in addition, he says, an incorporeal soul which stands to the corporeal soul as the latter stands to matter. Thus he sees the relation of soul and body in traditional fashion as a difference of degree throughout a hierarchy, rather than a difference in kind.[1] In his physical arguments also Gassend is

[1] *Syntagma Philosophicum*, *Opera*, II, Lyons, 1658, pp. 250ff.

conservative. It seems to him self-evident that there are only two principles of motion, namely impulse and attraction, and that attraction can take place only if the attracting body emits something material which actually lays hold of the object attracted. This, he thinks, is what must happen in the cases of sympathy and antipathy, and he likens the corpuscles emitted by electrical bodies to 'innumerable rays darted out like tongues', as the chameleon darts out its tongue to catch a fly.[1] But there is no mechanical account of how the 'grasping' action takes place, in fact motion towards a magnet is ascribed to a quasi-soul in the object which is stimulated into motion by the magnetic emanations.[2] Light is a substance transmitted like arrows or javelins, and fire warms by sending out a stream of heat atoms just as a fountain wets by dispersion of water-drops. Visual 'species' are, as in Lucretius, corporeal images peeled off the object of vision.[3]

In contrast with Gassend's version of atomism, Descartes wholly rejects the notions of moving souls and of the void. Rejection of the void follows directly from his identification of bodily substance with spatial extension, for wherever there is space, there is *ipso facto* body. Body, he holds, has essentially no qualities except shape, size, and motion, since these are the qualities that are clearly and distinctly perceived, and whatever is clearly and distinctly perceived is true. All the phenomena of physical nature, including, Descartes thinks, the life of animals, depend exclusively upon these qualities. Mental substance or soul is quite distinct from body, and as it is found only in man, the whole of physical science can proceed without reference to it, on purely geometrical principles, and then the secondary qualities or 'sensations' of 'light, colours, sounds, odours, taste, heat, hardness, and all other tactile qualities'[4] will be seen to be produced by the relation between body and mind. On this basis Descartes constructs the first mechanical physics which is both detailed and comprehensive, and for this reason it is worth describing here as much of it as is relevant to his theories of physical action.

Since there is no void, it follows that all the spaces between observable bodies are filled with a subtle, insensible, fluid substance which extends throughout the universe. This substance does not consist of indivisible atoms, but must be indefinitely divisible, since extension is indefinitely divisible, but Descartes asserts that we can

[1] ibid., I, p. 450 [2] ibid., II, pp. 128ff. [3] ibid., I, pp. 422ff.
[4] *Les Principes de la Philosophie* (French trans., 1647), Part I, 48; *Œuvres*, ed. Adam and Tannery, IX, p. 45

speak of particles of substance which temporarily move in one piece relative to other particles.[1] It is not clear at this point how a 'piece of extension' can be said to move at all, since there are no qualities which could mark out the boundaries of such a particle, and while it is intelligible to say that a body defined by certain qualities moves in extended space, it seems meaningless to say that extension itself moves. Descartes has after all slipped something other than extension into his conception of matter. Motion, in fact, has to be thought of as a mode of body as fundamental as shape or size, and it is this that leads him to his statement of the law of inertia, namely, that no more action is required for motion than for rest. Descartes, unlike Galileo and Hobbes, regards inertial motion as rectilinear, and realises that therefore, when a body moves in a circle, force has to be applied only along its radius and not in any other direction.

If the universe is a plenum it is clear that the motion of any particle must result in a complete ring of bodies moving together, as in the Platonic circular thrust.[2] Descartes pictures the whole universe as a system of interlocking vortices ('tourbillons') each containing rotating particles. The total amount of motion given to the system by God in the beginning remains constant, since God is immutable. The result of all these circular motions is that most of the large particles are rubbed into a spherical shape, but the spaces between the spheres cannot be empty, so they must be filled with other, perhaps smaller, particles, there being no lower limit to their size. Descartes distinguishes three different kinds of matter, according to the size of the particles and their origin :

1. Luminous particles which have been rubbed into a spherical shape, and which make up the sun and stars.
2. Very small particles derived from the rubbed-off corners of the luminous particles. These fill the heavens and are insensible, transparent, and offer no resistance to the motion of larger bodies. This Descartes calls *matière subtile*, and it corresponds to the aether of other physical theories.
3. Large opaque pieces of matter which constitute the earth and the planets, and which are derived from the adhesion of luminous particles.

[1] ibid., II, 16ff.
[2] It might be thought that motion also takes place by condensation and rarefaction, that is, by the same continuous body taking up more space by becoming less dense, but this possibility is not open to Descartes since for him density depends on weight, and weight is a secondary quality explained by the motions of surrounding matter.

Fig. 1 Descartes's Vortex-system. The line NCEVB is the path of a comet passing from the vortex *S* of the sun into neighbouring vortices.

Descartes goes on to make ingenious use of the vortex principle in explanations of the motions of planets and comets, terrestrial gravitation, and elementary properties of light, heat, chemical reactions, electric and magnetic attractions.

In his theory of light and of sight,[1] he rejects the conception of species flying from an object to the eye, conveying form without matter. In what is really a return from Aristotelianism to Aristotle himself, he asserts that forms or qualities are not substances, and cannot be separated from the substance in which they inhere. Objects are seen by light reflected from them to the eye, and the reflection causes modifications in the light which are perceived as different colours. As for the transmission of light itself, this is not a flight of corpuscles, but a pressure in the all-pervasive second element, which is caused by the centrifugal force of particles of the first element spinning round in the sun and other luminous bodies.

[1] *Principes*, III, 55–63 ; *Réponses aux Sixièmes Objections*, 7, 9, *Œuvres*, IX, pp. 234ff.

The transmission of this pressure is instantaneous, and Descartes compares it with the sensation conveyed along a blind man's stick and to the pressure of wine in a vat of grapes tending to make the small wine particles move, while the larger grapes remain at rest.[1] The latter illustration shows how light is transmitted through transparent solids. Heat is simply the agitation of particles of the second element produced along the line of the light pressure.[2]

If, as Descartes asserts, bodies are composed simply of pieces of extension in juxtaposition, the problem of the cohesion of solids becomes acute. There cannot of course be any ' cement ' between their parts, and Descartes's solution is that hard bodies are those whose parts are in relative rest, while fluids and soft bodies have their parts in greater or less relative motion. Thus he attempts to explain the fact that soft bodies cannot break hard ones, by application of his law of inertia and rules of impact. A perfectly hard body has all its parts in relative rest, and since their inertia will cause them to remain so unless acted upon by a force, a very large force will be required to separate millions of them from each other. But then the objection arises that according to his own fifth rule of impact, any moving body, however slow, can always move a body smaller than itself, so how is it that our hands cannot break stones smaller than themselves? This, Descartes replies, is because our hands are soft, that is, their parts are in relative motion, and so do not all act together upon the stone. Thus it is easier for the parts of our hands to be separated than the parts of the stone, and in general, hard bodies can only be broken by bodies harder than themselves.[3] The tendency of liquids to cohere in spherical drops Descartes explains by the impacts of particles of the second element on the liquid surface.[4]

Magnetic attraction [5] is explained by means of small ' striated ' or spiral particles which flow through channels in the magnet acquiring rotation as they go, like a bullet through a rifle-barrel. Outside the magnet they flow in vortices, following the lines of magnetic force, and free magnets which are placed in their way move to allow them the easiest possible passage. Thus a compass needle turns to the earth's pole so that its channels lie along the paths of the earth's magnetic particles. Most substances are non-magnetic because they have no channels of the required shape, but there are channels in iron and steel which can be orientated in one

[1] *La Dioptrique* (1637), *Œuvres*, VI, pp. 84ff. [2] *Principes*, IV, 29
[3] ibid., II, 54–63; *Traité de la Lumière* (posthumous, 1664), *Œuvres*, XI, p. 12
[4] *Principes*, IV, 19 [5] ibid., IV, 133ff.

direction under the influence of a magnet, or by being fired and cooled, so that iron and steel than behave like natural magnets. By means of ingenious applications of this theory, Descartes tries to account for all the observed properties of magnets without postulating any fundamental attraction or action at a distance.

The theory of electric attraction is more tentative. Here Descartes suggests that in substances such as amber, wax, and resin, attraction is due to rubbing off small branched particles which remain hooked together, and are drawn back to the body carrying small bits of straw with them, just as a drop of sticky liquid shaken at the end of a stick carries small straws into the main drop. But Descartes's theory of the composition of glass does not allow this explanation of its attractive powers, so he suggests an alternative : perhaps particles of the first (luminous) element collect together in long bands which fit the passages in glass through which it transmits light, and when the ends of these bands are rubbed off, they tend to return because the holes in the surrounding air are not the right shape to receive them. If it seems that this explanation is as *ad hoc* as anything the Aristotelians could produce, it must be remembered that Descartes admits its tentative nature, and that in any case the phenomenon of electric attraction was at this time a very minor one, confined to a weak effect upon neighbouring light objects, and therefore not an important part of any theoretical system.

The question of gravitational attraction was more important and here Descartes's theory was intuitively more acceptable. He remarks [1] that if the earth were turning in a void, bodies would fly off its surface by centrifugal action, but he believes this is prevented by the second element which surrounds the earth and which is tending to recede from the centre more strongly than the earthy particles. If one of the latter is displaced from the surface of the earth, the tendency of the surrounding second element to rise causes the earthy particle to fall, in order to preserve the vortical motion of the plenum. Gravitational theories of this kind were elaborated by Descartes's successors, and became an accepted part of Cartesian physics, remaining serious rivals to Newton's theory of attraction until well into the eighteenth century.

Huygens, for example, made serious efforts to commend Descartes's theory and to make it consistent with the developing science of mechanics. By stating for the first time a detailed theory of centrifugal force Huygens was able to clear up various confusions

[1] ibid., IV, 20ff.

which had arisen in the use of eddy theories as an explanatory device by the Greeks, and which had been perpetuated by Descartes in this account of gravitation. Huygens described before the Académie des Sciences [1] an illuminating experiment in which a cylindrical dish containing water and powdered sealing wax is rotated about a vertical axis. At first the particles of wax are driven to the circumference because they are denser than water. When the rotation is stopped, the particles at once move to the centre, because by friction with the bottom of the dish they tend to come to rest more quickly than the water, so that the centrifugal force upon them is less. Descartes had made the mistake of assuming that the mere rotation of the earth and its accompanying second element would be enough to make heavy bodies tend to its centre, but Huygens points out that the opposite is the case. Bodies denser than the medium will always move outwards if their rotational velocity is as great as or greater than that of the medium, and this would apply in the gravitational case, because surely gross bodies must be denser than the aether. Heavy bodies will move inwards only if their motion is slower than that of the medium, as in the second part of the experiment, and it is this effect which Huygens likens to gravitation. If it is to provide an explanation, then the subtle matter surrounding the earth must be rotating much faster than the earth itself (Huygens calculates that it must be 17 times faster) in order to produce the observed gravitational force. He also points out, as Descartes failed to do, that the rotation must take place, not about the earth's axis, but in great circles about the earth's centre, in order to produce the force in a radial direction at all points on its surface. This makes it very difficult to understand how the required motion of subtle matter is possible at all, but Huygens suggests that it is an effect of very rapid random motion of the particles which averages out to a resultant motion in great circles because it is confined by the neighbouring spatial vortices to a spherical region surrounding the earth.

Descartes's Method

These detailed applications of Descartes's ideas, however, belong to the latter half of the century, and if Descartes himself was the pioneer of such theories, he certainly did not fully grasp the method of experiment and hypothesis which was to carry science forward, and most of his own explanations did not survive. The blame for

[1] *Mémoire* of 1669, *Œuvres Complètes*, The Hague, XIX, p. 631; *Discours de la Cause de la Pesanteur* (1690), ibid., XXI, p. 453

this must be placed on the way in which he used his mechanical explanations. In practice his method was an unsystematic mixture of rationalism and empiricism, but in theory he intended it to be entirely rationalist, for he claimed that when the first principles or 'simple natures' of philosophy (including natural philosophy) are clearly and distinctly perceived, all other truths can be deduced from them. Verification of the simple natures by comparing the deductions with experience is then a mere formality, interesting only as a means of finding out which deductions are in fact realised in nature, and which might have been but are not. If experience appears to contradict the deductions that must be because limited experience is deceptive, and Descartes therefore sees no possibility of rigorous testing for his theories.[1]

On the other hand, the process by which to arrive at the simple natures does involve observation and even active experiment. There must first be an enumeration of all the observable properties of the phenomenon to be explained, and then an attempt to deduce from them the complex of simple natures which would produce those effects.[2] This is very like the inductive method as recommended by Aristotle and by Newton. As Descartes himself saw it, his chief innovation in physics, as compared with Aristotelian and scholastic science, lay in the type of explanatory principles or simple natures which he 'clearly and distinctly perceived'. He points out that his simple natures are analogies drawn from familiar experience, and are not the unintelligible principles of the Aristotelians. Speaking of the mechanical actions attributed in his theory to the particles of subtle matter, he says

> 'We do much better to judge of what takes place in small bodies . . . by what we see occurring in those that we do perceive . . . than, in order to explain certain given things, to invent all sorts of novelties, that have no relation to those that we perceive such as are first matter, substantial forms, and all the great array of qualities which many are in the habit of assuming, any of which it is more difficult to understand than all the things which we profess to explain by their means.'[3]

The most clear and distinct notions of material things are those of figure, magnitude, and motion, and the rules of geometry and

[1] cf. his reply to the complaint that experience contradicts his rules of impact (*Principes*, II, 53), and his disagreement with Harvey on the mechanism of circulation of the blood ('Description du Corps Humain', *Œuvres*, XI, p. 242).

[2] 'Regulae', *Œuvres*, X, p. 427.

[3] *Principes*, IV, 201

mechanics which describe them, therefore the behaviour of insensible bodies is best described by their means, ' and for this end the example of certain bodies made by art was of service to me ',[1] for machines are necessarily large enough to be sensible, but they may imitate the effects of insensible bodies.

In the application of his method to physics, Descartes makes far more use of such analogies from familiar observation than of any principles derived deductively from his metaphysics. The passage of particles of the ' second matter ' through transparent solids is likened to that of small lead shots through a closely packed barrel of apples ; the pressure of light is transmitted like that of wine through a vat of grapes ; the action of the second matter in liquids is described in terms of the behaviour of impurities in vats of wine ; the vortex theory itself is exemplified in many natural processes : eddies in streams of water, the circulation of the blood, the motion of water from rain to rivers to sea to atmospheric vapour, the suction of the vacuum (which he explains by a circular thrust), and so on.[2] There is also evidence of considerable observation of manufacturing techniques such as the preparation of metals and the making of glass. The easily picturable nature of the analogies was undoubtedly one of the reasons for the popularity of Cartesian physics, and for its later persistence in face of the success of Newtonianism.

In the end, however, Cartesian physics fails to be either convincingly rationalist, or adequately empirical. The cosmology does not follow necessarily from self-evident principles. It depends too much on Descartes's own intuitions of the simple natures of physics, and his intuitions are far from being self-evident to all rational minds, even in the sense in which Euclid's axioms might be said to be self-evident. This failure to be strictly rationalist would not affect Descartes's claim to have produced an adequate scientific theory, if it were not for his second failure, which was the absence from his method of systematic empirical testing. Explanatory analogies may be drawn from any source whatsoever, including metaphysical principles or intuitions guided by familiar observations, but the new philosophy was beginning to demand exact quantitative correspondence with the facts, and this Descartes failed to provide, for two reasons. The first was his self-confident assurance that what he conceived as clear and distinct ideas were necessarily more reliable than the deliverances of the senses, and the second resulted from the inadequacy of his mathematical equipment. He was attempting to explain physical processes in terms of action in a

[1] ibid., IV, 203 [2] ibid., III, 30 ; IV, 18, 65 ; *Traité de la Lumière, Œuvres*, XI, p. 20

continuous fluid medium filling all space, but he gave no precise mathematical account of the behaviour of such a medium, for in spite of his eminence as a geometer and his frequent insistence on the importance of mathematics for the clarification of ideas, he never attempted a systematic hydrodynamical theory such as that initiated later by Newton. Lacking this he could not make precise deductions from his theory to be checked against observations, even if his general position had not at the same time discouraged reliance on sense perception.

Descartes cannot, however, be dismissed so quickly from the history of science. His theory, like most of those of his predecessors, was initially a plausible description, but when made precise enough to be refutable, it was either refuted or became unduly cumbersome, and hardly any of its details passed into the structure of physics. But no-one who reads the works of his contemporaries, and even of many of his successors, can doubt that in stripping physics down to extension and motion to the exclusion of all the picturesque accretions of Greek and Renaissance science, he was the first of the moderns. The purgation turned out to be too thorough : even physics could not progress without the additional concepts of mass and of force, and his cleavage of mind and body was certainly too drastic, but his manner of reasoning is for the first time recognisable as that of theoretical physics as we understand it. There is no doubt that Descartes's greatest influence upon subsequent physics lay in his firm elimination from it of mental categories. Among his predecessors only Galileo had been consistent about this, and he was no system-builder, being content rather to begin with the simplest problems and refusing to speculate about the more complex. As a result his work was more lasting than Descartes's, but it is doubtful whether his influence alone would have accustomed the seventeenth century to a concept of nature without spirits. Galileo dealt with idealisations of nature, and even the correctness of what he regarded as his greatest triumph—that of showing that a projectile in vacuum describes a parabola—remained in doubt for many years, because the effects of air resistance on actual projectiles vitiated all experimental tests, and made the theory useless to practical artillerists. But Descartes took the real world as his province, and banished mind from the whole of it, with the single exception of human volition. Even in his physiology, all remnants of the quasi-animist pneuma of the medical schools were translated into 'a certain very subtle air or wind which is called the animal spirits', and 'what I here name spirits are nothing but material bodies, and their one peculiarity is that they are bodies of extreme minuteness and that they move very quickly like the particles of the flame which issues from a torch'. Heat and fire are still the primary agents in physiological processes, but now they are understood mechanically. Even human passions are the result of movements and disorders of the spirits, as when for example the spirits of wine are converted into 'stronger and more abundant' animal spirits in the brain, and move the body 'in many strange fashions'.[1]

The human mind is, however, able to control the movements of the body by volition. This must be, by definition, a non-mechanical process, since mind according to Descartes is not extended substance, but the nature of the connection between mind and body is something which he did not succeed in clarifying. One suggestion, made in his correspondence with Princess Elizabeth of Bohemia, shows him, as it were, returning the analogy of organism to its proper place, and reverting to the original experiences from which the analogy developed, both in primitive animism and in Greek philosophy. He is discussing the nature of gravitation, and objecting to the Aristotelian notion that it is an immaterial moving force, a substance somehow penetrating the moving body. All qualities such as gravity, hardness, heat, and attraction are, he says, inseparable from the substantial body in which they inhere, and are simply effects of motion and configuration, as he shows in his physics. But we may correctly think of *mind* as an immaterial moving force: it *is* a substance and can act at any point of a body just as gravity can; moreover it may be thought of as concentrated at one point, just as weight is concentrated at the centre of gravity of a body.[2]

Corpuscular and Medium Theories

Descartes does not take the possible next step and assert that mind, too, is nothing but motion and configuration, although this might have increased the consistency of his theory. It was inevitable that someone at this stage should take the step, and it was Thomas Hobbes who did so. Hobbes seems to have arrived independently of Descartes at a physical system involving only matter and motion, and for him mind was not a substance distinct from matter, but was a function of matter in motion just as were all other qualities of bodies.

Hobbes's physics compares unfavourably in its details with that of Descartes, but it illustrates the same mechanical principles

[1] *Passions de l'Âme*, Part I, vii, x, xv, *Œuvres*, XI, pp. 332ff.
[2] 'Letter to Princess Elizabeth, 21 May 1643', *Œuvres*, III, pp. 663ff.

of method. With regard to the alternative theories of transmission presented by a mechanical physics, namely transmission through a medium, or by emission of corpuscles, he eventually came down on the same side as Descartes. He is quite as certain as Aristotle or Descartes that change is produced only by contact of bodies, and this appears so self-evident to him that his statements about it are assertions rather than arguments, for example :

> ' There can be no cause of motion, except in a body contiguous and moved. . . . For if it shall be moved, the cause of that motion . . . will be some external body ; and, therefore, if between it and that external body there be nothing but empty space, then whatsoever the disposition be of that external body or of the patient itself, yet if it be supposed to be now at rest, we may conceive it will continue so till it be touched by some other body.' [1]

Hobbes had at various times three theories of the transmission of light, the first was a corpuscular development of the Aristotelian doctrine of species, while the other two, in his more mature physics, were medium theories. In the second theory, put forward in the *Tractatus Opticus* (1644), he likens light transmission to an alternate expansion and contraction of the medium caused by a motion of the luminous body similar to that of the heart. This makes it easier to conceive the instantaneous propagation of light to infinitely great distances, in which he believes in common with most of his contemporaries, and to explain the diminution of light intensity with distance. He is puzzled, however, about how light passes through a vacuum, for he still regards air as the medium of light, and even suggests that the light passes, not through empty space, but through the glass walls surrounding a vacuum, and he asks Mersenne to tell him about the shape of the images seen through a Torricellian tube [2] in order to test this part of his theory.

In his third and final theory, developed in *De Corpore* (1655), Hobbes replaces expansion and contraction by what he calls the ' simple circular motion '. This he seems to think is natural to bodies, in spite of what he says elsewhere about the law of inertia, for, like Galileo, he seems not to have clearly conceived that inertial motion is rectilinear and not circular. The simple circular motion is a translation of a body in a circle without rotation, like the motion

[1] *De Corpore* (1655), Chap. ix, para. 7

[2] Letter to Mersenne, 1648, described in Brandt, *Thomas Hobbes' Mechanical Conception of Nature*, Copenhagen and London, 1928, p. 206

used by a man sieving. In this theory Hobbes will not admit any vacua at all—the fluid aetherial medium which fills space and interpenetrates all grosser bodies does not consist of fine grains, but is perfectly homogeneous, like the interior of the Democritan atom, or the void. Experiments showing apparent contraction or evacuation of air, such as those with barometers and thermoscopes, Hobbes explains by assuming that air penetrates the mercury or water bounding the alleged vacuum, so that in fact the amount of air present remains the same.[1] It is never made clear whether the all-pervasive aetherial medium is air, or some more subtle fluid, and there is no attempt to test any of these assertions by experiment.

Hobbes goes on to give mechanical accounts of light, heat, sound, and cohesion, with an ingenuity as great as that of Descartes, but with ever greater carelessness about the correspondence of his hypotheses with fact. The simple circular motion of the sun causes ripples of varying pressure to press outwards through the aether, and this pressure upon the eye and the body produces the sensations of light and heat. Sound is generated in a similar way by the motion of the medium, but by perceptible vibrations rather than by pressure producing only imperceptible motions. Even smell is ascribed to air-motion rather than to emitted corpuscles. Gravity, Hobbes remarks, must be due to the external action of the earth and not to an internal appetite of the falling body, as the Aristotelians assert, and he explains it in terms of convection currents of air produced by the centrifugal force of the earth's rotation in a way similar to that of Descartes, and with the same vagueness about the direction of fall which such a theory involves. Hobbes sees that if his theory is correct, the velocity of fall should be less towards the poles because the centrifugal force is less there, but he does not seem to grasp, as Huygens did, that the direction of fall would be towards the earth's axis and not towards its centre unless more complicated air-motions are assumed.

Magnetic and electric attraction are treated together, and ascribed to a simple circular motion of small parts of the attracting body. The motion is transmitted through the air and produces movement of those bodies which 'resonate' with it: iron in the case of the magnet, straw etc. in the case of jet and amber. Hobbes compares the process to the resonance produced in strings having the right properties to receive the vibration, and to the raising of grains of sand from the bottom of a vessel full of liquid by stirring

[1] *De Corpore*, Chap. xxvi

the liquid. How microscopic motions of parts of the attracted body are converted into macroscopic motion of the whole body Hobbes does not explain.

Less reliance on systems and more attention to the philosophy underlying mechanical explanations are to be found in the writings of Robert Boyle. Boyle was a professed disciple of the Baconian method of induction, and on the whole averse to speculative theoretical explanations which could not be associated closely with experiment, but this tentative attitude towards hypotheses in general did not extend to the corpuscular theory itself. His grounds for acceptance of this were not however metaphysical but empirical and methodological. In a tract entitled *The Excellency and Grounds of the Mechanical Hypothesis* published in 1674, he states that his mechanical hypothesis is not Democritan nor Cartesian but simply the belief that the phenomena of the world are ‘ produced by the mechanical affections of the parts of matter and that they operate upon one another according to mechanical laws ’.[1] He gives five reasons for this belief. Firstly, mechanical principles are intelligible and clear, unlike the principles of the peripatetics and Paracelsian chemists, and by this he means that they provide explanations at which the mind rests, as in the mechanical explanations of eclipses and optical images. Secondly, no explanation can make do with fewer principles than two, and matter and motion are the two principles of mechanism. Thirdly, no principles can be more primary than these two. Fourthly, none can be more simple. Fifthly, these principles are comprehensive, for they can give rise to as many combinations as letters in a literature, and they can account for microscopic as well as macroscopic effects, for one sees, for example, in a microscope that a single grain of sand is as various as the rocks themselves.

The first four of these arguments might be said to be purely methodological ; Boyle is merely recommending economy of thought in the introduction of principles of explanation ; but the last argument brings in something else, for Boyle here says that whatever detailed hypothesis may be found to correspond with the facts, it must be reducible to matter-in-motion and cannot contradict it. If there are immaterial substances, he says, their way of working is unknown to us and they are no help in conceiving how things are effected ; furthermore, even ‘ if an angel himself should work a real change in the nature of a body, it is scarce conceivable to us men, how he could do it without the assistance of local motion ’.[2] This

[1] *Works*, IV, p. 69 [2] ibid., p. 73

appears to be a generalisation from experience and not an *a priori* argument to show the comprehensiveness of matter and motion, for why could not the angel change, for example, the colour of a body directly? But Boyle does not consider that hypotheses, including the mechanical hypothesis, need be shown to be true *a priori*, and he compares them to the key to a cipher, which is justified by its results, an analogy later taken up by Leibniz.[1]

Boyle's tract might be called the manifesto of the mechanical philosophy. It is noticeable that it is intelligibility and simplicity which are insisted upon rather than testability, and indeed if the matter-in-motion model is really presupposed by all kinds of change as Boyle's fifth argument suggests, then the model is irrefutable. But if this theoretical framework is filled in by detailed hypotheses, then Boyle is cautious and insists upon tests. He thinks matter is in some sense particulate, but comes to no firm conclusions about the exact nature of the corpuscles, that is, whether they are divisible or indivisible, and whether the spaces between bodies are void or filled with subtle matter. The Cartesians have not, he says, proved the existence of the subtle matter experimentally, and it is therefore a metaphysical question.[2]

Most of Boyle's voluminous writings in physics are devoted to the attempt to ascertain by experiment the modes of motion which would account for various physical processes. He uses his experiments on the vacuum to refute the theory of attraction by suction, and this leads him to suspect that all examples of alleged attraction are really examples of impulsion, just as when a man is drawing a chain behind him, the action on each link is really pulsion from behind. He is hopeful that magnetic attraction can be explained in this way, perhaps by means of screwed particles as Descartes had suggested, but he recognises the difficulty that magnetic attraction does not appear to be hindered by the medium.

> 'But if there were none of these [screwed particles], nor any other subtil agents, that cause this motion by a real, though unperceived pulsion; I should make a distinction betwixt other attractions and these, which I should then stile attractions by invisibles. But, whether there be really any such in nature, and why I scruple to admit things so hard to be conceived, may be elsewhere considered.'[3]

[1] ibid., p. 77; and Leibniz: *New Essays concerning Human Understanding*, Book IV, Chap. xii, 13. cf. also L. E. Loemker, 'Boyle and Leibniz', *Journ. Hist. Ideas*, XVI, 1955, p. 22
[2] *Experiments on the Spring of the Air* (1660), *Works*, I, p. 37
[3] *Of the Cause of Attraction by Suction* (1674), *Works*, IV, p. 130

Elsewhere he describes how he tested the hypothesis of magnetic particles by careful weighing, but cannot satisfy himself on the question, and in any case he thinks that a magnet may be replenished by a continuous stream of such particles from the earth, so that there is no change in weight due to their emission. And in general he looks upon the doctrines of ' effluvia, of pores and figures, and of unheeded motions, as the three principal keys to the philosophy of occult qualities '.[1] By experiments in a vacuum Boyle shows that electric attraction is not due to air motion, and he speaks favourably of the hypotheses of electrical effluvia of Gassend and others.[2] His experimental investigations into the elasticity of the air, for which he is best remembered, led him to a hypothesis of air atoms consisting of coiled springs which exert an outward pressure due to the centrifugal force of their rotation. Heat, he suggests, increases the rotatory motion, and therefore produces expansion of the air.[3]

Of all the new experiments of the seventeenth century, that of Torricelli was one of the most influential from a theoretical point of view, because it led to re-examination of the question of void by seeming to dispose of arguments against the void based on the *horror vacui*. But the Torricellian vacuum was of course fundamentally irrelevant to the controversy about void, because it showed only that there could be space void of air, not that the space was void of all matter whatsoever. There was no contradiction in asserting that the space above the mercury in a barometer tube contained ' subtle matter ', or aether, many times lighter than air. It was moreover convenient to adopt this interpretation of the experiments, because most philosophers came to think that influences such as light, heat, and magnetism passed from one body to another by means of modifications in a medium, rather than by corpuscles of some kind emitted from the body as in the original atomic theories, and in order to explain the transmission of these influences across the Torricellian vacuum it was necessary to postulate a subtle medium which remained when the air was removed. The question of whether this aetherial fluid was ultimately composed of indivisible parts surrounded by void spaces, or whether it was continuous, became less and less relevant to the physical theories in which it appeared, and natural philosophers became less and less willing

[1] *Of Effluviums* (1673), *Works*, III, p. 660

[2] *Experiments and Notes about the Mechanical Origin or Production of Electricity* (1675), *Works*, IV, p. 345

[3] *Experiments on the Spring of the Air*, *Works*, I, p. 54 ; *An Explication of Rarefaction*, ibid., p. 178

to speculate upon what had become, as Boyle said, a metaphysical question, incapable of test. But whether the corpuscularian philosophers favoured theories of transmission by means of a medium or by emission of corpuscles, they had no doubt that action must be transmitted by contact and not at a distance, and this was a direct result of the association of action at a distance with souls, sympathies and antipathies, and other organic analogues.

In various mechanical theories the analogy of 'resonance' became popular. We have seen that Hobbes used it, and it is also developed by Robert Hooke in a theory of 'congruence', by means of which he explains all transmission of action. In his lecture on springs, entitled *De Potentia Restitutiva* (1678),[1] he suggests that matter itself consists of parts in continual vibration, and that the apparent extension of a body is due to the amplitude of these vibrations, just as the apparent extension and impenetrability of a cube of side one foot might be due to the vibrations of an iron plate one foot square through a distance of one foot in a direction perpendicular to its plane. The vibrations, he thinks, are communicated from one particle to another by impact at the ends of their respective oscillations, and from this hypothesis he is able to derive the variation of force with extension in springs.

Earlier, in his *Micrographia* (1665), Hooke equated the vibratory motions with heat, and suggested that they are more violent in fluids than in solids, just as sand shows more fluid properties when it is shaken. If sands of various kinds are shaken together, it is observed that 'like goes to like', and in this Hooke sees an analogy for the formation of liquid drops and the cohesion of solids: parts which cohere are those which have 'congruous motions', incongruous motions tend to drive bodies apart. This he thinks may account even for gravitation: perhaps the incongruity of the vibrations of heavy bodies and those of the aether drives the heavy bodies towards the earth with which they are congruous.[2] Hooke criticises the Cartesian theory of gravitation by remarking that the moon probably exerts a gravitational force, and that this cannot be due to centrifuging, because the moon has no rotation on its axis.[3] Light, he thinks, is transmitted by very fast, short vibrations, in a medium which must be congruous with them and capable of transmitting them to great distances in a very short time, although not necessarily instantaneously. Light is propagated in straight lines in all directions and produces spherical wave fronts spreading

[1] Facsimile ed., R. T. Gunther, *Early Science in Oxford*, VIII, pp. 339ff.
[2] Gunther, XIII, pp. 12–22 [3] ibid., p. 246

out from the source like the ripples on the surface of water. Refraction is due to the fact that the propagation is less easy in some media than in others, perhaps because of differing degrees of congruity of transparent substances and rays of light.[1] The 'tenacious and attractive power' of congruity is also invoked to explain electric and magnetic attraction, which Hooke regards as forces similar to cohesion but exerted by congruous motions transmitted through a medium instead of by direct contact. The cohesive force of direct contact between bodies with congruous motions is however much stronger than that transmitted to a distance.[2]

On the whole the medium theories of transmission, at least of light, had ousted emanation theories by the second half of the century. In 1661 Joesph Glanvill wrote

> 'Whether sensation be made by corporeal emissions and material eidola [images], or by motions impressed on the Aethereal matter, and carried by the continuity thereof to the Common Sense, I'll not revive into a Dispute : the ingenuity of the latter hath already given it almost an absolute victory over its Rival.' [3]

The idea of vibration offered more possibilities of explaining the properties of light, by variation of frequency and amplitude, than the simple flight and collision of corpuscles, and the flexibility of the latter as an explanatory analogy was increased only when Newton postulated light corpuscles which were not simple Democritan atoms, but centres of force capable of attracting and repelling at a distance. The greatest triumph of the seventeenth-century plenum theories was undoubtedly the formulation of the wave theory of light by Huygens. Huygens was, on the whole, a follower of Descartes, but without his desire for universal explanation in terms of one all-powerful hypothesis, and with a far greater ability for applying mathematics to the details of the physical world. He had a clear conception of the method of mathematical hypothesis, and in the Preface to his *Traité de la Lumière* of 1690 he remarks that demonstrations in physics cannot be as certain as those of geometry, since the principles are verified by their conclusions. But, he continues,

> 'It is always possible to attain thereby to a degree of probability which very often is scarcely less than complete proof. To wit, when things which have been demonstrated by the Principles that have been assumed correspond perfectly to the phenomena

[1] ibid., pp. 54ff. [2] ibid., pp. 31, 32 [3] *Vanity of Dogmatizing*, p. 28

which experiment has brought under observation; especially when there are a great number of them, and further, principally, when one can imagine and foresee new phenomena which ought to follow from the hypotheses which one employs, and when one finds that therein the fact corresponds to our prevision.'[1]

Huygens's theory of wave motion in a plenum illustrates a general fact about the development of physics at this time, also evident in Newton's theory of gravitation, namely that once the right mathematical principles were arrived at, they could be exploited and made to yield new predictions even though the exact nature of the physical processes described by the mathematics remained obscure. Huygens did put forward various physical theories of aethers and subtle matters of the Cartesian kind, but their details were tentative. 'It is inconceivable to doubt', he says, 'that light consists in the motion of some sort of matter', and he believes that 'the true Philosophy' is that 'in which one conceives the causes of all natural effects in terms of mechanical motions'.[2] He rejects the explanation of light in terms of corpuscles emitted from the luminous object, because its speed of travel is very great, and its rays cross one another without mutual disturbance. Therefore, he thinks, it must be a transmission like sound which disturbs the air in concentric spherical wave-fronts, just as, in two dimensions, a stone thrown into water disturbs its surface in concentric circular ripples. He refutes Descartes's arguments supporting instantaneous transmission of light, and accepts the estimate of its speed made by Römer, in 1666, on the basis of observations of the eclipses of Jupiter's satellites. Huygens then goes on to show by experiments that air propagates sound but not light, and that light must therefore be propagated by a subtle aether which penetrates bodies impenetrable to air, and which must be very hard and elastic in order to transmit disturbance at such great speed. The detailed structure of this aether is not discussed, but he suggests that its individual particles must have springiness like that of billiard-balls which, when placed in line, will transmit an impulse from one end to the other, so that the last ball moves off while the others remain at rest, in other words, Huygens's theory is one of longitudinal waves. Various types of phenomena involving transmission through a medium seem to Huygens to require three different aethers interpenetrating one another: one to account for gravity and two for

[1] *Treatise on Light*, trans. S. P. Thompson (1912), p. vi; *Œuvres*, XIX, p. 454
[2] *Treatise on Light*, p. 3

the anisotropic refracting properties relating to Iceland spar.[1] The mathematical wave-theory does not however depend upon the nature of the medium, but only upon the assumption of a disturbance propagated in spherical wave-fronts, and from this Huygens is able to deduce, in demonstrations of great clarity and elegance, the laws of reflection and refraction, including those of Iceland spar.

Locke on the Mechanical Philosophy

The main contention of the corpuscularian philosophers was simple and often repeated. Most of the simple physical processes that we observe involve, in some way or other, motion of parts of matter. Heat is certainly engendered by the friction of bodies against each other, flame is mobile, and it does not need much imagination to compare the effects of burning with those of breaking bodies apart by impact. Where there is heat there is often light, and if one cannot conceive of action at a distance one is forced to the conclusion that light ‘ travels ’; in both respects light seems to require motion of some kind. It is obvious that sound depends on a motion of the sounding body, and its transmission can be connected with the motion of air by observing the effects upon it of wind and obstacles. Electrical effects are associated with frictional motion, and magnetic properties can be induced in iron by striking and heating. It is immediately clear that if all these effects are to be ascribed in detail to motion and impact, it must be the motion and impact of small, invisible parts of matter that is generally involved. It became easier to conceive these when the microscope became widely known, and gave fascinating evidence of the subtle structure of gross matter and the motions of its parts. The title of Hooke's *Micrographia* is instructive, for it contains fundamental discussions of the nature of matter and light as well as descriptions of various bodies seen under the microscope.

In addition there was the prestige acquired by mechanical explanations on more philosophical grounds. For Galileo, shape, size, and motion were the primary qualities of bodies upon which all others depended, and he and Descartes asserted that the other qualities are functions of the human sense organs rather than of the external stimuli, just as tickling is in the armpit and not in the feather. For Descartes, extension and motion were the clear and distinct ideas of body, and he thinks that the behaviour of the subtle parts of matter is best described by analogy with the behaviour of mechanical devices which are familiar to us. The science of

[1] ‘Letter to Papin’ (1690), *Œuvres*, x, p. 177

mechanics itself made steady progress among Galileo's disciples in Italy, Descartes and his contemporaries in France, and among the founders of the Royal Society in England, so that by the time Newton began his work the laws of motion and of impact were fairly well understood.

When Descartes, Hobbes, and others concluded that most if not all processes in the universe were reducible to matter-in-motion, there were naturally philosophical and theological protests, but there was in general no criticism of the programme as a method in physics, and many of its ablest exponents were far from being philosophical materialists. Doubts were, however, expressed on philosophical grounds by John Locke, not because he doubted that physical processes did ultimately depend on the motions of small parts of matter, but because he doubted whether these motions could ever be known.

In his *Essay concerning Human Understanding*, Locke makes the familiar distinction between primary qualities, that is, solidity, extension, figure, motion or rest, and number; and secondary qualities, which derive from the primary qualities of the small parts of matter. But he doubts whether we can have knowledge of these small parts. The mind, he thinks, is bound to ascribe the same primary qualities to the minutest parts of matter that it perceives in gross matter, but we cannot know what are the size, figure, and texture of these parts, upon which the secondary qualities depend, or even whether they depend upon them at all, rather than 'upon something yet more remote from our comprehension'.[1] It is therefore impossible, says Locke, to know which secondary qualities have a necessary connection with one another, and further, it is impossible to know any necessary connection between the primary qualities of small parts and the secondary qualities which they produce in us. Again, we treat things as if they were isolated from the universe, but we do not know how many properties of things may be due to their relations with other things:

> 'The great parts and wheels, as I may so say, of this stupendous structure of the universe, may, for ought we know, have such a connection and dependence in their influence and operations one upon another, that perhaps things in this our mansion would put on quite another face, and cease to be what they are, if some one of the stars or great bodies incomprehensibly remote from us, should cease to be or move as it does.'[2]

[1] *Essay*, Book IV, Chap. iii, 11 [2] ibid., IV, vi, 11

Locke is here comparing the situation in natural philosophy with that in mathematics and in ethics, where, he thinks, we can have direct knowledge of the real essences involved, because we can perceive directly the agreement and disagreement of ideas and can therefore be certain of their truth or falsity. If we could know the real essence of gold as we do of a triangle, we could be certain of all the properties of gold and of their interconnections. But we cannot, so 'natural philosophy is not capable of being made a science',[1] and our knowledge cannot reach beyond experience.

The conclusion that natural science cannot have the same kind of certainty as mathematics is one that no modern philosopher would wish to dispute, although we no longer share Locke's view about ethics. His pessimism about the potentialities of the mechanical philosophy turned out to be unjustified, but his rejection of essentialism in regard to 'the minutest parts of matter' has not. The pragmatic success of science has not depended upon infallible intuitions of clear and distinct ideas as in the rationalisms of Aristotle and Descartes. In his discussion of theories about hidden causes Locke outlines exactly the analogical method which science has used since the seventeenth century in dealing with those things which 'either for their smallness in themselves or remoteness from us, our senses cannot take notice of'. No better summary could be given of the corpuscular philosophy as an analogical method than this passage from Locke, and there is no doubt that he himself would have extended the method to include analogies other than those from matter and motion if the state of science had seemed to require it.[2] The passage is worth quoting in full as a statement of the method of the corpuscular philosophy as it was understood in the seventeenth century.

> 'Concerning the manner of operation in most parts of the works of nature: wherein though we see the sensible effects, yet their causes are unknown, and we perceive not the ways and manner how they are produced. We see animals are generated, nourished, and move; the loadstone draws iron; and the parts of a candle successively melting, turn into flame, and give us both light and heat. These and the like effects we see and know: but the causes that operate, and the manner they are produced in, we can only guess and probably conjecture. . . . *Analogy* in these matters is the only help we have, and it is from that alone we draw

[1] ibid., IV, xii, 10

[2] He was in fact prepared to adjust his views on the communication of motion after the publication of Newton's *Principia*; see *infra*, p. 166.

> all our grounds of probability. Thus observing that the bare rubbing of two bodies violently one upon another, produces heat, and very often fire itself, we have reason to think, that what we call *heat* and *fire*, consists in a violent agitation of the imperceptible minute parts of the burning matter. Observing likewise that the different refractions of pellucid bodies produce in our eyes the different appearances of several colours ; and also, that the different ranging and laying the superficial parts of several bodies, as of velvet, watered silk, etc. does the like, we think it probable that the *colour* and shining of bodies is in them nothing but the different arrangement and refraction of their minute and insensible parts. Thus finding in all parts of the creation, that fall under human observation, that there is a gradual connection of one with another, without any great or discernible gaps between, in all that great variety of things we see in the world, which are so closely linked together, that in the several ranks of beings, it is not easy to discover the bounds betwixt them ; we have reason to be persuaded that by such gentle steps things ascend upwards in degrees of perfection. . . . This sort of probability, which is the best conduct of rational experiments, and the rise of hypothesis, has also its use and influence ; and a wary reasoning from analogy leads us often into the discovery of truths and useful productions, which would otherwise lie concealed.[1]

In the later part of this passage, Locke is basing his method of analogy on belief in the ' great scale of being ', which was a commonplace in the sixteenth and seventeenth centuries, and according to which created beings are ordered in an almost continuous scale, from the lowest beings through animals and man and higher ranks of ' intelligent beings ', up to the perfection of the Creator. This theory is dead, but it provides Locke with that continuity and correspondence between the macroscopic and the microscopic which is always necessary if theories of the microcosm are to be testable. We have seen that so-called observation statements must be interpreted in the language of a theory before they can test that theory, and in the case of the seventeenth-century corpuscular theories, which speak in terms of matter endowed with the primary qualities, this condition is satisfied at once, for the model is a specialisation of ordinary language. Ordinary language contains the necessary concepts for describing matter in motion, and their

[1] *Essay*, IV, xvi, 12

reference is clear ; the corpuscular theories are therefore not only intelligible in themselves, but also have clear reference to empirical situations. To take Locke's example : if we are told that heat is produced by the violent motion of the small parts of matter, we can not only develop a model in which, for example, heat is associated with the kinetic energy of molecules, but we also know how to test the theory, for imperceptible motion of the parts of matter will clearly be produced by violent rubbing, by striking, by passing an electric current (itself thought of in terms of moving particles), and so on. The point may appear trivial, because we are so accustomed to presupposing the seventeenth-century world-model according to which all physical change is *really* produced by matter in motion, that we forget that this is in a sense a metaphysic; derived from the Greek metaphysical theory of atomism; established by the overthrow of opposing metaphysical systems, namely the Aristotelian, Stoic, and Neo-Platonic ; and justified by the Pythagoreanism of Copernicus, Kepler, and Galileo, and by the metaphysical arguments about primary qualities of Descartes, Galileo, and Locke. But it differs from some other metaphysical systems in being not only intelligible in itself, but also in having clear empirical reference, and here perhaps are some of the necessary conditions for such systems to be useful in science.

The transition to the seventeenth-century world-model is instructive, for, as we shall see later, an important question for modern physics is whether all future theories must be built upon the foundations of this model, or whether a new transformation of the fundamental analogy can take place, comparable with those from mythological animism to Aristotelianism, and from both to mechanism. Such a transformation would however be more fundamental than the dispute about contact-action and action-at-a-distance, for the classic instance of this dispute, as we shall now see, took place wholly within the terms of the analogy of matter-in-motion.

Chapter VI

THE THEORY OF GRAVITATION

Gravity as Internal Tendency or External Attraction

THE phenomena of magnetism and of falling bodies had always been among the most difficult to reconcile with action by contact, and to these should be added the tidal motion of the sea. The tides had been correlated with the path of the moon since before the time of Ptolemy, and he and Albumasar (the latter in the ninth century) had ascribed them to a special virtue exerted by the moon upon the oceans. The theory was rejected by strict Aristotelians, who would have nothing of occult virtues acting at a distance, and were therefore driven to ascribe the tides to the moon's light and heat. But this theory was untenable, for, as Albumasar showed, there is no correlation between the tides and the amount of light received from the moon. Grosseteste evolved an ingenious theory, which was repeated by Roger Bacon and Albertus Magnus, according to which the moon's power draws mist up from the sea-bed and through the surface of the sea, whereupon the sea collapses like a pricked bubble. Other medieval writers, including Aquinas,[1] admitted that there might be influences from the heavenly bodies other than their light and heat, and the action of the moon on the sea was widely compared with that of the magnet on iron. Astrologers had, of course, always asserted that special virtues emanated from the heavenly bodies, and for them there was no difficulty in assuming such virtues as the cause of the tides, especially as the moon was traditionally the 'humid' planet, and by the principle of the attraction of like for like, everything containing water on the earth was supposed to swell and shrink with the waxing and waning of the moon. Physicians, again, were used to the idea of lunar influences, for it had been a theory of Galen's that the critical days of diseases were correlated with the phases of the moon.

The Copernicans on the other hand, true in this to the spirit of Aristotle, rejected any such occult cause of the tides. Galileo in particular tried to show that they are due to the combined effects of the daily rotation and annual revolution of the earth, regarding them in fact as the best proof that the earth does rotate. It is

[1] *Summa Theologica*, I, cxv, 4

absurd, he says, to think that they are due to the moon 'For neither by the light of the moon or sun, nor by temperate heat, nor by differences of depth can we ever make the water contained in a motionless vessel run to and fro, or rise and fall but in a single place.'[1] He severely criticises Kepler because he has 'lent his ear and his assent to the moon's dominion over the waters, to occult properties, and to such puerilities'.[2]

The Copernican theory had, however, removed any plausibility there may have been in Aristotle's theory of falling bodies. It could no longer be maintained that bodies have a natural tendency to fall to the centre of the universe, since that centre no longer coincided with the centre of the earth, but there had already in the Middle Ages been some modification of the Aristotelian tradition here, in the suggestion that both magnetism and gravitation should be described as tendencies of kindred bodies to unite, as in the theory of the attraction of like for like, which, of course, Aristotle himself had strongly repudiated.[3] The medieval Aristotelians, including Averroes and Aquinas,[4] spoke of a substantial form of bodies which resisted attempts to separate them from their like, and which caused their return when separated, as iron returns to its kindred magnet.

After the Copernican revolution it was still possible to retain this version of the Aristotelian tradition. Copernicus himself suggests that all earthy bodies tend to get as near as possible to the centre of gravity of the earth, thus explaining the fall of bodies, and also the fact that the earth has a spherical shape. The other heavenly bodies too are observed to be spherical, and it is reasonable to suppose that there is a similar attraction of solar matter to the centre of the sun, Martian matter to the centre of Mars, and so on. Neither he nor his immediate disciples suggest that there is attraction between the matter of one heavenly body and that of another, and Galileo, who follows Copernicus closely in this theory of gravitation, agrees that the matter of each heavenly body tends only towards its own centre of gravity.[5] The point is made explicit by Gilbert:

[1] *Two Chief World Systems* (1632), trans. S. Drake, Berkeley and Los Angeles, 1953, p. 421 [2] ibid., p. 462

[3] The suggestion is at least as old as Plutarch. He considers that the moon is a body of the same kind as the earth, and that bodies tend, not to the centre of the universe as such, but to the earth to which they are akin. Why then may there not also be attraction to the sun and moon? (*On the face of the moon*, 924 E). Plutarch also likens the swing of the moon in her orbit to that of a stone in a sling (ibid., 923 D), an analogy taken up by Hooke. [4] *De Physico Auditu sive Physicorum Aristotelis*, §1812

[5] *Two Chief World Systems*, p. 33

'All that is earthly unites itself with the earth's globe; in the same way all that is of the same substance as the sun tends towards the sun, all the moon's substance towards the moon, and the same for the other bodies which make up the universe. . . . It is not a desire or inclination towards a place, or a part of space, or a terminus, but towards a body, the fount and source and origin, where all the parts are united, conserved, and where they can rest, safe from all peril.' [1]

As we have seen, Gilbert thinks that the tendency to unite is due to contact with the atmosphere of the parent body. Thus there is some ambiguity as to whether, like the Aristotelians, he locates the active agent in the moving body which 'desires' to move to its parent, or in the parent body which sends out its 'breath' and lays hold on bodies which are akin to it. In neither case, however, is there any suggestion of external attraction at a distance. This notion does not appear either in Galileo, who suggests,[2] not only that the natural motion of a falling body is internal, but that its upward motion after it has been thrown is equally internal, since when it has left the projector there is no external force acting upon it. He gives the example of a swinging pendulum to show that the same internal motion carries the bob alternately nearer to and farther from the centre of the earth. There are echoes here of the medieval theory of impetus, and the 'internal motion' he speaks of is nearer to our 'momentum' than to the Aristotelian 'innate gravity', but there is at any rate no suggestion that gravity may be due to an external attraction. Galileo prefers to remain silent on hypothetical questions of that kind, and is not prepared to say more about gravity than that it manifests itself in an impetus similar to that shown by a rising body or a pendulum.

For Galileo's contemporaries, however, the difference between internal tendency and external attraction remained important, because the two explanations presupposed quite different types of physical causation. The first suggestions linking gravitational effects with external attraction came, curiously enough, not from the 'progressive' schools of Copernicus and Galileo, but from the traditions of astrology. In Kepler an ability to discern the mathematical implications of the new astronomy was combined with a not unsympathetic attitude towards astrology, and it was Kepler who asserted explicitly that 'Gravity is not an "action", it is a

[1] *De Mundo nostro Sublunari Philosophia Nova* (posthumous), London, 1651, pp. 115, 116. cf. *De Magnete*, p. 229

[2] *Two Chief World Systems*, pp. 234ff.

" passion " of the stone which is drawn ', and ' it is the earth which draws the stone, rather than the stone which seeks the earth '.[1] He insists that attraction can be exerted only by bodies, not by a mathematical point such as the centre of the universe, but he does not take the final step of suggesting that *all* bodies exert mutual attractions. In this he agrees with the Copernicans and Gilbert: bodies are attracted only to those to which they are akin. If two stones were placed near each other at a great distance from all kindred bodies, they would move to meet each other, but there is no suggestion that they would move to meet any other body, for example the sun, or one of the planets.

Kepler describes magnetic attraction in a similar way and unlike Gilbert saw in gravitation an influence of magnetic character. The tides are due to a magnetic attraction of the moon, not for the *waters* of the sea, but for its *earthy* parts, for Kepler considers that the substance of the earth and moon are akin and exert a reciprocal attraction.

The Planetary Orbits

A new problem was now coming to the fore, concerning the forces which kept the planets in their orbits, although at first there was no apparent connection between this and the problem of falling bodies. The Copernican theory had not greatly affected the situation here, because the planets were still conceived to revolve in circular paths, as in the Aristotelian universe, the difference being that the centre of revolution was shifted from the earth to the sun. Copernicus himself seems to have thought, in Aristotelian fashion, that the natural motion of the planets is circular, and he therefore required no special force to keep them in their orbits. Similarly, Galileo suggests that the planets were at first given their uniform motions by the Creator, and that a circular orbit is appropriate to uniform motion, which can continue indefinitely. His nearest approach to the law of inertia for bodies moving on the earth was the statement that once they had acquired motion in a *horizontal line*, that is, along the earth's surface, they would continue in it perpetually and uniformly, and in a similar way he seems to have thought that the inertial motion of the planets is uniform circular motion about the sun.[2]

Kepler, however, discovered that the planetary orbits are not

[1] ' Letter to Herwart von Hohenburg, 28 March 1605 ', trans. from *Gesammelte Werke*, ed. Caspar, xv, p. 184; *Astronomia Nova* (1609), ibid., III, p. 25

[2] *Two Chief World Systems*, pp. 21, 28

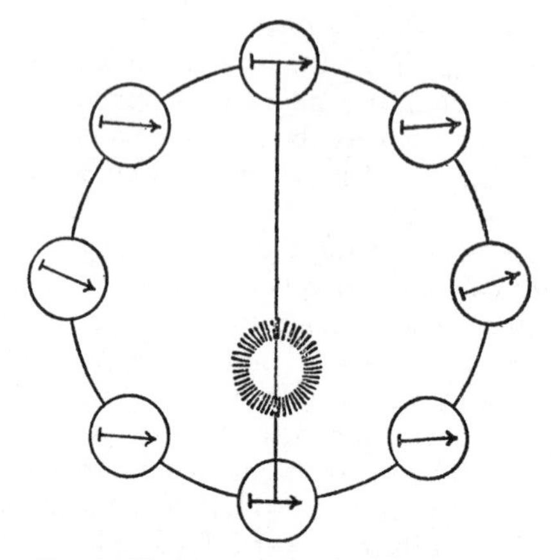

Fig. 2 Kepler's eccentric orbit produced by magnetic attraction of the sun

circular but elliptical, and as it was implausible to suggest that the 'natural' orbit of a body is anything as complex as an ellipse, the fact seemed to require explanation in terms of forces acting at a distance between the sun and the planets. The medieval theory of multiplication of species lay to hand as a means of expressing such action, and Kepler uses its terminology in describing a *species motus* emitted by the sun and passing through space to the planets.[1] This species has two distinct effects on a planet. In the first place it has a component of velocity resulting from the natural revolution of the sun on its own axis,[2] and which is in a direction tangential to the planet's orbit and therefore provides the motive power necessary for maintaining revolution in the orbit. Kepler has here dropped the Copernican assumption of natural circular motion, and uses the Aristotelian axiom that all motion requires a moving force. Secondly, the species acts like a magnetic force in relation to the planet, which is itself a natural magnet similar to the earth. There are certain magnetic fibres in the planet lying parallel to each other, and when these are tilted so that one end is nearer the sun than the other, the species of the sun either attracts or repels the planet, depending on which end is nearer. Thus the

[1] *Astronomia Nova*, loc. cit., pp. 240ff.; *Epitomes Astronomiae* (1618), loc. cit., VII, pp. 298ff., 333ff.

[2] Gilbert had suggested that a spherical lodestone perfectly balanced would rotate naturally upon its axis (*De Magnete*, pp. 220ff.) The suggestion is criticised by Galileo (*Two Chief World Systems*, p. 413).

distance of the planet from the sun is varied, and an eccentric orbit results.

Owing to his initial assumption that a tangential force is required to move the planet along its orbit, Kepler had to assume that the *species motus* of the sun was emitted only in the plane of the planetary orbits, so that it diminished in the inverse ratio of the distance. This was the only supposition which was consistent with the observed relation between the periods and mean radii of the orbits, namely the relation $T^2 \propto a^3$, which is well known as Kepler's third law, but the assumption meant that he departed from the usual theory of multiplication of species, in which the species were emitted in all directions, and hence diminished in power in the ratio of the inverse square of the distance. The inverse-square law had long been recognised in the case of light emission. Thus Kepler's force is inconsistent with the only conception of the action of emanations then accepted, and, apart from connecting the earth's magnetism with its orbit, it is *ad hoc*.

This conception of action at a distance by the sun upon the planets was not acceptable to Kepler's immediate successors. Galileo and Descartes, as we have seen, rejected it. Attractive forces next appear in a work by Roberval,[1] in which, afraid to appear openly as a Copernican, he pretended merely to be editing a manuscript of Aristarchus of Samos. Here Roberval supposes that all the parts of the fluid aether filling the universe have a mutual attractive property, so that they tend to move to the centre of gravity of the whole. All parts of the earth, the sun, and the planets have a similar property, but as with Kepler, there is no mutual attraction between parts of one system and those of another. Roberval does not discuss the variation of this attractive force with distance, except to say that the effect of each planet is felt only in its own neighbourhood. This limitation on the range of the attractive forces was generally thought to be necessary to account for the fact that the planets and the stars do not fall in upon one another, and even Newton has recourse to it.[2] Roberval attempts to derive the relations between the planets and the sun in terms of ordinary hydrostatical principles : the planets 'float' in a fluid aether whose density varies at different distances from the sun. His idea of attractive forces is severely criticised by Descartes, who says it assumes that every piece of matter in the universe is ensouled, and

[1] *Aristarchii Samii de Mundi Systemate*, Paris, 1644

[2] General Scholium to the *Philosophiae Naturalis Principia Mathematica*, 2nd ed., 1713 ; 3rd ed., Motte's trans. revised by Cajori, Cambridge, 1934, p. 544. Unless otherwise stated, references to this work will be to the latter edition.

that the souls are 'intelligent and even divine to know what is going on in places very distant from them without any messenger, and even to exert their power there'.[1]

In a memoir of 1669 Roberval clearly states the alternative theories of weight then being debated and indicates the kind of evidence which would help to decide between them. There are three opinions: first, the Aristotelian thesis that weight is a quality of the heavy body which carries it downwards; second, that it is a reciprocal attraction between all parts of a body causing them to unite as closely as possible; third, the Cartesian theory that surrounding subtle matter pushes heavy bodies downwards. But Roberval continues:

> '. . . although there is manifest contradiction between these opinions, they have nevertheless this in common, that they are founded only on the pure thought and imagination of their authors, who have no clear and distinct principle; and so they have no certain proof of what they say on the subject.'[2]

The second opinion, however, seems to Roberval the most likely, if one is allowed to postulate occult qualities: 'that is those for which we have not proper and specific senses'. It can be tested, for if it is true, one would expect that the same body will weigh less near the centre of the earth than upon its surface, and less also on the top of a mountain. Thus, metaphysical objections to attraction at a distance are less important to Roberval than the methodological requirements of 'clarity and distinctness' and the possibility of experimental test.

The next theory of interest was that of Borelli, published in Florence in 1666.[3] Borelli thinks that three forces are required to maintain a planet in its orbit, namely a 'natural instinct' to approach the sun, which he compares with the 'inborn force of gravity' of heavy bodies and with iron in the presence of a magnet; a force along the tangent to the orbit produced by light emanating from the rotating sun; and a centrifugal force due to the orbital motion, which balances the tendency towards the sun and prevents the planet falling into it. The second of these forces is reminiscent of Kepler's theory of species, but Borelli has so far understood the law of inertia as to realise that such a tangential force would produce continual increase of speed, although he does not explain what sets

[1] Letter to Mersenne, 1646, *Œuvres*, IV, p. 401

[2] *Registres de l'Académie des Sciences*, V, 1669, p. 129; quoted in R. Dugas, *La Mécanique au XVIIe Siècle*, Neuchâtel, 1954, p. 313

[3] *Theoricae Mediceorum Planetarum*, Florence, 1666

a limit to this speed or why the force is necessary at all. In explaining the force towards the sun, Borelli rejects the idea of external attraction and returns to the traditional innate tendency. The difference at this stage in the argument was important, because, as Roberval saw, there is more reason to expect the force to vary with distance if it is an external and mutual attraction, and Borelli does not in fact suggest that his ' natural instinct ' varies with the distance of the planet from the sun.

During the whole of this development, in which the elements of Newton's gravitational theory gradually emerged from Aristotelian physics, Neo-Platonic light metaphysics, astrology, and the mathematical idealisations of Galileo, it was not only that the concepts of mechanics were becoming clearer and more precise, but also, and perhaps more important, modes of thought were changing. The same search for quantitative, testable, theories is in evidence here as in the development of the corpuscular theories, but it is remarkable that hardly any influence from corpuscular theories can be found in the direct ancestry of Newton's theory. Those who, like Descartes and Hobbes, suggested corpuscular explanations of gravitation, did not contribute to the mathematical theory, and those who tried merely to describe the observed motions in terms of the new mechanics were content to remain agnostic about the physical causes of the motions. Even in the case of Hooke, who belonged to both streams of thought, there was no apparent connection between his corpuscular speculations and his planetary theory, and Huygens was too wedded to Cartesian ideas to exploit his theory of centrifugal force in the direction of a mathematical theory of gravitation.

The essential elements of Newton's theory were : an acceptance of Kepler's laws, a consistent application of the rectilinear law of inertia, a clear understanding of centrifugal force, and a doctrine of *universal* gravitation in which all matter in the universe is considered equivalent, and in which every part attracts every other part with a force proportional to the quantity of matter and inversely proportional to the square of the distance between the attracting parts. One or more of these elements appear in several of Newton's predecessors, and all of them appear in the work of Hooke, who stated the doctrine of universal gravitation in 1670 in these words :

> ' All Celestial Bodies whatsoever, have an attraction or gravitating power towards their own Centres, whereby they attract not only

> their own parts, and keep them from flying from them, as we may observe the Earth to do, but that they do also attract all the other Celestial Bodies that are within the sphere of their activity ; and consequently that not only the Sun and Moon have an influence upon the body and motion of the Earth, and the Earth upon them, but that Mercury, also Venus, Mars, Jupiter, and Saturn, by their attractive powers, have a considerable influence upon its motion as in the same manner the corresponding attractive power of the Earth hath a considerable influence upon every one of their motions also.' [1]

As early as the 1660s, Hooke had been experimenting on high buildings and in deep wells in order to find out how gravity varied with distance from the centre of the earth, and although these experiments were inconclusive, Hooke decided, certainly before 1680, that the attractive force of universal gravitation is proportional to the inverse square of the distance. He tried to find a model for the planetary orbits in the conical pendulum, the bob of which can be made to execute circles or ellipses about a centre under the influence of a force directed towards that centre. But the sun is not in the centre of the planetary ellipses, but in one focus, and Hooke did not realise that the law of force in the two cases is different.

This last point illustrates the fact that one thing was still lacking, and that was the mathematical technique required to interpret all the hitherto unrelated factors and to exhibit the logical relations between them. It was Newton who presented the complete theory in all its mathematical inevitability in his *Principia Mathematica* of 1687. This is not the place for a detailed historical account of this theory, but it is important to examine its logical shape in some detail in order to show how Kepler's laws, together with Newton's second and third laws of motion and a few additional assumptions, led ' deductively ' to the law of gravitation, and hence to judge how far Newton was justified in claiming that the theory involved no hypotheses.

An Analysis of Newton's Laws of Motion

Newton's first law of motion (the law of inertia) states that ' Every body continues in its state of rest, or of uniform motion in a right line, unless it is compelled to change that state by forces impressed upon it.' [2]

[1] ' An attempt to prove the motion of the earth ', London, 1674 ; in Gunther, op. cit., VIII, pp. 27, 28

[2] *Principia*, p. 13

From the historical standpoint this law arises from experiences with projectiles and bodies moving on smooth surfaces in which the ideal conditions of motion under no resistance are approached, and in which velocity is observed to remain nearly uniform. Force then comes to be understood as something which disturbs uniform motion or which is necessary to overcome resistance, itself thought of as resisting *force*. From this point of view the law can be regarded as an empirical generalisation, involving extrapolation to ideal experimental conditions, and presupposing that a space-time framework is given within which there are recognised methods of measuring time and distance, and that we can recognise situations in which force is acting.

From a logical point of view on the other hand, the law can be regarded as a definition of inertial motion, that is, if we do not assume that we have independent ways of recognising force-situations, it provides the means of testing whether or not there are forces acting upon the body. No doubt the form of the law is suggested by the fact that when what we normally call 'forces' are missing, bodies do tend, on the whole, to continue in their state of rest or uniform motion, but they do not always do so, as the example of falling bodies shows. If the law of inertia is regarded as a definition, we are then bound to *postulate* a gravitational force to account for the acceleration of fall, in spite of the fact that no obvious physical cause seems to be acting, whereas, if the law is regarded as an empirical generalisation, we shall be inclined to say that we have *discovered* a gravitational force, and, if we hold the view that all physical forces must act by contact, we are then bound to look for a hidden contact-mechanism. In view of Newton's known desire not to be bound to identify the physical mechanism involved, at least in the *Principia*, it is significant that the law of inertia is already implied in his definitions of 'inertial force' and 'impressed force'. Definition III states that 'The *vis insita* [later called *vis inertiae*] of matter, is a power of resisting by which every body, as much as in it lies, continues in its present state, whether it be of rest, or of moving uniformly forwards in a right line', and Definition IV states that 'An impressed force is an action exerted upon a body, in order to change its state, either of rest, or of uniform motion in a right line'.[1]

The second law of motion states that 'The change of motion is proportional to the motive force impressed; and is made in the direction of the right line in which that force is impressed'.[2] This

[1] ibid., p. 2 [2] ibid., p. 13

again may be interpreted empirically or logically: either as the natural generalisation of the fact that when what appear to be two equal forces act upon a body, they produce twice the acceleration produced by one force acting alone; or, if we assume with Newton (Definition I) that a statical definition of mass is already given, the law may be taken as the definition of the measure and direction of the force acting, the constant of proportionality between force and acceleration being the inertial mass of the body.[1]

Thus from the first and second laws we have two possible interpretations, depending upon whether force is assumed to be recognisable by its muscular, tensional, pressureal, centrifugal, and other effects. The status of the third law depends essentially upon which interpretation is adopted, and as this is not at all clear in the *Principia* it is important to consider it in detail. The third law states that 'To every action there is always opposed an equal reaction' and it is introduced by Newton in relation to the forces exerted on each other by bodies in contact, for example when a horse draws a load, and as a statement of the conservation of momentum (which he calls 'motion') in collisions. Let us suppose first that the empirical interpretation of force is adopted, namely that force can be recognised apart from the acceleration it produces. Now in force-situations where the forces are acting between parts of the same body, it is clear that the third law follows from the first, for if action and reaction between any two parts of the body were unequal, there would be a resultant force between them and hence a resultant acceleration of the body not due to any force external to it, and this is contrary to the first law. If, on the other hand, the forces are acting between different bodies, the third law appears to be independent of the first, for the first law does not mention systems of bodies, and the conservation of total motion of a system must therefore be established empirically. Thus with an empirical interpretation of force, the status of the third law depends on the kind of force involved. But, secondly, suppose the logical interpretation of force is accepted, then acceleration of a body may take place at any time with no apparent external cause, and nothing can follow

[1] What is here called the logical interpretation of the first and second laws is not the same as that in Mach's reconstruction, for although Mach does not assume force as an independent concept, he wishes also to circumvent Newton's definition of mass, and so he already assumes in his first 'Experimental Proposition' (*Science of Mechanics* (1883), trans. McCormack, La Salle, Ill., 2nd ed., 1902, p. 243) that it is only *bodies* that initiate accelerations in each other and that these are along the line joining the bodies. In the seventeenth century this would have been a dubious hypothesis of just the kind Newton wanted to avoid, and it is in fact Newton's conclusion at the end of a long deductive argument. It will be my contention that my 'logical interpretation' is one which Newton did hold, in a sense to be explained below, pp. 144ff.

from the law of inertia (which is now a mere definition) about the change of motion produced by parts of a body acting upon one another. It is then an empirical fact that single bodies and systems of bodies whose parts are interacting, satisfy the law of conservation of momentum. Thus with the logical interpretation of force, the third law is simply the law of conservation of momentum, and must in all cases be regarded as empirical.

Newton's own discussion of the third law is not quite satisfactory. He regards it as empirical in the case of collisions, and describes experiments by which he has established it, but he gives what appears to be a logical proof from the first law in the case of attractions. The proof is as follows : suppose an obstacle is placed so as to hinder any two mutually attracting bodies A and B from coming into contact. Then if A is more attracted to B than B to A there will be a resultant pressure on the obstacle towards B, and the whole system will ' in free spaces . . . go forwards *in infinitum* with a motion continually accelerated ; which is absurd and contrary to the first Law '.[1] He shows that for a similar reason the parts of an attracting body, for example the earth, must exert on each other equal and opposite forces. Now in the first place, this proof only applies when the attracting parts are in contact or separated by another rigid body, and so does not apply to bodies attracting at a distance as in the case of the sun and earth. Here the law must be empirical. Furthermore, Newton assumes without comment that when bodies at a distance are moving relative to one another, the law still holds,[2] and this implies that action between them takes place instantaneously, for if the transmission takes time, the action of *A* on *B* may not be simultaneous with that of *B* on *A* and therefore in general not equal to it at all times.[3] Secondly, the supposed proof of the third law in the case of attraction between contiguous bodies, explicitly appeals to the first law as an empirical law, not a definition, for if it were a mere definition, no behaviour of bodies could be ' absurd and contrary ' to it.

We now come to the question, which has been much debated, of whether Newton intended force to be regarded as an independent and absolute concept or not, and indeed he is not wholly consistent about this. Since, however, there is no doubt that he wished to

[1] *Principia*, p. 25 [2] ibid., Book I, Section XI

[3] It is clear that Newton is not thinking of power emanating in all directions with finite velocity, as in the light-model which had previously led to postulation of the inverse-square law. In any case Newton's considerations are quite general, applying not only to inverse-square attractions, but also to bodies attracting with forces of any kind, and he recognises for example that magnetic force does not obey the inverse-square law (ibid., p. 414).

eliminate hypothetical elements from his theory of gravitation, it seems that we should seek as far as possible to interpret the *Principia* in such a way that an independent concept of force is not required. I shall now try to show how far Newton's statements will bear this interpretation, and that if his view is carried through consistently, there are some consequences which he failed to notice and which are sometimes not clear in contemporary discussions of the relation of force to space-time reference frames.

If the first and second laws are to be regarded as far as possible as definitions, as in the logical interpretation described above, then we have seen that the law of conservation of momentum is empirical, and although Newton describes experiments to demonstrate it in the case of collisions, it must remain hypothetical in the case of bodies attracting at a distance upon which we can perform no direct experiments. Again, the logical interpretation presupposes a method of measuring distance and time and hence acceleration, so it may be significant that Newton's account of space and time comes, almost like an afterthought, in his Scholium to the Definitions, that is, *before* the Laws of Motion. Here he says ' I do not define time, space, place, and motion as being well known to all ', and is prompted to speak of them only because ' the vulgar conceive these quantities under no other notions but from the relation they bear to sensible objects ', whereas Newton considers that the time and space of philosophy are absolute, infinite, and independent of our relative measures of them. He goes on to support these assertions by showing how true accelerations may be distinguished from relative, and hence the framework of absolute space identified, at least approximately, with that of the fixed stars. The famous rotating-bucket experiment, or an experiment with two globes connected by a cord and rotated about their common centre of gravity, would show by the concave surface of the water or the tension in the cord whether the bodies are absolutely rotating or not ' even in an immense vacuum, where there was nothing external or sensible with which the globes could be compared '.[1] Furthermore, Newton suggests that we could test the direction of rotation of the globes by trying the effect of equal and opposite impressed forces, one on each globe, then if the tension in the cord increases, the direction of rotation is the same as the sense of the torque, if it decreases, it is opposite. Newton does not discuss the detection of translational acceleration, but his method might be extended to test for true accelerations produced by non-gravitational forces. Suppose we

[1] ibid., p. 12

came across our two globes connected by a tense cord in otherwise empty space, and on applying a torque we found that the tension increased for *both* senses of the torque. We should then conclude that one or other body was accelerating and drawing the other behind it, as for example a piece of iron acted upon by an invisible magnet and drawing a piece of wood. This method will not however do for forces such as gravitation which act upon all bodies alike, for as Einstein pointed out in his well-known lift example, if a lift is falling freely in a gravitational field, everything inside it has the same acceleration and therefore no internal forces are observed and no acceleration detected.

There is, however, a logical circle in these tests, in that the notion of force is presupposed in testing for the absolute motions that are to determine absolute force, for although the surface of the water and tension in the cord are observable apart from the notion of force, we need also some kind of absolute impressed force which is recognisable apart from absolute space, for example, muscular tension. Thus Newton's arguments establish only absolute *rotations*, and not absolute lateral accelerations, independently of the notion of force, and a half-realisation of this no doubt partly explains his silence in this Scholium about lateral accelerations, although another explanation will be suggested below.[1]

Let us for the moment accept the fixed-star framework of measurement, and consider some of Newton's statements about forces. His commentary on Definition IV (of impressed force) is ambiguous : 'But impressed forces are of different origins, as from percussion, from pressure, from centripetal force'. This may be taken to refer to forces known independently of accelerations, but may also be read simply as the assertion that these situations have been found empirically to produce accelerations, and hence that they are, by the definition, forces. The second interpretation is supported by Definition V, which states 'A centripetal force is that by which bodies are drawn or impelled, or in any way tend, towards a point as to a centre', and by the commentary on Definition VIII, where Newton remarks that the cause of a centripetal force may be

[1] It may also be remarked that Newton could have introduced the absolute space of linear accelerations by means of what he calls (in the second and third editions) Hypothesis I : '*That the centre of the system of the world is immovable.* This is acknowledged by all, while some contend that the earth, others that the sun, is fixed in that centre' (ibid., p. 419). From the theorem which follows, it is clear that Newton only intended this hypothesis to be used to distinguish among reference frames moving uniformly relative to absolute space, but, at the cost of introducing an explicit hypothesis earlier in the deduction, it might have provided a more logically satisfying account of absolute space itself.

'some central body (such as is the magnet in the centre of the magnetic force, or the earth in the centre of the gravitating force), or *anything else that does not yet appear*', and he goes on

> 'For I here design only to give a mathematical notion of those forces, without considering their physical causes and seats . . . wherefore the reader is not to imagine that . . . I anywhere take upon me to define the kind, or the manner of any action, the causes or the physical reason thereof, or that I attribute forces, in a true and physical sense, to certain centres (which are only mathematical points); when at any time I happen to speak of centres as attracting, or as endued with attractive powers.'[1]

Two more passages are relevant to this distinction between the mathematical and physical consideration of forces. First, in Section XI of Book I, where, having developed the mathematical theory of central forces acting on bodies from a fixed geometrical point, Newton considers its application to forces acting between movable bodies:

> 'I have hitherto been treating of the attractions of bodies towards an immoveable centre; though very probably there are no such things existent in nature. For attractions are made towards bodies . . . I shall therefore at present go on to treat of the motion of bodies attracting each other; considering the centripetal forces as attractions: though perhaps if I spoke physically they would more truly be called impulses. But these Propositions are to be considered as purely mathematical . . .'[2]

Finally, in the Scholium which concludes this section, where Newton remarks on the 'analogy' between the mathematics of centripetal forces there developed and 'the central bodies to which those forces are usually directed:

> 'I here use the word *attraction* in general for any endeavour whatever, made by bodies to approach to each other, whether that endeavour arise from the action of the bodies themselves, as tending to each other or agitating each other by spirits emitted; or whether it arises from the action of the ether or of the air, or of any medium whatever, whether corporeal or incorporeal, in any manner impelling bodies placed therein towards each other. In the same general sense I use the word

[1] ibid., p. 5 (my italics—M. B. H.) [2] ibid., p. 164

impulse, not defining in this treatise the species or physical qualities of forces, but investigating the quantities and mathematical proportions of them; as I observed before in the Definitions. In mathematics we are to investigate the quantities of forces with their proportions consequent upon any conditions supposed; then, when we enter upon physics, we compare those proportions with the phenomena of Nature, that we may know what conditions of those forces answer to the several kinds of attractive bodies. And this preparation being made, we argue more safely concerning the physical species, causes, and proportions of the forces.' [1]

What then emerges about Newton's view of force 'in this treatise'? It is particularly important to notice the distinction he continually makes between the mathematical properties of force and its physical causes.[2] As far as mathematics is concerned, it would be quite sufficient for his deduction of the law of gravitation that he is able to identify relative accelerations between the heavenly bodies, without assuming that the fixed stars correspond with absolute space, and this can be done in terms of what we have called the logical interpretation of force. If, however, we are concerned with physically absolute motions, we have to assume that we can detect absolute space, and this, as we have seen, means assuming that we recognise at least *some* absolute forces directly. The choice between the empirical and logical interpretations of force now depends upon whether or not we wish to restrict the use of the word 'force' to situations where the cause is recognisable apart from its effect on motion.

It is clear that, in principle, Newton does wish to use 'force' in this restricted sense. He continually speaks in terms of cause and effect, and his implicit view is that every effect must have an assignable physical cause and that even if we cannot recognise that cause directly, we infer it from the laws we find in the effects: 'the whole burden of philosophy seems to consist in this—from the phenomena of motions to investigate the forces of nature, and then from these forces to demonstrate the other phenomena'.[3] That Newton assumes in this way that there are causes for all effects lends some plausibility to the criticism later levelled against him

[1] ibid., p. 192

[2] For an interpretation of Newton's absolute space and time in terms of this distinction, see S. E. Toulmin, 'Criticism in the History of Science: Newton on Absolute Space, Time, and Motion', *Phil. Rev.*, LXVIII, 1959, pp. 1 and 203.

[3] *Principia*, Preface to the First Edition, p. xvii

and his less cautious disciples, namely that he made absolute space the cause of centrifugal effects, although Newton himself never uses this form of words. But it is unfortunate for him that the dynamical difference between uniform rotation and all other kinds of acceleration means that physical causes are assignable to the latter as long as they continue, since an external force is always required, but an external force is not required to maintain a uniform rotation. This fact, the conservation of angular momentum, was not clearly realised in seventeenth-century mechanics,[1] but some notion of the uniqueness of circular motion in this respect may have been an additional reason why Newton does not explicitly refer to the effects of linear acceleration in his discussion of absolute space. When we observe linear accelerations we always observe a source of power, either as an ' engine ' in one of the bodies, or in some other body acting either at a distance or by contact. And this is where the possible ambiguity in Newton's notion of force arises, for the question is, does he wish to say that absolute space may be a ' cause ' of observed linear accelerations as in a sense it is of rotational effects ? Or does he think that some physical cause must always be present in linear acceleration even though, as can be seen from the above quotations, he is not sure whether the list of possible kinds of cause has yet been closed ? The quotations indicate that the second alternative was almost certainly his view, but he wished to set no limits on the new sorts of physical forces which might be discovered, and does not even disqualify *a priori* possible attractions towards points of space, although ' very probably there are no such things existent in nature '. We must conclude, then, that Newton does not regard the law of inertia, considered physically, as a pure definition, for we are not to ascribe observed accelerations to the arbitrary behaviour of absolute space, about which in itself we can know nothing, but, on the other hand, observed accelerations are not evidence of any particular kind of physical force. The first law has empirical content only to this extent : *observed accelerations do not take place without some assignable cause, which it is the business of natural philosophy to investigate, and which generally turns out to have its source in some body, whether at a distance or not, and to be describable by simple mathematical laws.*

Why then does Newton not look for external physical causes of centrifugal effects, such as the fixed stars themselves ? This must surely be ascribed to the ' accident ' that the centrifugal

[1] Although everyday examples of this law, such as spinning wheels and tops, no doubt account in part for the long-persistent belief in the uniqueness of circular motion.

effects we normally observe are very regular and easily related to the fixed stars, whereas all translational accelerations that we observe are very irregular and only simply relatable to definite, local, physical causes. If we were used to seeing the effects of accelerations which appeared more or less regular but could be assigned to no local cause, we should perhaps ascribe them to the acceleration past us of absolute space, and if on the other hand such accelerations were erratic, we could still ascribe them to the capricious behaviour of absolute space, but there would then be no science of dynamics. It is worth pointing out that even the general theory of relativity does not alter the fundamental position with regard to force. Einstein noticed that we always have the choice of describing a system either as accelerating or as at rest under the action of a gravitational force, but no-one seriously suggests that gravitational forces should be invoked indiscriminately and without sources in observable matter. It has, however, been asserted in a recent controversy that the principle of general relativity entails that if a rocket is shot from the earth into space and then returns, one can either say that the earth remains at rest while the rocket travels, or that the rocket remains at rest while the earth travels, and that the two descriptions are entirely equivalent. But this is not so, for in order to change direction and return to the earth, supposed to be moving freely in space, the rocket requires an engine. The alternative description implies either that the braking of the rocket's engine and its subsequent acceleration in the opposite direction causes the *earth* to change direction at the same instant, which is very implausible, or that a gravitational field is suddenly switched on in space, which accelerates the earth, and that the rocket requires its engine to overcome this field and enable it to stay in the same place. But, like Newton, we do not countenance the postulation of *ad hoc* fields of force having no physical cause, and therefore the two suggested descriptions are not equivalent. In the sense of eliminating *ad hoc* forces, and in this sense only, our, and Newton's, notion of force remains absolute.[1]

[1] We should not be misled by Newton's anthropomorphic illustrations in Book I into thinking that they are necessary for his mathematical concept of force, for he warns us against this interpretation in the preface to Book III : 'In the preceding books I have laid down the principles of philosophy ; principles not philosophical but mathematical : . . . but, lest they should have appeared of themselves dry and barren, I have illustrated them here and there with some philosophical scholiums, giving an account of such things as are of more general nature . . . such as the density and the resistance of bodies, spaces void of all bodies, and the motion of light and sounds' (*Principia*, p. 397). Thus, I cannot think that Jammer does full justice to the subtlety of Newton's thought when he says that force is 'a concept given *a priori*, intuitively and ultimately in analogy to human muscular force' (*Concepts of Force*, p. 124).

Universal Gravitation as a Mathematical Law

Turning now to Newton's derivation of the law of gravitation, we shall inquire how far he was justified in claiming that he had avoided hypotheses. He begins his deduction in Book III by accepting two of Kepler's laws as 'Phenomena',[1] that is, in modern terminology, as empirical generalisations from direct observation :

1. The radius from the sun to each planet (or from each planet to its own moons) describes equal areas in equal times.

2. The squares of the periodic times of each planet (or moon) in its orbit vary with the cubes of their mean distances from the sun (or its own planet).

The law usually called Kepler's first law, namely, that the planets move round the sun in ellipses with the sun at one focus, is not assumed as a Phenomenon by Newton ; what he does assume initially is that the orbits are circular, and he then derives the correction to ellipses as a Proposition [2] of which he says, 'We have discoursed above on these motions from the Phenomena. Now that we know the principles on which they depend, from those principles we deduce the motions of the heavens *a priori*.' This logical pattern, of assuming phenomena as first approximations, deducing their consequences, and then using these more generally to derive corrections to the original phenomena, is repeated in relation to Kepler's third law (2 above). This law is accurate only if the centre of force (e.g. the sun) is at rest, but since the mass of the sun is very great compared to that of a planet, it is a good approximation even though strictly false, and can be corrected when the law of gravitation has been found. Similar remarks apply to corrections which arise from the (small) actions of the planets upon each other.

This logical pattern is of some interest, as it is an indication of part of what Newton means by 'deducing' theories from phenomena. The rest of his deduction of the law of gravitation involves only mathematics together with the laws of motion and some explicit general assumptions which Newton calls Rules of Reasoning in Philosophy. These Rules are as follows :

1. We are to admit no more causes of natural things than such as are both true and sufficient to explain their appearances.

[1] *Principia*, pp .401–5. These 'Phenomena' were called 'Hypotheses' in the first edition, see *infra*, p. 147*n*. [2] ibid., Book III, Prop. XIII, p. 420

II. Therefore to the same natural effects we must, as far as possible, assign the same causes.

III. The qualities of bodies, which admit neither intensification nor remission of degrees, and which are found to belong to all bodies within reach of our experiments, are to be esteemed the universal qualities of all bodies whatsoever.

IV. In experimental philosophy we are to look upon propositions inferred by general induction from phenomena as accurately or very nearly true, notwithstanding any contrary hypotheses that may be imagined, till such time as other phenomena occur, by which they may either be made more accurate or liable to exceptions.[1]

It is significant that Newton places these Rules just before the sections of the *Principia* dealing with universal gravitation, for with their help the argument from Phenomena can be rendered deductive. He shows that a planet P, which satisfies Kepler's laws, has an acceleration towards the sun S of amount μ/SP^2 (in modern notation), where the constant of proportionality μ is the same for each planet. Since force is determined by the second law of motion, there must therefore be an attractive force acting towards the sun of amount $\mu m/SP^2$, where m is the inertial mass of P. Also the satellites of Jupiter and Saturn and the earth's moon are observed to move relative to their own planets according to laws similar to Kepler's. It must therefore be concluded, according to the second Rule of Reasoning, that each planet is the centre of an attractive force acting on its own satellites, and similar to that between the sun and the planets. In particular the earth exerts such a force upon the moon, and the acceleration of the moon towards the earth can be calculated from the moon's orbit. Assuming now that the force acts at all distances from the earth's centre with magnitude given by the inverse-square law, it is possible to calculate the acceleration which would be produced in a body at the earth's surface. Newton performs this calculation,[2] using a recent estimate of the earth's radius, and finds the acceleration to be the same as the measured acceleration g of freely falling bodies of any mass at the earth's surface. Therefore, in conformity with his Rules I and II, he identifies the force exerted upon falling bodies with the force which maintains the planets and satellites in their orbits.

At this point in the argument Newton knows that the sun exerts a gravitational force upon the planets, and the planets upon their

[1] ibid., pp. 398–400 [2] *Principia*, Book III, Prop. IV, p. 407

own satellites. Following Rule III, he must assert that a planet exerts a reciprocal force upon the sun, and a satellite upon its planet. The planet's force upon the sun would be $\nu M/SP^2$, where M is the mass of the sun, and ν a constant characteristic of the planet. But by the law of equality of action and reaction, this must be equal to the sun's force upon the planet, namely $\mu m/SP^2$. Thus $\nu M = \mu m$, where ν is a constant depending only upon the planet, and μ a constant depending only upon the sun. Hence $\nu = \lambda m$ and $\mu = \lambda M$, where λ is constant for all gravitating bodies whatever.

Rule III now leads to the law of universal gravitation : 'That there is a power of gravity pertaining to all bodies, proportional to the several quantities of matter which they contain', and 'The force of gravity towards the several equal particles of any body is inversely as the square of the distance of places from the particles'.[1] In modern notation, $F = \lambda m_1 m_2/r^2$, where λ is a universal constant; m_1 and m_2 are the *inertial* masses of the two bodies; and r is the distance between them. It should be noticed that the fact that gravitation depends on inertial mass is not logically necessary in Newton's system, but is a deduction from phenomena, either from Kepler's laws as above, or from the equality of acceleration of all freely falling bodies or pendulums at the earth's surface, irrespective of their mass.[2] Thus in Newton's theory the equality of inertial and gravitational mass is an empirical fact, and one that may appear surprising and inessential to the gravitational theory. In relativity theory, as we shall see, this fact is taken to be fundamental.

In view of later controversy about Newton's claim to have made no hypotheses, it is important to see how far the theory of gravitation can be regarded as a pure deduction from empirical generalisations. The attempt just made to set out the theory in this way indicates that the argument is deductive, as deduction is generally understood in science, if and only if the conservation of momentum as well as Kepler's laws is regarded as a 'phenomenon'; if relative forces are defined 'operationally' in terms of the fixed-star framework; and if the Rules of Reasoning are assumed as methodological axioms. Then from the mathematical point of view the forces mentioned in the theory are pure constructs having a possible empirical, but not a logically necessary, relation to physical efforts and tensions. This is clear if we consider what

[1] ibid., Book III, Prop. VII and Corr. II, p. 414

[2] In Book II, Prop. XXIV (ibid., p. 304) Newton remarks that 'by experiments made with the greatest accuracy, I have always found the quantity of matter in bodies to be proportional to their weight'. The experiments are described in Book III, Prop. VI.

interpretation Aristotle would have put upon the theory. If he had had the mathematical genius of, say, Archimedes, and the records available to Ptolemy, he could in principle have arrived at a description of the velocity and acceleration of the planets identical with that of Newton, but he would have used the word ' force ' in a different sense, because his notion of force was different from Newton's. He would have accepted uniform circular motion of a planet about the earth as the natural motion requiring no forces, and would have looked for forces which disturbed this circular motion to give the orbit the form of an ellipse with the sun in one focus, as we have seen that Kepler in fact did. These forces would have had no resemblance to Newton's gravitational attraction, but would have provided a valid explanation in the context of Aristotle's general principles. As soon as Newton's forces cease to be mathematical definitions and become physical attractions in an absolute space, they are no less hypothetical than those of Aristotle and Kepler.

As for the Rules of Reasoning,[1] they are put forward in the second and third editions of the *Principia*, very plausibly, not as hypotheses, but as the simplest modes of procedure to be followed unless and until evidence is found which compels their modification. Their reasonableness and conformity with the spirit of modern science should not, however, blind us to the total difference between the world picture here presented, and that of, for example, Aristotle, to whom it appeared obvious that there are great qualitative differences between different parts of the universe. Newton's absolute space and time, together with the uniformity and continuity postulated in these Rules, do in fact present us with a world-model whose hypothetical nature has only become clear since a great part of it has had to be abandoned.

It should also be remarked that the theory of gravitation can be looked at as a hypothetico-deductive system in which the statement of universal gravitation is the hypothesis from which the phenomena of Kepler's laws, the laws of falling bodies, of comets and tides and so on can be deduced. This represents the order of discovery quite as well, if not better, than the operational form, for the law was certainly known to or guessed by Hooke and Newton

[1] Professor I. B. Cohen has pointed out (*Franklin and Newton*, Philadelphia, 1956, pp. 127ff.) that in the second edition (1713) Rules I to III replace what are called Hypotheses I, II and III in the first edition. Rule IV was added in the third edition (1726). In the first edition Hypothesis IV is the statement that ' the centre of the system of the world is at rest ', and Hypotheses V to IX concern the observed orbits and periodic times of the planets and their satellites. Thus Newton did not at first distinguish between the statuses of these nine hypotheses. In the second and third editions, however, Hypotheses V to IX become Phenomena, and Hypothesis IV becomes Hypothesis I.

before its deduction from Kepler's laws, and it is only an accident that in this instance it happens to be possible to prove the converse of many of the propositions in the deductive part of the theory, so that the argument may proceed from empirical laws to hypothesis as well as the other way round. Even though this is possible, Newton nowhere claims that the inverse-square law is nothing but a deduction from phenomena, as is clear from his use of the law to *correct* phenomenal laws, which are then regarded as first approximations, as has been described above. But he was entitled to claim a close deductive relation from phenomena to hypothesis, and could not be expected to foresee that such a relation would not always be possible as mathematical physics developed. For this reason gravitation has appeared to later positivists and operationalists to be an ideal form of theory, and extreme positivism has, unjustifiably, been read into Newton's own words.

Universal Gravitation as a Physical Hypothesis

There is, however, another sense in which it may be said that the law of gravitation was not a hypothesis. It is not clear that Newton distinguished the two senses, for when he says that a hypothesis is anything 'not deduced from the phenomena'[1] he may or may not be claiming that the law as a mathematical expression is deducible from Kepler's laws in the way discussed above, but he certainly is claiming that he is not making any hypothesis about physical causation. In modern terminology, the law of gravitation may be seen either as an operational law, or as the hypothesis in a hypothetico-deductive system, but in either case, Newton was justified in replying to his critics that he had made no *special* physical hypotheses, but had merely described the phenomena in the simplest possible way. This is borne out by his definition of centripetal force[2] as 'that by which bodies are drawn or impelled, or in any way tend, towards a point as to a centre', which may apply, as he says, to gravity, or to magnetism, or to the force on a stone whirled in a sling. It deliberately abstracts from the physical means whereby the force is exerted, and concerns itself only with the resulting tendency of the body to move towards the centre.

The theory would in fact have been consistent with a multiplicity of hypotheses about the 'cause' of gravitation, of which at least four were held at the time or had been held earlier, for instance :

[1] *Principia*, General Scholium (written for the second edition, 1713), p. 547
[2] ibid., Def. v, p. 2

1. Angels propel the planets in their orbits, exerting a force which according to Newton's definitions must be towards the sun and proportional to the inverse-square of the sun's distance. This hypothesis is perfectly compatible with the phenomena and involves no action at a distance, but it offends against the seventeenth-century view of science in general and Newton's first and second Rules of Reasoning in particular.

2. Bodies have internal natural tendencies towards one another which vary with mass and distance according to the inverse-square law. These tendencies may be unconscious or conscious or even intelligent. This was the Aristotelian theory of heavy bodies as understood in the Middle Ages, and it was still possible to hold it, but would have been implausible, seeing that the attraction was now known to be reciprocal, to diminish with distance, and to depend on the magnitude of both attracting masses, so that a body would have to ' know ' the properties of other bodies at a distance. It became more natural from an Aristotelian point of view to suggest the next hypothesis.

3. Bodies have an innate quality of gravitational attraction which acts at a distance according to the required law. This is to replace the *tendency to move* of the last hypothesis, for which the moving body is *active*, by a quality of attraction within the attracting body, so that the moving body is *passive*. This was the Aristotelian-type theory which Newton's critics thought he was putting forward, and which led to the accusations that gravity was an ' occult quality ' and had no place in true philosophy. Newton strongly repudiated this suggestion. Attraction was not in any sense an ' occult quality ' like the qualities postulated by the Aristotelians, because the gravitational forces of his theory could be used to derive the motions of bodies, whereas the Aristotelian qualities were mere names which explained nothing. This point is made very clearly in the Query which Newton added to the second edition of his Optics in 1717 ' to show that I do not take Gravity for an essential Property of Bodies ' :

> ' These Principles [gravity, fermentation or chemical actions, and cohesion] I consider, not as occult Qualities, supposed to result from the specific Forms of Things, but as general Laws of Nature, by which the Things themselves are formed ; their Truth appearing to us by Phenomena, though their Causes be not yet discovered. For these are manifest Qualities, and their Causes

only are occult. And the *Aristotelians* gave the Name of occult Qualities, not to manifest Qualities, but to such Qualities only as they supposed to lie hid in Bodies, and to be the unknown Causes of manifest Effects : Such as would be the Causes of Gravity, and of magnetic and electric Attractions, and of Fermentations, if we should suppose that these Forces or Actions arose from Qualities unknown to us, and uncapable of being discovered and made manifest. Such occult Qualities put a stop to the Improvement of natural Philosophy, and therefore of late Years have been rejected. To tell us that every Species of Things is endowed with an occult specific Quality by which it acts and produces manifest Effects, is to tell us nothing : But to derive two or three general Principles of Motion from Phenomena, and afterwards to tell us how the Properties and Actions of all corporeal Things follow from those manifest Principles, would be a very great step in Philosophy, though the Causes of those Principles were not yet discovered : And therefore I scruple not to propose the Principles of Motion above-mentioned, they being of very general Extent, and leave their Causes to be found out.' [1]

The interpretation of gravity as an innate quality of bodies like extension or mass is perhaps the one which creeps in most easily, and indeed the use of language about a body exerting attractive force lends itself to misunderstanding, for it evokes the notion of invisible power streaming out and laying hold on distant bodies. Newton explicitly guards against this interpretation of his language in his definition of centripetal force, but even his friends and supporters had to be gently chided for their misunderstandings. For instance he writes to Richard Bentley :

> 'You sometimes speak of gravity as essential and inherent to matter. Pray, do not ascribe that notion to me ; for the cause of gravity is what I do not pretend to know, and therefore would take more time to consider it.' [2]

But this is exactly the doctrine that appears in the Preface written by Roger Cotes for the second edition of the *Principia* in 1713. Here Cotes denies that gravity is an occult quality, because its effects show that it really exists, but he asserts that it is a primary quality of bodies as are extension, mobility, and impenetrability, and

[1] *Opticks*, London, 2nd ed., 1718, Query 31 ; 4th ed., reprinted 1931, p. 401. Page references to this work will be to this reprint.

[2] Newton's 'Second Letter to Bentley' (1692–3), *The Works of Richard Bentley*, III, London, 1838, p. 210

suggests that since it is simple, no further explanation of it can be expected.[1]

4. The fourth possible hypothesis is that there is some corpuscular mechanism which produces the required force. Newton himself hoped for this, although he could not see any satisfactory way of demonstrating it. In a letter to Oldenburg of 1675 [2] he allows himself to speculate upon an aether hypothesis in relation to light propagation because, as he says, some people find it difficult to understand his theories without the aid of an illustrative hypothesis. He is not, however, concerned with the truth or otherwise of this hypothesis and does not wish to enter into controversy over it. His first supposition is that there is a subtle elastic aether, not necessarily homogeneous, because it may contain the electric and magnetic effluvia and the ' gravitating principle '. He suggests that parts of this aether may account for electric and gravitational effects by a kind of vaporisation and condensation like that of water from the surface of bodies. When glass is rubbed, the aether contained in its pores is vaporised, creating a wind in the neighbourhood, and this causes the erratic motions of small pieces of paper which he has observed in an experiment. The aether condenses on to the glass, drawing the paper with it. Perhaps gravitation is due to a ' gummy tenacious and springy ' part of the aether which continually condenses in the pores of the earth, its place being taken by air, exhalations, and vapours rising from the earth, for nature is ' a perpetual circulatory worker '. Gravitation between the sun and the planets might be explained similarly : the sun ' feeds ' on this aetherial spirit, which conserves its shining, and whose sunward motion draws the planets with an attractive force. Newton goes on to suggest explanations of light refraction, cohesion, and the contraction of muscles in terms of differential pressures due to the varying degrees of density of the aether in different bodies.

In a letter to Boyle (1679),[3] Newton repeats some of these speculations, but adds a different and more truly mechanical suggestion about gravity, which, he says, ' came into my mind now as I was writing this letter '. This is a hydrostatic theory depending on variations in aether density, somewhat like the earlier theory of Roberval. But these speculations are tossed out in a remarkably casual fashion, and Newton never confuses them with mathematical

[1] *Principia*, pp. xxvi ff.
[2] Reprinted in D. Brewster, *Memoirs of the Life of Sir Isaac Newton*, I, 1855, pp. 390–409
[3] ibid., pp. 409-19

demonstrations, in which 'hypotheses non fingo'.[1] At the end of the letter to Boyle, he remarks 'For my own part, I have so little fancy to things of this nature, that had not your encouragement moved me to it, I should never, I think, have thus far set pen to paper about them'.

It is clear, however, that Newton thought there must be some physical cause of gravity still to be found, and it was quite inconceivable to him that 'inanimate brute matter should, without the mediation of something else, which is not material, operate upon and affect other matter without mutual contact.'[2] Again, 'It is absurd to suppose that gravity is innate and acts without a medium, either material or immaterial.' The mention of an 'immaterial medium' is reminiscent of the spirits, vapours, emanations, and the like, which still lingered from early seventeenth-century physics, and shows that Newton was not quite an orthodox corpuscularian. He is even prepared to suggest in the Queries to the *Optics* that some physical actions may be the result of the direct intervention of an omnipresent God who 'being in all places, is more able [than we are] by his will to move the bodies'. In this way God prevents the fixed stars falling on each other, and corrects disturbances in the solar system due to perturbations.[3] And Newton's friend and editor of the third edition of the *Principia*, Henry Pemberton, wrote that Newton had often complained to him that he had been misunderstood on attraction, for he had not intended to give a philosophical explanation of any appearances, but to point to a power which is 'worthy of diligent inquiry'. Pemberton adds:

> 'To acquiesce in the explanation of any appearance by asserting it to be a general power of attraction, is not to improve our knowledge in philosophy, but rather to put a stop to our farther search.'[4]

This Newton certainly did not intend to do. At this stage however the suggestion that the medium of gravitation might not have the properties of ordinary matter could not be taken into an increasingly mathematical physics. Before this could be done, the notion of

[1] This famous phrase occurs in the General Scholium at the end of the *Principia* (2nd ed.) in connection with the cause of gravity. It is followed by a paragraph (the last in the book) where it is suggested that 'a certain most subtle spirit' may be responsible for all physical phenomena, and even for the movement of animal bodies at the command of the will—a clear echo of Descartes.

[2] 'Third Letter to Bentley', *Works of Richard Bentley*, III, p. 211

[3] Queries 28 and 31

[4] *A View of Sir Isaac Newton's Philosophy*, London, 1728, p. 407

'immaterial' had to be divested of its association with animism and given mathematical expression, and this had to wait until the nineteenth century.

Newton's Atomism and Active Principles

Meanwhile Newton was content to assume that the small particles of bodies have 'certain powers, virtues, or forces, by which they act at a distance',[1] exhibiting both attractions and repulsions. Ultimately these might be explained by impulses or some other means, but at least, Newton thought, the phenomena compel the conclusion that certain gross bodies behave *as if* there are attractions and repulsions between them, and these are demonstrated not only in gravitation, electricity, and magnetism, but also in the refraction of light and in the phenomena of chemical affinity.

In his theory of light, Newton was led to postulate corpuscles as centres of force, rather than waves in the aether, for various empirical reasons.[2] The first was the apparent absence of diffraction which, on a wave theory, would be expected to cause light to bend round obstacles into their 'shadow' as in the case of water or sound waves. Again, the explanation of double refraction in Iceland Spar implies that light rays are polarised, that is, have different properties on different sides, and Newton cannot conceive how this can be the case with wave motion, which he (and Huygens) thought of as longitudinal pulses. Light corpuscles, on the other hand, can easily be imagined to be polarised. Another reason for rejecting the wave theory was that it implied the existence of a fluid medium extending throughout space, and it was difficult to see how this could be reconciled with the fact that the motions of the planets show no tendency to slow down.

If the light corpuscles are simply Democritan atoms, it is natural to suppose that reflection takes place when they impinge upon the solid parts of bodies. Newton, however, describes[3] several experiments to show that this cannot be the case. For instance, light is reflected as strongly at the surface between glass and air when it passes out of the glass as when it passes out of the air. Again, why should there be partial reflection and partial refraction at some angles of incidence, while there is total reflection at others? It seems unreasonable to suppose that the light finds pores in the air at one angle of incidence and none at another. And there were

[1] Query 31, *Optics*, p. 375 [2] Query 28, ibid., pp. 362ff.
[3] ibid., Book II, Part III, Prop. VIII, pp. 262ff.

the experiments with the colours of thin plates, in which Newton was a pioneer, which could not be accounted for by means of simple Democritan atoms. Newton therefore concludes that ' the Reflexion of a Ray is effected, not by a single point of the reflecting Body, but by some power of the Body which is evenly diffused all over its Surface, and by which it acts upon the Ray without immediate Contact.'

When a light corpuscle approaches a reflecting surface it suffers a repulsive force acting at right angles to the surface, and is reflected in such a way that the angles of incidence and reflection are equal. When it crosses the interface between a rarer and a denser transparent medium, it is attracted along the line perpendicular to the interface until it has passed a small distance beyond it. This causes refraction according to the sine law, and causes the velocity of the corpuscle to be increased in the denser medium.[1] The occurrence of both reflection and transmission at the surface of transparent media is explained by ' fits of easy reflection and easy transmission ' : the rays of light are put alternately into a state to be easily transmitted and easily reflected, and this also explains the coloured rings seen in experiments with thin plates. So far, the ' fits of easy reflection and transmission ' may be said to be a mere description summarising the results of several sorts of experiments, but in the Queries Newton goes further and suggests a mechanism for their production in terms of the vibrations of an aether in the neighbourhood of the refracting surface.[2] This suggestion is however as inconclusive as his other speculations about aether as the cause of gravity.

In Query 31 Newton suggests that many chemical reactions are due to attractions and repulsions between particles, the violence of some of them and the heat engendered being due to the high velocities produced by these forces. Cohesion is also best explained by ' some Force, which in immediate Contact is exceeding strong, at small distances performs the chemical Operations abovementioned, and reaches not far from the Particles with any sensible Effect '. The same explanation will serve for capillary effects. Newton concludes these speculations with a picture of the universe consisting of ' solid, massy, hard, impenetrable, moveable Particles ' which have, as well as inertia, ' certain active Principles, such as is that of Gravity, and that which causes Fermentation [chemical

[1] Newton's theory here contradicts the wave theory of light, but the crucial experiments were not made until 1850, when Foucault and Fizeau showed that the velocity of light in water is less than that in air.

[2] Queries 17–20

activity in general], and the Cohesion of Bodies '.[1] For if there were no active principles, Newton thinks that motion would eventually decay, since it is lost in all inelastic collisions and in the motion of viscous fluids. The principles, however, conserve and recruit motion, by causing bodies to acquire motion by falling into centres of force, and by causing heat in chemical reactions :

> ' And if it were not for these Principles, the Bodies of the Earth, Planets, Comets, Sun, and all things in them, would grow cold and freeze, and become inactive Masses ; and all Putrefaction, Generation, Vegetation and Life would cease, and the Planets and Comets would not remain in their Orbs.' [2]

For these reasons it seems that Newton has become doubtful about the possibility of finding explanations of attraction and repulsion in terms of impulse, although he nowhere definitely excludes it. Since he has no other kind of mechanical explanation to suggest, and since he cannot conceive action without any kind of medium, he falls back on the suggestion that the medium is the ' sensorium of God ',[3] to whom, ultimately, the order of the world must be ascribed.[4] But the teleology implicit in this belief was not the only alternative to mechanism, for the central-force theory was able to act as a model without help from either of these types of explanation. It had been rendered ' clear and distinct ' by mathematical expression whose adequacy was self-evident to anyone equipped to understand it, and in this respect it was far preferable at that stage to any attempted corpuscular explanation which could only have made it more complex and obscure. The change of emphasis from corpuscular pictures to mathematical description was in one sense more important than that from animistic to mechanical science earlier in the century, because for the first time an abstract theory came to be regarded as explanatory. In a sense, the less familiar was in this case an explanation of the more familiar and concrete, for its intelligibility was that of a mathematical system, but at the same time it retained the clear empirical reference and hence testability of mechanism, because its concepts were still those of matter and motion. From this point of view

[1] Query 31, ibid., pp. 389, 400, 401 [2] ibid., p. 399

[3] H. G. Alexander points out that this phrase expresses Newton's extreme representative theory of perception : according to that theory, we perceive images *inside our heads*, whither they have been conveyed by the sense-organs, but Newton believes God to perceive them directly without mediation by sense-organs, that is to say, external space *is* God's sensorium (*The Leibniz-Clarke Correspondence*, Manchester, 1956, pp. xv ff.).

[4] *Optics*, pp. 369, 402 ; General Scholium, *Principia*, p. 544

Newton's avoidance of hypotheses and occult qualities meant that no new interpretations of his concepts into empirical terms were required—he had only explored the interrelations of those terms more exactly. But having done so, the theory of gravitation itself became a useful analogue for magnetism, cohesion, elasticity, and electricity, and could be exploited in connection with the attractive and repulsive properties of theoretical entities, for example gas molecules and light corpuscles. It is proper to speak of the theory as a 'model', for it has the characteristics of a model in being generalisable in such a way that its resulting new properties can be compared experimentally with the phenomena: thus having used the equation $F = mf$ in the particular case $F = Mm/r^2$, the mathematical apparatus associated with it can be exploited and tested in connection with other forms of F. Even when the mathematical details of the theory were not used in other physical problems, its qualitative features were presupposed, for example in the firmly entrenched assumptions that distance forces could only be central, that is, act along the line joining bodies, and that they are propagated instantaneously.

The wide applicability of the theory of central forces shows that it is not necessary that modes of action should have everyday analogues such as were provided by the phenomena of vacuum suction or impacts of particles. The theory was the first mathematical model, intelligible because it was a coherent logical system, and not because it was a familiar process or event. Mechanical models also continued to make good analogues, but in the subsequent history of physics this ceased to be primarily because they are objects of a familiar sort that could be imagined or built; it was rather because their structure and functioning was easily susceptible of mathematical expression. This lesson had to be learnt again at the end of the nineteenth century, but it should have been learnt during the preceding two hundred years, when the theory of central forces was a clear example of an abstract theory acting as a good analogue in several branches of physics, but not interpretable in terms of mechanical action by contact.

Chapter VII

ACTION AT A DISTANCE

Leibniz's Attack on Action at a Distance

NEWTON'S theory of gravitation won immediate acceptance in England, where it was expounded and defended by a number of his disciples and admirers, but on the Continent the situation was very different. There the Cartesians regarded the notion of attraction as an occult quality, a surrender to Aristotelianism, and a retrogression to the immaterial influences and sympathies which had been banished from physics so recently and with such difficulty.[1]

Even so astute a mathematician as Huygens was unable to accept the new theory as it stood. He made generous acknowledgment of Newton's work and accepted his proof that the planets exhibit a centripetal force proportional to the inverse square of the distance from the sun, realising that this had made Descartes's vortex theory untenable,[2] but he continued to seek a modified vortex theory which would accord with the facts which Newton had demonstrated. As late as 1730, the Swiss mathematician John Bernoulli won a prize of the Académie des Sciences for a paper in which he derived Kepler's third law from a vortex hypothesis,[3] and four years later he concluded that in his theory the law of attraction ought to be an inverse-cube, not an inverse-square law. Later still, Euler was refusing to accept the inverse-square attraction as a universal property of bodies, although he admitted it empirically as a description of the motion of the planets.

One of Newton's most outspoken critics was Leibniz, who held a theory of gravitation similar to that of Huygens. Although he was no Cartesian, for his own work in dynamics had led him to criticise the identification of matter with extension, Leibniz retained Cartesian principles in his denunciations of action at a distance. He was the last great philosopher of the eighteenth century seriously

[1] cf. the French writer Saurin: '[Newton] likes to think of weight as a quality inherent in bodies, and to revive the discredited ideas of occult quality and of attraction . . . let us not abandon philosophizing always on the basis of the clear principles of mechanics; if we abandon them, all the light we can get is extinguished, and we are plunged back into the old darkness of peripateticism, from which heaven would preserve us' (*Hist. de l'Acad. Roy. des Sci.*, 1709, p. 148).

[2] *Discours de la Cause de la Pesanteur* (1690), *Œuvres*, XXI, p. 472

[3] *Sur la Système de M. Descartes*, *Opera Omnia*, III, 1742, p. 134

to contest the theoretical basis of the new science on purely metaphysical grounds, for most of his arguments in physics depend in the end, not on empirical facts, but on the conviction that theoretical explanation of the facts must conform to certain principles which appeared to him self-evident, especially the principle of sufficient reason and the principle of continuity. According to the former, nothing occurs in the contingent world unless there is sufficient reason for that particular event to occur rather than another, and even God is subject to the principle in his acts of creation : his mere will to create is not enough, there must be a sufficient reason why he creates one thing rather than another. And according to the principle of continuity, all created substances form series in which every possible quantitative and qualitative gradation is realised once and only once. Leibniz maintained that these principles enable one to distinguish *a priori* between contingent facts and logically possible facts, that is, to decide which of the many logically self-consistent states of affairs are actually realised in the world.

Leibniz's exploration of these principles led him to put his finger on some important inconsistencies in the current mechanical theories of atomism and Cartesianism.[1] Against atomism, he argues that the true units of matter cannot be hard, extended, indivisible atoms, since this would involve a breach in the continuity of nature, and there would be no sufficient reason why there should be an end to the process of subdivision at one point rather than any other. The atoms would be indistinguishable from each other, and this would imply that God had no sufficient reason for originally creating them in one spatial order rather than in any other. Other arguments depend on more purely physical considerations. Leibniz, rejecting attractive forces, thinks that cohesion between atoms would require a perpetual miracle. Again, if all substances are made up of extended atoms, what does the substance of the atoms themselves consist of? Another argument comes from the theory of elastic impact, which had been clarified by Huygens and others : atoms are supposed to be perfectly rigid, but in the collision of perfectly rigid bodies, all motion is lost, so that collision of bodies which are ultimately supposed to be made up of atoms could never be elastic. If, on the other hand, atoms were regarded as perfectly elastic they would be deformable and therefore not truly atomic. The modern reader is tempted to reply to these arguments

[1] See, for example, his *New Essays concerning Human Understanding*, trans. Langley, New York, 1896, Appendices III, IV, V, VII.

that in objecting to notions such as the substance of the atom and its elasticity, Leibniz is asking for too much. From the scientific point of view there is no reason why atoms should not be ultimately and inexplicably substantial or elastic, if they provide thereby useful models for physical interactions. But it must be remembered that Leibniz cannot accept these arguments, for he is committed to the belief that rationality, that is, intelligibility in detail to the human mind, is part of the essence of things, and he therefore requires that descriptions of the ultimate structure of matter should be understood literally and not metaphorically, although, as we shall see, he was prepared to admit corpuscular descriptions for practical purposes.[1]

Leibniz's arguments against the void are purely metaphysical : all possible perfection has been imparted to things, and space which is full of matter is more perfect than that which is empty, therefore all space is full of matter. Again he argues on the basis of the principle of sufficient reason that if there were matter and void, there could be no reason for any particular proportion of matter to void, or for any particular arrangement of matter in the void. If the answer to a question is indeterminate, the question cannot arise in reality, therefore he concludes there is no void.

So far Leibniz agrees with the Cartesians, but he has other objections to their doctine, and his arguments here are more telling. If matter were mere extension, as Descartes seemed to hold, it would have no determinate figure, and there would be no meaning in saying that it moved. Again, the laws of motion show that bodies have inertia, as Descartes himself admitted, and this cannot be derived from pure extension. Above all, Leibniz corrects Descartes's principle of conservation of motion by pointing out that quantity of matter multiplied by speed is conserved only in a given direction in mechanical interactions, and that it is the quantity mv^2, which Leibniz calls *vis viva,* which is conserved absolutely in elastic collisions.

That bodies do not act at a distance may be suggested by the absence of void, but cannot be deduced from it. Kant, for example, as we shall see later, made action at a distance fundamental, but was

[1] He describes his intellectual pilgrimage as follows : ' After I had left the elementary school, I fell in with the modern philosophers, and I recall that I was walking alone in a little grove called the Rosenthal, not far from Leipzig, at the age of fifteen years, in order to make up my mind whether I should adopt the doctrine of substantial forms. The theory of mechanism, however, won finally the upper hand with me and led me to mathematics. Yet, in my search for the ultimate grounds of mechanism and of the laws of motion, I came back to metaphysics and the doctrine of entelechies' (*Opera Philosophica,* ed. Erdman, Berlin, 1840, p. 701 ; quoted in Jammer, *Concepts of Force,* p. 159).

very dubious about the existence of void. Leibniz, however, puts forward no arguments against action at a distance other than the principle of continuity and the arguments which lead him to reject the void. Even in his metaphysical theory of monads, which he regards as a more satisfactory description of the ultimate units of reality than physical atomism or Cartesianism, he never makes clear how the principles of dynamics are implied in the real relations between monads of which they describe the phenomenal appearances. Monads themselves are non-spatial, have no parts, have no causal interactions one with another, and are more analogous to souls than to atoms, but Leibniz is not concerned to show that this metaphysical theory entails any particular mode of physical (phenomenal) interaction. It is indeed difficult to see how a theory of casually independent, non-spatial monads could do so, for although there is a pre-established harmony between them, there is no reason why this should give rise to one kind of apparent interaction rather than another. Leibniz does not seem to think that any explanation of this is necessary, or perhaps even possible, for he insists that his metaphysical principles are no use for detailed theories of physics, and that

> '. . . all the phenomena of bodies can be explained mechanically, or by the corpuscular philosophy, according to certain principles of mechanics, which are laid down without taking into consideration whether there are souls or not.' [1]

But on the phenomenal level Leibniz appeals to the principle of continuity to show that there must be continuity of cause and effect in space and time. Hence his polemic against attraction in his letters to Clarke, and many explicit assertions that action involves contact :

> 'A body is never moved naturally, except by another body which touches it and pushes it ; after that it continues until it is prevented by another body which touches it. Any other kind of operation on bodies is either miraculous or imaginary.' [2]
> 'As we maintain that it [attraction] can only happen in an explicable manner, i.e. by an impulsion of subtler bodies, we cannot admit that attraction is a primitive quality essential to matter.' [3]

[1] Letter to Arnaud (1686), *Die Philosophischen Schriften von G. W. Leibniz*, ed. Gerhardt, Berlin, 1875–90, II, p. 78

[2] *The Leibniz-Clarke Correspondence*, ed. H. G. Alexander, Manchester, 1956, p. 66

[3] Letter to Bourguet (1715), *Die Philosophischen Schriften*, III, p. 580

Leibniz does not, however, envisage ultimate contact between hard extended atoms, for he has rejected atoms of this kind for the reasons mentioned above. It is rather that every collision, however far analysed in terms of the small parts of bodies, is elastic like the collision of two inflated balloons : 'there are no elements of bodies, . . . analysis proceeds to infinity'.[1] This explains how Leibniz could maintain the conservation of *vis viva* even in non-elastic collisions. He believed that motion is not entirely lost when two inelastic bodies bring each other to rest, but that it is always transferred to smaller and smaller parts, however far analysis into these parts is carried.

The best account of the points at issue between Leibniz and the Newtonians is in the Leibniz-Clarke correspondence. This was initiated by Leibniz, only a year before his death in 1716, in a letter to Caroline, then Princess of Wales, in which he asserted that the philosophy of Sir Isaac Newton was contributing to the decay of natural religion in England. This provoked Clarke to reply, as a churchman and disciple of Newton, and the correspondence continued through Princess Caroline until Leibniz's death, ranging widely over Leibniz's metaphysical principles and over scientific topics such as the nature of space and time, the existence of vacuum, and the theory of gravitation.

Leibniz repeats his arguments against the void, and asserts that the vacua produced in Torricelli's tube and in Guericke's air pump are void only of gross matter and not absolutely empty. The space contains rays of light 'which are not devoid of some subtle matter', and 'the effluvia of the load-stone, and other very thin fluids may go through' the glass.[2] Clarke objects that if these were material they would exert resistance, and there is no resistance to bodies moving in a vacuum.[3] But he is prepared to admit that immaterial substances may be present in void space : 'God is certainly present, and possibly many other substances which are not matter ; being neither tangible, nor objects of any of our senses'. Leibniz replies that if the subtle matter were non-viscous it would exert no resistance, and ridicules Clarke's notion of immaterial substances :

> 'If the space (which the author fancies) void of all bodies, is not altogether empty ; what is it then full of? Is it full of extended spirits perhaps, or immaterial substances, capable of extending

[1] *New Essays*, p. 687 [2] *Leibniz-Clarke Correspondence*, pp. 65, 195

[3] ibid., p. 47. Boyle had conducted experiments on mechanical resistance in vacua : *Experiments touching the Spring and Weight of the Air* (1669), *Works*, III, pp. 250–9.

and contracting themselves; which move therein, and penetrate each other without any inconveniency, as the shadows of two bodies penetrate one another upon the surface of a wall? Methinks I see the revival of the odd imaginations of Dr. Henry More. . . . Is it not overthrowing all our notions of things, to make God have parts, to make spirits have extension?'[1]

On the question of action at a distance, Clarke agrees with Leibniz that it is impossible. In reply to Leibniz's assertion that an attraction causing a free body to move in a curved line would be a miracle, he says:

'That one body should attract another without any intermediate means, is indeed not a miracle, but a contradiction: for 'tis supposing something to act where it is not. But the means by which two bodies attract each other, may be invisible and intangible, and of a different nature from mechanism; and yet, acting regularly and constantly, may well be called natural; being much less wonderful than animal-motion; which yet is never called a miracle.'[2]

But Leibniz will not have any talk of

'. . . some immaterial substances, or some spiritual rays, or some accident without a substance, or some kind of *species intentionalis*, or some other I know not what. . . . Of which sort of things, the author seems to have still a good stock in his head, without explaining himself sufficiently. That means of communication (says he) is invisible, intangible, not mechanical. He might as well have added, inexplicable, unintelligible, precarious, groundless and unexampled. . . . 'Tis a chimerical thing, a scholastic occult quality.'[3]

Clarke replies that attraction is only the name of an empirical fact, whatever may be its explanation, and therefore it should not be called occult, even if its efficient cause is not yet discovered. If Leibniz or anyone else can explain these phenomena by the laws of mechanism he will 'have the abundant thanks of the learned world'.[4] But elsewhere Clarke is doubtful of the possibility of

[1] *Leibniz-Clarke Correspondence*, p. 72

[2] ibid., p. 53. Also p. 21: 'Nothing can any more act, or be acted upon, where it is not present, than it can be where it is not.'

[3] ibid., p. 94 [4] ibid., p. 119

explaining gravitation by impulse, since it depends on quantity and not surface area of matter; hence he thinks it must be due to something 'immaterial' which penetrates matter.[1]

The argument is clearly the same as that we have seen running through the whole history of theories of interaction. If there is continuity of causes transmitting actions from one part of space to another, what is the nature of these causes? Leibniz holds the comparatively 'modern' view of Cartesian mechanism, asserting that matter can act upon matter only by contact and according to the laws of mechanics, and that soul or spirit cannot act upon matter at all, for that would violate the natural conservation of *vis viva*.[2] Clarke holds the traditional view that immaterial spirit may act upon body, remarking that this is as easy to conceive as cohesion 'which no mechanism can account for',[3] and he maintains in particular that God is substantially present everywhere and may intervene in the universe by acting directly upon matter.[4] For Leibniz the activity of God is of another order, and though he does not deny God's omnipresence, he regards his action upon matter as a continual production, not a spasmodic intervention.[5] In this emphasis on the distinction between theological and scientific explanation Leibniz is not, like many who followed him, dismissing God as an unnecessary hypothesis, but is concerned rather to safeguard his uniqueness and transcendence. And from the scientific point of view he was no doubt right, in 1716, to question Clarke's immaterial substances and non-mechanical explanations, for the time had not come when these could be made sufficiently precise to be acceptable in a scientific theory, but on the other hand, Clarke shows the more empirical spirit. There is no disagreement between them about the facts, the disagreement lies in the question of what kinds of theory are to be admitted in explanation of the facts, and Leibniz is certainly too rigid in insisting that theories must conform to certain preconceived metaphysical ideas.

Philosophical Justifications of Action at a Distance

The vague and unsatisfactory nature of Leibniz's arguments from the principle of continuity to action by contact is indicated by the use of the same principle by Boscovich in reaching the opposite

[1] Clarke's notes to *Rohault's System of Natural Philosophy*, 1st English ed., London, 1723, p. 50; cf. M. A. Hoskin, 'Mining all within,' *The Thomist*, xxiv, 1961.

[2] *Leibniz-Clarke Correspondence*, p. 86 [3] ibid., p. 117

[4] ibid., p. 47. He quotes the General Scholium of the *Principia*: 'He is omnipresent not virtually only but also substantially; for virtue cannot subsist without substance.'

[5] *Leibniz-Clarke Correspondence*, p. 17

conclusion. In his *Theoria Philosophiae Naturalis* of 1758,[1] Boscovich derives his metaphysical principles from Leibniz, but integrates them with Newtonian physics. He suggests that matter consists of identical points of no extent, having no essential properties except inertia and the capacity of exerting forces on one another of magnitudes depending on their mutual distances. These point centres of force are spatial versions of Leibniz's monads, and Boscovich claims that Leibniz, as well as Newton, requires every particle to be connected with every other at a distance. As we have seen, however, this cannot be understood in physical terms in Leibniz.

Boscovich invokes a Leibnizian law of continuity to show that even the impact of bodies must ultimately involve forces at a distance. He agrees with Leibniz that if the ultimate particles of matter are finite, there would be a discontinuous change of density at their boundaries, and that if they came into absolute contact, their velocities would change discontinuously and an infinite force would be required. But instead of invoking an infinite regression of elastic parts, he concludes that the primary elements of matter must be simple points. The force exerted between two point particles is a continuous function of the distance between them, tending to infinite repulsion at very short distances, becoming alternately repulsive and attractive as the distance increases, and eventually tending to an inverse-square law attraction at distances comparable with the sizes of ordinary bodies. In this way Boscovich tries to subsume under one continuous force-function the gravitational attraction at finite distances, the repulsive forces producing deflections which are ordinarily called impacts, the cohesion and stability of aggregates of point masses which are in mutual equilibrium, and the attractive and repulsive forces of electricity and magnetism, claiming that the theory unifies Newton's three principles of gravity, cohesion, and fermentation.[2]

Boscovich's concept of force is a development in physical terms of the mathematical concept we have seen in Newton, for he asserts like Newton that 'force' does not denote any particular mode of action nor any mysterious quality, but only the propensity of masses to approach and recede :

> 'The various motions that arise from forces of this kind, such as when one body collides with another body, when one part of a solid is seized and another part follows it, when the particles

[1] *Theory of Natural Philosophy*, trans. Child, Chicago, 1922. Russell suggests (*Philosophy of Leibniz*, p. 91) that Boscovich has here a more logical development of the monadology than Leibniz himself.

[2] *Theory of Natural Philosophy*, p. 15

of gases, and of springs, repel one another, when heavy bodies descend, these motions, I say, are of everyday occurrence before our eyes. . . . In all of these there is nothing mysterious ; on the contrary they all tend to make the law of forces of this kind perfectly plain.'[1]

He does not intend to examine the ultimate nature of either force or inertia, and considers that these probably cannot be discovered, but he claims that the mere idea of change of motion does not

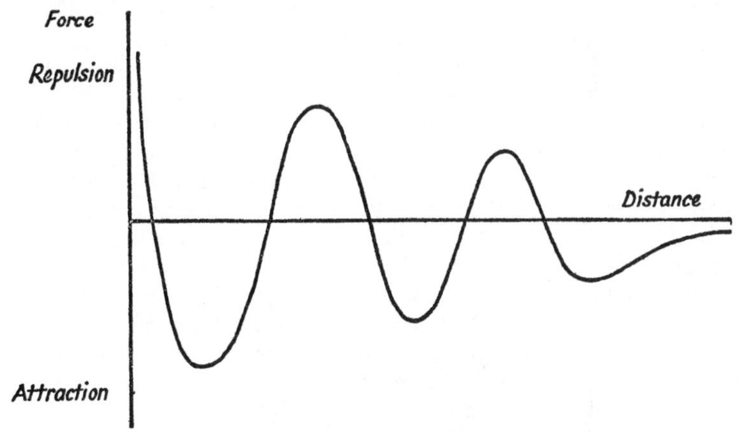

FIG. 3 The force-function of Boscovich

commit the philosopher even to action at a distance, although 'this method of explaining phenomena . . . is strongly recommended'[2] and since it is concerned with change of motion it is just as mechanical as explanation by impulse. In general, however, the theory of the force-function as applied to physics can be interpreted consistently with various philosophies : Cartesian occasional causes, Aristotelian substance and attribute, or the impenetrability and active forces of Leibniz and Newton. In any of these forms the theory would produce the same sensible effects.[3]

The theory of the force-function itself may be criticised on grounds of being *ad hoc*, but Boscovich was misled by his data rather than his method. Considering the confused state of the theory of matter and of chemical interaction at the time Boscovich wrote, it was not to be hoped that the details of his theory would survive, but his method was in the tradition of mathematical physics

[1] ibid., p. 95 [2] ibid., p. 109 [3] ibid., p. 365

leading from Newton to the present time, indeed the method of deriving a force-function *ad hoc* from the phenomena is very similar to that used at present in postulating short-range nuclear forces. What was influential however was Boscovich's fundamental idea of replacing the notion of extended particles by mathematical points having inertia, for in spite of his explicit positivism in defining his forces solely in terms of change of motion, his comparison of the forces between points with the action between the ends of a spring suggests a physical picture in which force is more fundamental than matter. His conception was developed in this sense by Kant, Faraday, and the nineteenth-century field theorists.

Prejudice against Newton's attraction could not last. The obvious success of the theory in astronomy, together with the difficulty of explaining such physical phenomena as cohesion, elasticity, and magnetism, in terms of ultimate action by contact were sufficient to ensure eventual acceptance of attractive forces by physicists, and the growing influence of empiricism in philosophy helped to modify the dogmatic mechanism of the seventeenth-century outlook. Arguments like those of Leibniz and Boscovich showed that even impact was not as intelligible as an ultimate mode of interaction as had been assumed by the corpuscularians, and this line of argument was reinforced by the empiricist philosophers who had, as might be expected, few of Leibniz's difficulties in accepting action at a distance.

Locke, for instance, although in sympathy with the corpuscular philosophy, allowed himself to be convinced by the *Principia* that the power of attraction was as real as the power of impulse. In the first edition of his *Essay Concerning Human Understanding* (1690), in speaking of ' how bodies operate one upon another ', he says :

> ' . . . that is manifestly by impulse and nothing else. It being impossible to conceive that body should operate on what it does not touch (which is all one as to imagine it can operate where it is not).' [1]

In the fourth edition (1700), however, the words ' nothing else ' are omitted, and Locke contents himself with remarking that, when bodies produce ideas *in us*, it is by impulse, which is ' the only way which we can conceive bodies operate in '. He writes to the Bishop of Worcester that he still cannot conceive any way of operation of bodies other than impulse,

[1] *Essay*, Book II, Chap. viii, 11

'But I am since convinced by the judicious Mr. Newton's incomparable book, that it is too bold a presumption to limit God's power, in this point, by my narrow conceptions. The gravitation of matter towards matter by ways inconceivable to me, is not only a demonstration that God can, if he pleases, put into bodies, powers and ways of operation, above what can be derived from our idea of body, or can be explained by what we know of matter, but also an unquestionable and every where visible instance, that he has done so. And therefore in the next edition of my book, I shall take care to have that passage rectified.' [1]

In any case the idea of attraction was not difficult to assimilate into Locke's empiricism. He had already maintained that bodies of themselves give us no idea of their power to initiate motion; we only observe the motions produced by, for example, the magnet, and infer the idea of power from our knowledge of the power of our own minds to move our bodies.[2] The power of communicating motion by impulse is a primary idea we have of bodies, and we observe its effects in daily experience, but how this power is exercised is quite obscure; no more intelligible in fact than the manner in which thought interacts with matter, of which we also have daily experience but cannot clearly understand. Cohesion is similarly a primary idea of body, and yet it cannot be understood, for any explanation in terms of the pressure of surrounding bodies involves an infinite regress. Locke here uses the difficulties inherent in understanding impulse and cohesion to show that the action of thought is no more difficult to conceive than the fundamental ideas of mechanics, in fact that spirit is no more mysterious than body, but the argument might just as well have been used in regard to attraction. He does so use it in his letter to the Bishop of Worcester: we cannot say that God may not give powers to matter apart from what we conceive to be its essence. He may give thought, reason, volition, sense, and spontaneous motion, and has certainly given attractive powers; they are inexplicable and inconceivable, but so is the manner in which bodies cohere: 'The omnipotent creator advised not with us in the making of the world, and his ways are not the less excellent because they are past our finding out.' [3]

Locke would therefore seem to have none of Newton's certainty that action at a distance must ultimately be explicable in terms of the impacts of corpuscles. He is prepared to accept attraction

[1] 'Second Reply to the Bishop of Worcester' (1698), *Works of John Locke*, 1768, I, p. 754 [2] *Essay*, Book II, Chap. xxi, 4; Chap. xxiii, 9 [3] *Works*, I, p. 750

as an empirical fact, and in any case, as we have already seen, he doubts whether there can be any certain knowledge of explanations beyond the empirical facts. Berkeley is even more definitely of the opinion that the business of physics is to establish laws of motion, and not to inquire into efficient causes. Attraction, he says, ' is only a mathematical hypothesis, and not a physical quality '.[1] It and other such general terms are useful for reasoning about motion and for instruction, but should not be mistaken for real things ; only the sensible and the concrete are the proper objects of physics. Berkeley's discussion of the method of science is not however as wholly phenomenalist as these remarks might suggest. He speaks also of natural science as a skill in seeing analogies in nature and predicting by means of them,[2] and regards Newton's analogy of attraction as a better explanation in chemistry and in mechanics than Descartes's principles of size, figure, and motion. But these analogies can only suggest hypotheses, and although the consequences of the hypotheses may be realised in nature, it does not follow that the hypotheses themselves are true of nature. To determine this is not in any case, Berkeley thinks, the business of science but of metaphysics. Like Locke, Berkeley compares the notion of elasticity with that of attraction, and finds the latter no more obscure than the former.[3]

Hume, in his *Treatise of Human Nature* (1739), so far assumes the existence of gravitational attraction as to use it to argue by analogy to the cohesion and attraction between the simple ideas of our minds, whose connection in complex ideas we experience but cannot understand :

> ' Here is a kind of *attraction*, which in the mental world will be found to have as extraordinary effects as in the natural, and to show itself in as many and as varied forms. Its effects are everywhere conspicuous ; but, as to its causes, they are mostly unknown, . . . Nothing is more requisite for a true philosopher, than to restrain the intemperate desire of searching into causes.' [4]

Hume regards gravity as an undoubted property of matter, and agrees with Locke that it is not possible to go further than the sequence of sense-impressions in order to have knowledge of ultimate

[1] *De Motu* (1721), *Works of Berkeley*, ed. Luce and Jessop, IV, p. 38

[2] *Siris* (1744), *Works*, v, §252, and *Principles of Human Knowledge* (1710), *Works*, II, §105, where Berkeley holds that, in regard to their knowledge of phenomena, natural philosophers are distinguished from other men ' only in a greater largeness of comprehension, whereby analogies, harmonies and agreements are discovered in the works of nature, and the particular effects explained, that is, reduced to general rules.'

[3] *Siris*, §§228, 243, 247

[4] *Treatise*, Book I, Part I, Section IV

causes. The problem of causality with which Hume wrestled was the problem of finding reasons for our belief in necessary connections between cause and effect, and this problem arises as much from the experience of seeing one billiard-ball move another, as from any other physical or psychical sequence of events, including gravitational interactions.[1] But Hume remarks that we usually require to perceive contiguity in time and place between two events before we assert a causal relation :

> 'Though distant objects may sometimes seem productive of each other, they are commonly found upon examination to be linked by a chain of causes, which are contiguous among themselves, and to the distant objects ; and when in any particular instance we cannot discover this connection, we still presume it to exist.' [2]

But, like all generalisations according to Hume, this cannot be grounded in more than habit.

The cumulative effect of philosophical criticism from Leibniz to Kant was undoubtedly to encourage the view that science could progress independently of metaphysics, for where philosophers after Leibniz touched upon scientific matters it was rather to draw conclusions from existing and accepted science than to dictate to science its proper methods and categories. Again, the empiricists' discussions of substance and causality meant that no necessary connection could be seen between any of the properties of matter hitherto regarded as essential, such as extension, impenetrability, and communication of motion by impact. And if these properties were known only by constant repetition of experiences, there was no reason why attraction should not equally be regarded as a property of matter, since it was repeatedly exhibited in indubitable experiences.

These were the arguments used by Maupertuis when, in 1732, he pioneered the acceptance of Newton's theory in the less empirical climate of France : 'La manière dont les propriétés résident dans un sujet, est toujours inconcevable pour nous '.[3] Voltaire also enlisted in this cause, and in a series of letters and popular books, argued the merits of Newton as against Descartes. The Frenchman in London, he writes in 1728,

[1] Commenting on Euler's rejection of attractions in his *Letters to a German Princess*, his English editor remarks that some properties of matter must remain arbitrary ; if not gravitation, then impulse, and adds 'No metaphysical work has ever done so much service to philosophy as Mr Hume's admirable essay on "Necessary Connexion"' (*Letters*, trans. Hunter, London, 1795, I, p. 46*n*).

[2] *Treatise*, Book I, Part III, Section II

[3] *Œuvres*, 1752, p. 65

> '. . . left the World a *plenum*, and he now finds it a *vacuum*. . . . According to your *Cartesians*, everything is performed by an Impulsion, of which we have very little Notion ; and according to *Sir Isaac Newton*, 'tis by an Attraction, the Cause of which is as much unknown to us.' [1]

Eventually the mathematics of the *Principia* were accepted in France, but many remained unconvinced of its physical adequacy. The language of vortices was retained, and accusations of occultism were made against Newton throughout the eighteenth century. The Encyclopedist Diderot, for example, found it impossible to understand how attractive force varies with distance if it acts across a void, and he doubted in general the exact applicability of mathematics to nature.

Kant : ' The Metaphysical Foundations of Natural Science '

There were no such doubts in Kant's mind about the acceptability of Newton's theory. In his early ' Thoughts on the true estimation of living forces ' (1747) he regards the inverse-square law of action between bodies as so fundamental that he suggests that the three-dimensionality of space is derived from it.[2] The inverse-square law, he says, is entirely arbitrary, in the sense that God might have chosen any other law of interaction, and the dimensionality of space would then have been correspondingly different. Kant's best-known work of a scientific character is the *Universal Natural History and Theory of the Heavens*, subtitled ' An essay on the constitution and mechanical origin of the whole universe treated according to Newton's Principles ' (1755), in which he puts forward the first hypothesis of the origin of the solar and stellar systems, and here, as the subtitle indicates, Newton's theory is accepted without question. Kant remarks that he is avoiding all arbitrary hypotheses and postulating only an initial chaos of material elements together with their mutual attractive and repulsive forces which ' are both equally certain, equally simple, and, at the same time, equally original and universal '.[3] Both are borrowed from Newton's natural philosophy : attractive force is established beyond doubt ' by geometry ' and ' by means of indisputable observations ',[4]

[1] *Letters Concerning the English Nation*, London, 1733, pp. 109, 110

[2] §10 in *Kant's Inaugural Dissertation and Early Writings on Space*, trans. Handyside, Chicago, 1929, p. 11. Later, in the *Anfangsgrunde*, where the doctrine of the *Critique* is presupposed, and three-dimensional space is regarded as an *a priori* form of perception of matter, the inverse-square law is derived from three-dimensionality.

[3] *Universal Natural History*, trans. Hastie, *Kant's Cosmogony*, Glasgow, 1900, p. 35

[4] ibid., pp. 48, 49

and repulsive force, although not directly demonstrated by Newton, is accepted without question as the force between the parts of an elastic fluid. Kant proceeds to describe the origin of the solar system out of the initial chaos : matter began to congregate round the points which exerted greatest attractive force, but was disturbed and deflected by the contrary elastic repulsion, consequently part came together to form the sun, and part acquired a motion round the sun in orbits depending on the equilibrium set up between attractive and centrifugal forces. The mechanics of Kant's system violates the principle of conservation of angular momentum, since his original chaos is at rest, but the mechanics of rigid bodies were by no means clear at this time. The work leaves no doubt that Kant, from his earliest writings onwards, accepted Newton's theory in principle as the true system of the world.

He never modified his acceptance of forces acting at a distance. When, 'roused from his dogmatic slumbers' by Hume's attack upon causality, he embarked upon his great metaphysical works, he nowhere demanded, as Hume had come near to doing, that causes should act contiguously through space. In the *Critique of Pure Reason* (1781), the question of action at a distance is not raised explicitly, and Kant's general insistence on the continuity of space and time, motion and change, might lead one to expect a rejection of non-contiguous causes. But he is concerned here only to insist that *observable* states of affairs cannot succeed one another discontinuously, like the discrete images on the film-screen, and it does not follow from this that things in themselves cannot act upon one another at a distance, for causal action is not a phenomenon to be observed.

One might put the point in modern terms as follows : for Kant the very conditions of perception imply that observable phenomena should change continuously, but there is nothing to prevent theoretical explanations introducing discontinuity. Kant would clearly prefer that they did not, as his criticisms of atomism show, but for him explanations going beyond phenomena can never be certain and can never literally describe things in themselves, so in general he rightly refrained from limiting the types of explanation that might be given. In spite of his preference for continuity he accepted action at a distance, and would presumably have accepted even the quantum jumps of modern atomic physics, because to postulate discontinuity in a *theory* does not necessarily imply any discontinuity of phenomena. He would, of course, have denied that what we experience as continuous is 'really' a smoothed-out perception of discrete events in the world, because he denied that we can have any knowledge of ' the world ' at all apart from perception. It is plausible to interpret his attitude to theoretical science, as does G. Martin,[1] in terms of possible models which are ways in which we think about the world, but which may not and sometimes cannot be literal descriptions of the world, and in any case can never be known to be such descriptions. This attitude leaves the scientist free, as Kant remarks in connection with theories of void or plenum, to adopt the theory which is convenient.

It is in the *Metaphysische Anfangsgrunde der Naturwissenschaft* of 1786 that Kant explicitly accepts the possibility of forces acting at a distance, and even asserts their metaphysical necessity, although one suspects that without empirical knowledge of gravitational attraction he would not have been led to regard their existence as necessary *a priori*. The purpose of the *Anfangsgrunde* is to apply the principles of the *Critique* to physics. Kant begins by distinguishing metaphysics and natural science proper from merely empirical studies. Metaphysics contains only those *a priori* principles discussed in the *Critique*—those forms of perception such as space and time, substance and causality, without which there can be no knowledge, and which are therefore, according to Kant, independent of experience and absolutely certain. But in addition to this transcendental metaphysics of nature in general, metaphysics may concern itself with particular things of which the empirical concepts are given, for example, matter or thinking substance, and its task is then to discover the *a priori* forms under which such empirical concepts must be known. This is natural science proper, and it contains only basic empirical concepts together with what can be said about them *a priori*. Other properties of the world must be discovered empirically, and are not part of natural science. Kant remarks that chemistry must remain a ' systematic art or experimental doctrine ' unless the laws of the effects of substances upon one another can be shown to be *a priori*, ' a demand that will scarcely ever be fulfilled '.[2]

The empirical concept which, together with *a priori* principles, constitutes the science of physics, is the concept of matter. Kant considers matter as the *movable* under four aspects, corresponding to his fourfold division of categories in the *Critique*. Of these, the second, which he calls dynamics, is of greatest concern to us ; the first treats of matter as simply movable, and corresponds roughly

[1] *Kant's Metaphysics and Theory of Science*, trans. Lucas, 1955, pp. 58–64, 94–7

[2] *Metaphysiche Anfangsgrunde der Naturwissenschaft* (1786), *Sammtliche Werke*, Leipzig, 1839, v, p. 310

to our kinematics; the third treats of matter as movable by forces, that is kinetics. Dynamics treats of matter in so far as it fills space and is impenetrable, and from this conception Kant tries to derive the necessity of attractive and repulsive forces.

Kant: Infinite Divisibility of Matter as a Regulative Principle

He begins by asserting that impenetrability is the fundamental property of matter in its dynamical aspect, and that this is a conception which could not be formed *a priori* but is given in sensation. The question now is, how is this impenetrability to be understood? The atomists believe that it is an ultimate quality of the atoms and that compression of bodies is only possible if there is closer packing of the atoms into the empty space between them. Kant, however, prefers to regard impenetrability as due to a primitive repulsive force by which matter extends itself over a given space and hinders the entry of any other matter into that space. The status of this conception is not entirely clear in the *Anfangsgrunde*, for although Kant mentions several reasons for preferring the conception of relative impenetrability due to repulsive force to the absolute impenetrability of atoms, none of these are strong enough to show that impenetrability *must* be conceived as relative, and this cannot therefore have the status of an *a priori* principle. Its status will become clearer after we have considered the discussion of infinite divisibility in the *Critique*.

The objections which Kant suggests in the *Anfangsgrunde* to the hypothesis of absolutely impenetrable atoms are as follows:

1. It implies that each atom is absolutely homogeneous and cohesive and that no finite force can penetrate it, and Kant agrees with Leibniz that there should be continuity of degrees of a property —it may be indefinitely small or indefinitely large, but not zero or infinite.

2. Absolute impenetrability is an occult quality which gives no reason for impenetrability but only names it, while Kant asserts, like Newton in connection with gravity, that repulsive force is a real explanation, since it can be shown to be subject to laws of variation.

3. Ultimate atoms cannot be discovered experimentally and are therefore hypotheses, whereas Kant seems to think that infinite divisibility is not a hypothesis, or at least not such an objectionable one.

4. The hypothesis of atoms and the void is not necessary for the explanation of varying density, which can equally well be explained by repulsive forces, therefore the hypothesis ought not to be made.

Repulsive force is thus an original and fundamental property of matter and cannot be derived from any other property, indeed, it *is* matter in its dynamical aspect. The force can have any finite degree of intensity, so that impenetrability is only relative, and any body, however hard, can be compressed by an external force sufficiently great. Since repulsive force is equated with matter as the impenetrable, it follows that where force is, matter is, and since the space filled by repulsive force is divisible to infinity, matter is divisible to infinity. Kant uses this argument to refute any hypothesis, such as that of Boscovich, of *point* monads exerting repulsive force.[1] Matter is not to be identified with point singularities in the field of force, but with the continuous field itself. Thus matter in its dynamical aspect is understood simply as repulsive force, but in the next section on mechanics, it also has to be endowed with inertial mass, which Kant calls ' quantity of matter '.

The conclusion that matter is infinitely divisible leads Kant to mention the antinomy which is implied by this conception and which is discussed more fully in the *Critique*. There [2] he derives four antinomies of the pure reason resulting from extrapolation beyond all possible experience of phenomena originally given in experience. They are: the finiteness or infinity of the world in space and time, the analysis of substance into simple parts or its infinite divisibility, causality or freedom, and the existence or non-existence of a necessary being. These antinomies are all of the same form : beginning with empirical data, the extension of the world in space and time, the relative divisibility of substance, the observed relations of cause and effect, and of necessary and contingent being, the reason is inevitably led to consider the indefinite extrapolation of the series involved in these phenomena. Does the world extend to infinity? Is substance infinitely divisible? Do causal sequences extend backwards indefinitely or are there first causes? Is all being contingently dependent or is there an independent, necessary being? Kant shows that both possible answers to each of these questions lead to contradictions if the world outside experience is thought of as a ' thing-in-itself '. The solution to the resulting antinomies lies in his doctrine of *phenomena* as the only objects of knowledge.

[1] ibid., p. 352

[2] *Critique of Pure Reason*, Transcendental Dialectic, Book II, Chap. ii

Since all these questions are about conditions which are not possible objects of experience, the contradictions to which they lead are only apparent, and can be avoided by refusing to go beyond empirical phenomena. There is no transcendental answer to the questions, and if the world is not a thing-in-itself there is no contradiction in saying, for example, that it is neither finite nor infinite, because ' the world ' is entirely unknown, and may not be determined at all in regard to quantity.

Something more can, however, be said from the empirical point of view. To concentrate attention on the second antinomy, since this is the one with which Kant's dynamics is concerned, one can lay down as a *regulative principle* that no absolute empirical limits can be conceived to the process of division of substance, although there may be limits to the organisation met with during the subdivision, and these limits could be found only by experience.[1] In other words, whatever may at any stage be regarded as an ultimate constituent of matter can always be conceived to be further divisible, and may later be shown to be so empirically.

Thus both atomicity and infinite divisibility are possible empirical hypotheses, but neither can be true descriptions of matter as it is in itself. Neither alternative is falsifiable. The choice of a regulative principle must then depend on criteria other than truth or falsity, and the arguments so far adduced by Kant for preferring infinite divisibility to atomicity (where the size of the atoms is left indefinite, otherwise this alternative would be falsifiable) are the Leibnizian arguments against atomism, which, while not conclusive, suggest which is the most convenient model. This is all that can be hoped for in relation to a regulative principle of natural science, as opposed on the one hand to an *a priori* principle of metaphysics, or a synthetic statement which is empirically testable. The reason for the choice may be put like this : at any particular stage of empirical investigation it is conceivable that matter may be further divided ; to assume that it is infinitely divisible is therefore more economical than gratuitously to assert that at some point, which can never be specified, divisibility ceases.

Another criterion for choice between regulative principles is ease of mathematical manipulation, and if this were the only consideration in this case, Kant would undoubtedly prefer atomism. He says that as far as the mathematics is concerned absolute impenetrability may be assumed for some purposes, and this is done in what he calls the *mechanical mode* of explanation, in which bodies

[1] ibid., Sections 8 and 9

are regarded as mere tools of external moving forces, as in elementary Newtonian mechanics. This mode of explanation is, he says, 'the most amenable to mathematics',[1] and he treats it in the third division of the *Anfangsgrunde*.

Kant: Attraction and Repulsion

Returning to the dynamics, after his discussion of the repulsive force of expansion, Kant proceeds to derive attractive forces as equally fundamental to the possibility of matter: 'All motion which one matter can impress upon another, as in this respect each of them is only considered as a point, must always be regarded as distributed in the straight line between two points'.[2] Thus, apart from repulsive force, only attractive force is possible. Attractive force moreover is necessary, otherwise any given impenetrable body would disperse to infinity under the action of its own repulsive forces; thus attractive forces also are original and fundamental to the possibility of matter. Kant pauses to account for the fact that attractive forces seem to be more difficult to conceive than repulsive: it is because extension and impenetrability are the manifestations of repulsive force, and these are at once obvious to perception:

> 'This is undoubtedly the reason why, in spite of the clearest proofs on the other hand that attraction must be a fundamental force of matter just as repulsion is, one is so unwilling to admit it, or to concede any other moving forces but those of impact and pressure (both by means of impenetrability). . . . But as substance only reveals its existence to us by sense, whereby we perceive its impenetrability, namely by feeling, and therefore only in reference to contact . . . it seems as though the immediate effect of one matter on another could never be anything but pressure or impact, the only two we can immediately feel; while attraction, which can give us either no feeling at all, or at least no definite object of it, becomes difficult for us to conceive as a fundamental force.' [3]

Kant now proceeds to distinguish between the modes of action of the repulsive and attractive forces. Since physical contact is simply the direct action of repulsive forces, these forces necessarily

[1] *Anfangsgrunde*, p. 392

[2] ibid., p. 345. To consider matter as occupying *points* here appears to contradict the identification of matter with the field of repulsive force. If matter can be replaced by force, there is no reason why these forces should be central.

[3] ibid., p. 361

act where they are (that is, where matter is) and not at a distance. They act at surfaces of contact, and only upon distant matter by means of the matter that lies between. Attractive force, on the other hand, is ' possible without a medium '[1] ; it acts immediately at a distance as if through empty space. It is a penetrative force, acting directly upon the interior of bodies, and it is therefore proportional to the quantity of matter, and unlike repulsive force, it penetrates space without filling it. These two kinds of force, repulsive or elastic, and attractive or gravitational, are the only *a priori* universal characteristics of matter ; other forces such as cohesion ' when explained as the reciprocal attraction of matter, limited simply to the condition of contact '[2] is not an *a priori* characteristic, and is therefore part of empirical physics, not metaphysics.

From this account of attractive force it follows that, unlike repulsive force, it is not substantial. If it were, it would be subject to the same antinomy as that concerned with infinite divisibility of matter, and Kant would have to regard the question of action at a distance or continuous action as one to be settled by regulative principle and not as factual. But he nowhere suggests this, and appears to regard the existence of attractive forces at a distance as an *a priori* truth of the metaphysics of matter. He has, however, clearly imported into ' metaphysics ' empirical considerations other than the notion of matter as impenetrable. If he wished only to show that impenetrability implies attractive as well as repulsive forces, one would expect him to make the cohesive force of attraction fundamental, and this could be similar to the repulsive force in acting only where matter is and in filling the space where it acts. Kant argues that one cannot demand that attraction should act only between surfaces in contact, because the existence of attraction is the condition of the existence of determinate material bodies and therefore prior to the possibility of contact, but it does not follow that the attractive force acts at a distance outside matter, only that there is a cohesive force exactly balancing the repulsive force in the interior of matter, and a resultant repulsive force acting at its surface. But Kant wishes to make gravitational attraction at a distance fundamental even at the cost of asymmetry between his fundamental forces, and the characteristics with which he endows his attraction are clearly not *a priori*, but can come only from experience. He is, of course, perfectly entitled to take repulsive contact forces and attractive forces at a distance as fundamental

[1] ibid., p. 363 [2] ibid., p. 372

if he wishes, but their status must then be that of empirical hypotheses, not *a priori* metaphysics.

Kant next considers objections to the conception of forces acting at a distance. He has earlier remarked that 'Newton's system of universal gravitation is established, although it carries with it the difficulty that one cannot explain how attraction at a distance is possible ; but *difficulties are not doubts*'.[1] Now he admits that the possibility of the fundamental forces cannot be conceived, precisely because they cannot be deduced from any others ; all one can do is to show that they are not self-contradictory. 'The commonest objection to immediate effect at a distance is that a matter cannot directly operate *where it is not*.'[2] But this is not self-contradictory : everything acts upon something outside itself, and therefore acts where it is not. A point of contact is not part of one or the other body, for the mathematical contact of spaces must be distinguished from physical contact by repulsive forces. Kant seems to be suggesting here that two repulsive forces do not meet in a point, but his conception is far from clear, and seems to presuppose a model of surface-to-surface contact which is cruder than his own repulsive force model and contradicted by it. A sounder argument is that attractive forces acting at a distance follow definite laws which are independent of the repulsive forces, and can therefore be regarded as equally fundamental. Again, says Kant, if one admits with Newton that even aether has weight, how can weight itself be explained as due to the impulse and pressure of the aether ?

Finally comes the question of the existence of void. Kant has given reasons for objecting to the hypothesis of atoms, but the infinite divisibility of matter does not by itself imply that matter fills all space, and this question has to be discussed separately. Those who postulate a void often maintain that it is necessary in order to explain the different densities of bodies as a consequence of the proportion of empty space they contain. But on Kant's view, differences of density can be conceived as differences in the fundamental repulsive force and its relation to the fundamental attractive force. Thus density is not the proportion of homogeneous matter to empty space, as in atomism, but is an original property of every point of continuous matter, like heat, which 'can diminish in its degree *in infinitum*, without leaving the smallest part of this space in the least empty'. This is not however metaphysically certain :

[1] ibid., p. 314*n*   [2] ibid., p. 365

'I do not at all intend to assert that this is what actually occurs when material bodies differ in specific gravity, but only to establish from a principle of pure understanding that the nature of our perceptions allows of such a mode of explanation.' [1]

Thus Kant does not claim to refute the void *a priori*, but he points out that it is not necessary to assume it for the explanation of varying densities; moreover it is a hypothesis which could never be proved directly from experience, for this gives us only evidence of comparatively empty spaces. One function of metaphysics in relation to science is to point out possible alternative explanations, and thus free science from slavish adherence to one, such as atomism, which might otherwise be thought to be necessary. But in this instance metaphysics cannot go further and absolutely deny the possibility of void, although Kant himself would clearly prefer to deny it, and suggests that it may be *physically* though not *logically* impossible. For if cohesive forces are, as he suggests in the *Anfangsgrunde*, not real attractive forces, but due to the external pressure of the aether, then empty space within matter, and that outside all matter, would be impossible, because at short distances the inverse-square force of attraction would not be sufficient to prevent matter expanding into these spaces under the influence of its repulsive force (which Kant thinks is proportional to the inverse cube of the distance). The gravitational force by itself, he asserts, is sufficient only to ensure that matter does not disperse to infinity with a vanishingly small density everywhere.

Kant's remark that atomic explanations are more amenable to mathematical treatment than those involving a continuous medium, indicates a general weakness of physical theories of the continuum during the eighteenth century, for until mathematical analysis was applied to physics by the French mathematicians towards the end of the century, there were no general methods of dealing with continuous distributions of force. Kant does not even seem to be fully conversant with such methods as Newton had used. He suggests that his fundamental repulsive forces vary as the inverse cube of the distance because they are penetrative and have to spread themselves throughout a volume instead of over a surface as is the case with the attractive forces.[2] But Newton had shown [3] that if attractive (and therefore also repulsive) forces vary in inverse proportion to an integral power of the distance greater

[1] *Critique*, A 174 (trans. N. K. Smith); cf. Aristotle's conception of density

[2] *Anfangsgrunde*, p. 375

[3] *Principia*, Book I, Props. LXXXV, LXXXVI

than the second, the attraction (or repulsion) of contiguous particles is infinite. In 1771 a paper by Cavendish [1] was published, giving a more satisfactory proof of the same result in connection with the forces exerted in continuous distributions of electric charge, and hence showing that these forces must be proportional to an inverse power of the distance less than the third. In the case of Kant's continuous distribution of repulsive force, Newton's theorems mean that an inverse-cube law would give infinite repulsion between contiguous particles everywhere, and therefore that no such distribution could exist.

Elastic Fluid Theories in Physics and Chemistry

It remains to describe briefly the applications which were increasingly made during the eighteenth century of the concepts of attraction and repulsion in all branches of physics and in the still immature science of chemistry.

The phenomena of cohesion, capillarity, and electric and magnetic attraction, had been the most intractable from the point of view of seventeenth-century mechanism. Experiments showed that the rise of liquids in capillary tubes was not the result of unequal air pressures as had been suggested, and was therefore not reducible to action by impact, and the suggestion was soon made that it was due to mutual attraction of particles. The same suggestion was obvious as an explanation of cohesion and elasticity, although at this stage there was little reason to associate these attractive forces with electrical attraction.

The elasticity of gases received two explanations, both consistent with Boyle's law, but conceptually very different from each other. The first was Newton's proof [2] that if the particles of a gas mutually repel with forces inversely proportional to the first power of their distances, the gas will behave according to Boyle's law, and the second was the proof by Daniel Bernoulli (1738) [3] that the same behaviour results from a collection of small elastic particles in random motion, in which the effects of their mutual collisions, which are rare, are negligible, and the pressure of the gas is interpreted in terms of the collisions of the particles with the walls of the vessel. These alternative explanations thus utilise the two kinds of force : repulsions at a distance and impacts. But in the second case the problem of elasticity may be said to be merely postponed by

[1] *Phil. Trans.*, 1771, p. 584
[2] *Principia*, Book II, Prop. XXIII, and Scholium
[3] *Hydrodynamica*, pp. 200ff.

transferring the elasticity of the gas as a whole to its constituent particles, and this explanation seems to have had little influence until it was revived as the foundation of the statistical theory of gases in the nineteenth century. During the eighteenth century, once forces at a distance were accepted, it seemed more satisfactory and economical to think of all fundamental forces in this way, since impact could be reduced to forces at a distance, whereas gravitation, magnetism, and cohesion could not be reduced to elastic impact.[1]

The introduction of repulsive as well as attractive forces was thought by some writers to need special justification. Newton had merely remarked [2] that attractive and repulsive forces are analogous to positive and negative numbers in algebra. The status of negative numbers had only recently been accepted as similar to that of positive numbers, so the implication of Newton's analogy was that the introduction of repulsion as well as attraction among the fundamental forces did not involve any new physical principle. Berkeley makes the same comparison with positive and negative numbers.[3] Gowin Knight lists the experimental evidence for repulsive forces : there are electric and magnetic repulsions ; Newton had shown that the surfaces of two convex glasses repel ; he had also explained the elasticity of air and the reflection of light by repulsive forces ; and it seemed probable that particles of light repel each other.[4] Franklin remarks that the hypothesis of electrical repulsion is not a multiplication of causes without necessity, but is made inevitable by the experimental results, and further justifies himself by pointing to evidences of repulsion elsewhere, as in the vaporisation of water, the explosion of gunpowder, and the repulsion of magnets.[5] In *A Philosophical Inquiry into the Cause of Animal Heat*, in 1778, P. D. Leslie ascribes forces of mutual repulsion to phlogiston and to light, remarking that ' *a priori* it is certainly as easy to conceive that a body should have a tendency from, as to the center '.[6]

So it became common to describe physical phenomena in terms of gross matter and subtle fluids whose particles exerted mutual attractions and repulsions. These fluids were the direct descendants of the vapours and emanations of the early seventeenth century, stripped of all quasi-animist properties except the attractive and

[1] This point is made by Berkeley, *Siris*, *Works*, v, §§225, 243 ; and Desaguliers, *A Course of Experimental Philosophy*, London, 1744, II, p. 36.
[2] *Optics*, Query 31, p. 395 [3] *Siris*, §236
[4] Proposition XIV in Gowin Knight's work referred to below
[5] *Experiments and Observations on Electricity* (1774), ed. Cohen, Cambridge, Mass., 1941, p. 366 [6] p. 120

repulsive powers which could be given exact mathematical description. In a book revealingly, if wordily, entitled *An Attempt to demonstrate that all the phenomena in nature may be explained by two simple active principles, attraction and repulsion : wherein the attractions of cohesion, gravity and magnetism are shown to be one and the same ; and the phenomena of the latter are more particularly explained,*[1] Gowin Knight adopts a mathematical form clearly modelled on the *Principia*, and his physical reasoning shows no trace of the animism which Leibniz suspected on reading of Newton's 'active principles', but has in fact marked affinity with some of the qualitative methods of particle physics of much more recent date. Knight's world-model is based on two kinds of fluid matter consisting of very small, equal, spherical particles, together filling space, but allowing for interspersed vacua. The particles of one kind of matter are mutually repulsive, the repulsion being proportional to the inverse power of the distance, while those of the other are mutually attractive. He goes on to show that if the repelling fluid, which he identifies with light, is excluded from any part of space, for example by one or more particles of the attracting fluid, then the repelling particles appear to be attracted to the centre of the 'hole' with a force inversely proportional to the distance. Different kinds of corpuscles can then be constructed by combining repelling and attracting particles in various equilibrium configurations, and Knight builds up solid cohesive bodies and liquids, and provides explanations of the inverse square law of gravitation, and of light and heat phenomena, and magnetism. A characteristic feature of his system is that atmospheres of repelling particles are built up around all collections of attracting particles, that is, around solid bodies. A version of this theory appeared later in atomic theories of chemistry, in which each atom was supposed to be surrounded by an elastic atmosphere of heat.

Comprehensive theories of this kind were premature, however, and progress was made much more surely during this period by the patient accumulation and piece-meal interpretation of experimental results. This was particularly true in the case of electricity. Knowledge of the most elementary facts of electricity had been scanty during the seventeenth century, and it is noteworthy that even Guericke's classic experiments with the sulphur globe were undertaken not in order to discover the properties of electricity, which he did not recognise at all as a distinct kind of physical force, but to show that a rubbed globe has, among other virtues, that of

[1] London, 1754

attracting bodies to itself, and is therefore an illustration of the same power belonging to the earth ;[1] one of the few occasions on which gravitation was compared to what were really electrical rather than magnetic effects. During the eighteenth century, however, interest in electricity itself was widespread, and an accumulation of experimental results followed the invention of electrical machines and the Leyden jar. There were two theories as to the nature of the electricity so produced. Franklin suggested that there is a single electric fluid whose parts are mutually repulsive, that is, the fluid is elastic, and that its parts are attracted by ordinary matter 'as a sponge takes up water'.[2] He introduced the terms 'positive' and 'negative' to describe the excess or deficiency of electric fluid in an electrified body. The theory accounted for the repulsion of two positively charged bodies, since their excess fluid is mutually repulsive, and for the attraction between two bodies when one is negatively and the other positively charged, but not for the repulsion between two negatively charged bodies, because in that case it ought to follow that the only force between them is the gravitation of their ordinary matter. Aepinus and Cavendish[3] suggested that the force between particles of matter (presumably at short distances only) is in fact *repulsive* ; but a more satisfactory theory to explain this effect was that of *two* electric fluids, each of whose particles repel particles of like kind and attract those of unlike kind.

Magnetism was also generally ascribed to a fluid which was thought to be, as Franklin put it, 'in many respects analogous to the electric fluid'.[4] There were one- and two-fluid theories of magnetism as of electricity. In the hands of Coulomb and Poisson the two-fluid theory began to approximate to modern views of the nature of magnetic substances, for it was assumed that particles of each kind of magnetism are associated in pairs so that the smallest parts of the substance are themselves small magnets, and then magnetisation of the body as a whole consists of the orientation of the small magnets in the same direction. Coulomb, in 1785,[5] described in the terminology of the fluid theories the torsion experiments by which he showed that the attractive and repulsive forces of electricity and magnetism are both proportional to the inverse square of the distance between the centres of force.

[1] *Experimenta Nova Magdeburgica*, Amsterdam, 1672, Book IV, 'On the Mundane Virtues'

[2] *Experiments and Observations*, ed. Cohen, p. 213

[3] See Cavendish, 'An attempt to explain some of the principal phaenomena of electricity, by means of an elastic fluid', *Phil. Trans.*, 1771, p. 584

[4] *Experiments and Observations*, p. 366

[5] 'Ou l'on détermine suivant quelles lois le fluide magnétique ainsi que le fluide électrique agissent soit par répulsion, soit par attraction', *Collection de Mémoires*, I, p. 116.

Theories of heat added to the rapidly growing number of elastic fluids required by physics. During the seventeenth century heat had most often been explained in terms of vibrations of the small parts of matter or aether, and this was Newton's view in the *Optics*. But Newton also speaks there of the radiation of heat through the aether, which is itself an elastic fluid, and it was an easy transition for his eighteenth-century followers to begin to speak of the aether itself as the matter of heat. Berkeley for instance speaks of fire as

> ' the most elastic and expansive of all bodies. It communicates this quality to moist vapours and dry exhalations, when it heats and agitates their parts, cohering closely with them, overcoming their former mutual attraction, and causing them, instead thereof, reciprocally to repel each other and fly asunder.' [1]

And after quoting Heraclitus and the Stoics in his support, he concludes that an ' animated heterogeneous fire should seem a more adequate cause, whereby to explain the phenomena of nature, than one uniform aetherial medium.' [2]

Newton does himself speak of the power of heat to overcome the attractive forces which normally cause bodies to cohere, and ascribes this both to the violent motion of vibration and also to the mutual repulsion of parts of the aether. The ambiguity of the Newtonian tradition here is illustrated by an English translation of a work of Boerhaave on chemistry.[3] Boerhaave accepts the view that fire is a very subtle atomic fluid, whose presence in bodies causes their particles perpetual agitation, and whose complete absence results in a state of absolute rest between the particles. His translator, Shaw, however, supplements this by notes which put forward the ' orthodox ' Newtonian view that heat is simply a form of motion. But it is clear that both acknowledge Newton as their master, and that Boerhaave's attempted reduction of chemistry to the mutual attractions and repulsions of corpuscles is equally in the Newtonian tradition.

The tendency to replace vibrations by the matter of the aether itself as the explanation of heat was encouraged by the traditional inclusion of fire as one of the elements, and by the contemporary version of the ancient theory of elements, namely the phlogiston theory of Becher and Stahl. Here the traditional four elements were replaced by three, of which one was the ' combustible ' element phlogiston, which was supposed to be released from metals on

[1] *Siris*, §149 [2] *Siris*, §229

[3] *A New Method of Chemistry*, London, 1727, Part II, p. 220

ignition. The phlogiston theory was able to provide satisfactory explanations of many chemical reactions and was widely accepted until the end of the eighteenth century, but there were some phenomena which were better explained by the rival theory of heat-fluid or 'caloric'.[1] For example, one difficulty about the phlogiston theory of combustion of metals was the fact that the resulting metallic oxide weighed more than the original metal, so that if the oxide had been produced by driving off phlogiston, phlogiston would have to have negative weight. In the caloric theory on the other hand it was assumed that caloric was added to the metal during the heating which produced oxidation, and the gain in weight was then simply the weight of caloric added. Many experiments were carried out to confirm that caloric had weight, but most gave negative results.

Caloric was usually supposed to be an elastic fluid whose parts mutually repelled but were attracted by ordinary matter. This hypothesis accounted well for the expansion of hot bodies by the repulsive force of the extra caloric contained in them, and the passage of heat from hot to cold bodies was explained by the same repulsive effect. The theory also accounted better for Joseph Black's discoveries relating to heat capacity than the dynamical theory of heat. In his *Lectures on the Elements of Chemistry* [2] Black discusses this point, and remarks that if heat is a motion of small parts of bodies, then the densest bodies ought to have the greatest heat capacity, but he found that the reverse is often the case. He speaks favourably of the theory of Cleghorn,[3] who had suggested that the differing heat capacities of different substances are due to their different attractions for fire. Black suggests that the fact that water vapour, for example, has a greater heat capacity than water, might then be explained by assuming that when the particles of water are farther apart, more of the particles of fire are able to cluster round each to give an equilibrium between all the attracting and repelling forces. He objects, however, that this is to put the cart before the horse : it is not vaporisation that causes water to acquire heat, but the addition of heat that causes water to vaporise.

The notion of 'heat atmospheres' consisting of repelling particles of caloric surrounding particles of matter was developed to account for the properties of gases and chemical combinations. Brian

[1] The name was first used by Lavoisier (*Méthode de Nomenclature Chimique*, Paris, 1787), to replace the description 'matter of heat'. [2] Edinburgh, 1803, I

[3] *Disputatio Physica inauguralis Theoriam Ignis complectens*, Edinburgh, 1779 ; Black, *Lectures*, I, pp. 33ff., 195ff.

Higgins [1] postulated hard spherical atoms of matter attracting each other inversely as some power of the distance, and a pervasive fire, or caloric, which exerts repulsive forces. Various types of chemical combination were then ascribed to the equilibrium set up between these forces, and Higgins was able in this way to explain combination of substances in definite proportions. Dalton's theory of mixed gases also made use of the idea that their elasticity is due to the atmospheres of caloric surrounding each gas particle. In his *Researches into the laws of Chemical Affinity* (1801), Berthollet [2] lists among the forces affecting chemical combination : cohesion, gravity, and elasticity of the substance itself or of caloric, but he thinks that all are probably gravitational in origin, and ascribes the differences between astronomical and chemical attraction to the different distances at which they act. Gravitation is exerted at distances where the particular forms and arrangements of molecules have no effect, whereas chemical affinity depends to a great extent upon local conditions. The forces of chemical affinity can therefore be discovered only by observation, but Berthollet thinks it probable that the more general the theory of affinity becomes, the more analogy it will have with the mechanics of gravitation.

The fluid theory of heat began to lose its plausibility as a result of the experiments of Rumford at the end of the century on the heat produced by friction. The apparently inexhaustible amount of heat thus generated led him to believe that it could not possibly be a material substance, and he says he is unable

> ' to form any distinct idea of anything capable of being excited, and communicated, in the manner the heat was excited and communicated in these experiments, except it be motion.' [3]

Lavoisier seems to have thought that caloric need not be assumed to be a real substance, but only ' some kind of repulsive cause which keeps the molecules of matter apart ',[4] whose effects can be considered mathematically. But the fluid theory had its adherents until well into the nineteenth century, and they included Carnot, whose work later became the basis of the science of thermodynamics. Fourier, in his *Théorie Analytique de la Chaleur* (1822), points out that the assumption that the rate of flow of heat, however produced, is proportional to the difference of temperature is sufficient to found a theory of heat independently of any hypothesis as to its cause :

[1] cf. J. R. Partington, ' Origins of the Atomic Theory ', *Annals of Science*, IV, 1939, pp. 269ff.
[2] Paris, 1801, trans. Farrell, Baltimore, 1804
[3] *Phil. Trans.*, 1798, p. 99
[4] *Traité Élémentaire de Chimie*, Paris, 1789, I, p. 6

> 'In whatever manner we please to imagine the nature of this element, whether we regard it as a distinct material thing which passes from one part of space to another, or whether we make heat consist simply in the transfer of motion, we shall always arrive at the same equations.' [1]

The dynamical theory did not entirely supersede the caloric theory until the equivalence of mechanical and heat energy was established in the mid-nineteenth century by Clausius, Helmholtz and Kelvin. The hypothesis of repulsions at a distance in the case of gas particles then became unnecessary, and was replaced by Bernoulli's hypothesis of random impacts. Forces at a distance were however still required to explain deviations from Boyle's law (Van der Waal's equation), and in the case of the heat energy of liquids and solids, for this was interpreted as the energy of random motion of their molecules in oscillations about positions of equilibrium under their mutual attractive forces.

During the second half of the eighteenth century, all these physical and chemical theories required a large number of various kinds of elastic fluid.[2] Sometimes aether, phlogiston, fire, light, and electric fluid are all identified; sometimes they are thought to be distinct. They have in common that they are subtle, highly elastic and usually imponderable. Most of these theories have left no trace on the subsequent progress of science, for they belonged to a period when the chemical properties of matter were little understood, before the interchangeability of forms of energy had been discovered, and before mathematical theories of heat, electricity, magnetism, and radiation had been developed. The fact that they could be suggested at all, however, indicates how radically the climate of thought had changed in regard to attractions and repulsions since the seventeenth century. Such forces were now intelligible models, not requiring to be introduced under cover of a mathematical positivism like that of the *Principia* or Cotes or Berkeley, in fact the influence of positivism is next found in the tendency to *abandon* elastic fluids and other particle models and to replace them by abstract mathematical theories, as in the theory of heat of Fourier and the theories of elasticity of Stokes and Green. There were still signs, though, even in the eighteenth century, that the idea of action exerted over very great distances was not much more acceptable than it had been, for the aether was still often

[1] *The Analytic Theory of Heat*, trans. Freeman, Cambridge, 1878, p. 464

[2] cf. J. R. Partington and D. McKie, 'Historical Studies on the Phlogiston Theory', *Annals of Science* II, 1937, p. 361 ; III, 1938, pp. 1, 337 ; IV, 1939, p. 113

thought of as an intermediary in the case of gravitation between distant bodies. Gowin Knight, for example, while accepting attractions and repulsions between particles of a fluid, explains the gravitation of planets towards the sun by assuming that the ' repellent atmospheres ' of the sun and the planets extend far enough to penetrate each other.[1] And J. Elliot, writing in 1780, compares the action of aether in causing gravitation with that of fire causing repulsion,[2] that is, aether is thought of as an ' atmosphere ' of the attracting body, carrying its attractive force to a distance.

The physics of the imponderable elastic fluids was largely qualitative, and their disappearance was accelerated by the increasingly mathematical aspect of theories from the beginning of the nineteenth century onwards. The demand for precision and falsifiability which had been heard in the seventeenth century in regard to Aristotle and Descartes, was now heard in connection with the multiplicity of theories of elastic fluids into which almost any combination of phenomena could be fitted by repeated modifications. Some remarks of Lavoisier about phlogiston might have been applied to them all :

> ' Chemists have turned phlogiston into a vague principle which is not at all precisely defined, and which consequently adapts itself to all the explanations in which it may be required : sometimes it has weight, and sometimes it has not ; sometimes it is free fire and sometimes it is fire combined with the earthy element ; sometimes it passes through the pores of vessels, sometimes these are impervious to it : it explains both causticity and non-causticity, transparency and opacity, colours and their absence. It is a veritable Proteus changing in form at each instant.' [3]

[1] op. cit., Prop. LXI, Corr. I
[2] Quoted in Partington and McKie, *Annals*, III, 1938, p. 352
[3] *Mém. de l'Acad. Roy. des Sciences*, 1783, p. 523

Chapter VIII

THE FIELD THEORIES

Euler's Hydrodynamics

THE principles of theoretical mechanics are stated in Newton's three laws of motion, but as they stand there they can be applied only to point masses subject to external forces. Their application to continuous solid or fluid bodies requires further physical hypotheses, and here Newton was not successful. In Book II of the *Principia* he attempted a theory of motion for fluid media, in connection with the resisted motion of bodies, the propagation of sound, and the motion of vortices. This was partly intended to refute the Cartesian theory of vortices as applied to the solar system, and to clear up difficulties about possible resistance to the planetary motions arising from an aether,[1] but as a contribution to mechanical theory it was less satisfactory than the mechanics of point masses and the gravitational theory itself in the other two books of the *Principia*.

Euler must be regarded as the real founder of the mechanics of continuous media. In his *Mechanica* of 1736 [2] he sets out a programme for the development of the mechanics of various kinds of systems: infinitely small bodies, already treated by Newton; finite rigid bodies; flexible bodies; bodies which extend and contract; several bodies with mutual interactions; fluids. Of these, Euler initiated the theory of rigid dynamics; was less successful with the next three topics; but eventually completed the elementary theory of motions of fluids.

One difficulty hindering the application of Newtonian mechanics to continuous media was that which is common to all problems of mathematical physics, namely that of knowing how far to abstract from the physical situation in order to set up a simple and yet useful mathematical model with a minimum of physical assumptions. A similar problem had been solved by Galileo in the case of falling bodies and by Newton himself in the case of the inverse-square law, but Newton was unable to find the right mathematical model

[1] *Principia*, Book II, Prop. LIII; Book III, Prop. X

[2] 'Mechanica sive motus scientia analytice exposita', §98, *Opera Omnia*, 2nd series, I, Leipzig and Berlin, 1912, p. 38

in the case of fluid motion, and alternated between the corpuscular and the continuous, introducing various unsatisfactory *ad hoc* assumptions. Euler, however, was able to apply Newton's own mathematical method to the solution of this difficulty. There is in his work the same combination of corpuscular speculation and mathematical positivism which is to be found in Newton, for on the one hand he rejects attraction at a distance, in language reminiscent of Leibniz, as an unnecessary multiplication of the fundamental powers of bodies, as a 'perpetual miracle', and as unintelligible and occult, and he endeavours to explain elasticity, cohesion, gravitation, and electric and magnetic forces in terms of differential aether pressures.[1] His theories of these phenomena are as indecisive as any of the corpuscular theories of the seventeenth century. On the other hand, his mathematical attack upon the mechanics of fluids was independent of all such speculative hypotheses, and here he rejected all forms of atomism and of action at a distance, and tried to show that the phenomena result from the properties of continuous media, mathematically described.

Previous work on fluids by Newton, the Bernoullis, and D'Alembert had been characterised both by dubious assumptions about their material constitution and by too much concentration on particular cases. Euler, however, hit upon one simple principle, namely, that a perfect fluid can sustain no shearing force, and this was sufficient to generate the theory of pure hydrostatics and hydrodynamics, and was at the same time a sufficiently close description of the properties of real fluids to provide a useful first approximation. At times Euler seems chiefly interested in developing his mathematical model, with little reference to physical actuality; for example, in his treatise on fluid mechanics of 1766, he writes:

> 'I am about to inquire into the laws of equilibrium and motion of this sort of bodies, in which the definition of fluidity given here holds [that is, those that have no resistance to shear], little troubled whether or not such bodies exist in the world.' [2]

And elsewhere he rejects the molecular theories of fluids, then much in vogue, as 'absolutely sterile' in providing the foundations

[1] See, for example, *Letters to a German Princess*, trans. Hunter, London, 1795, I, letters xi, lxviii ff, lxxvii (on gravity); II, letters xxiv ff. (on electric attraction), lxi ff. (on magnetism).

[2] *Opera Omnia*, XIII, p. 9; cf. in a modern work by J. L. Synge: '. . . a theory in mathematical physics is to be judged primarily on the basis of its logical consistency and only secondarily on the basis of the truth of its physical predictions' (*Relativity: the Special Theory*, Amsterdam, 1956, p. 165).

of fluid mechanics.[1] In the Editor's Introduction to the volume of Euler's works dealing with fluid mechanics, C. A. Truesdell compares the abstract simplicity of his mathematical model with the molecular models and experimental or operational methods of his predecessors. According to Euler

> 'ideal materials are to be *defined*; their consequent properties are then to be explored mathematically; and their range of appropriateness to specific physical materials is to be established later by comparison of the detailed predictions of theory with the results of measurements.'[2]

In contrast, Bernoulli's *Hydrodynamica*

> 'suffers from too close adherence to quantities which can be measured: it lacks imagination. Daniel Bernouilli's definition of pressure . . . is 'operational' . . ., while Euler's is not. But Daniel Bernoulli's concept of pressure cannot be extended to compressible fluids in motion. It is the intellectual and artistic theory of Euler, who did not waste his time on experiments, which led ultimately not only to the other field theories but also to the practical fluid mechanics of today: the theory of the turbine, the ship, and the airplane.'[3]

There are few better examples of the relation between operationalism and physical and mathematical models.

A second difficulty in the way of applying Newtonian mechanics to continuous media was that of mathematical technique, namely, to find the right translation of the equations of motion for point masses into equations for continua. The notion of dividing a continuous body into differential elements in order to find its volume or centre of gravity or resultant gravitational attraction had long been familiar, although it appears that the application of this method in finding the resultant attraction of a sphere had given Newton considerable trouble.[4] In these cases, however, the method was hardly more than geometrical, and the idea of writing down equations of motion for the elements of a continuous body, as if they were point particles, did not emerge until John

[1] *Opera Omnia*, 2nd series, xii, p. 3 [2] ibid., xiii, p. xi

[3] ibid., xii, p. xliv *n*

[4] *Principia*, Book i, Props. lxx–lxxvi; Book iii, Prop. viii. It was apparently this problem which held up the completion of Newton's theory for nearly twenty years, from about 1665, when he was convinced of the law of gravitation, until about 1684, when he proved that a sphere acts upon external masses as if its mass were concentrated at its centre.

Bernoulli used it in his *Hydraulica*,[1] and even then its validity did not appear obvious. Euler, however, at once saw that it provided the necessary new method for dealing with the continuum, and applied it in the dynamics of rigid bodies and in hydrodynamics.

It was in Euler's hydrodynamics, too, that the notion of a

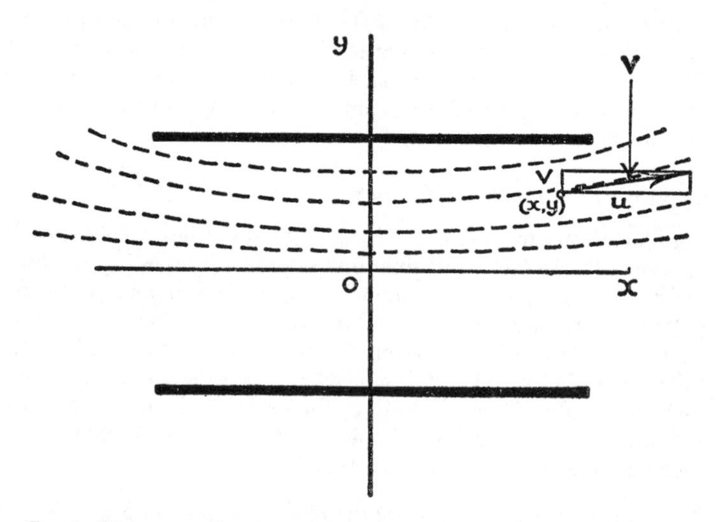

FIG. 4 Diagram to illustrate the two-dimensional field of flow in a slab of infinite width. The partial differential equation (in modern notation) to be satisfied by the vector velocity **v** at each point is

$$\left(\frac{\partial}{\partial t} + u\frac{\partial}{\partial x} + v\frac{\partial}{\partial y}\right)\mathbf{v} = \mathbf{F} - \frac{1}{\rho}\operatorname{grad} p$$

where **F** is the external force, ρ is the density, and p the pressure at the point (x, y).

field theory first took shape. A field in mathematical physics is generally taken to be a region of space in which each point (with possibly isolated exceptions) is characterised by some quantity or quantities which are functions of the space coordinates and of time, the nature of the quantities depending on the physical theory in which they occur. The properties of the field are described by partial differential equations in which these quantities are dependent variables, and the space and time coordinates are independent variables. In Euler's hands, hydrodynamics became a field theory, the field of motion of a fluid being characterised by the velocity

[1] Dated 1732, published 1743

of the fluid at each point, and the motion being described by partial differential equations involving the velocity components u, v, w, at the point (x, y, z) and time t.

The role of mathematical models as opposed to physical hypotheses can be illustrated again from Euler's theory of sound, which, like the theory of motion in liquids, had been confused by the difficulty of reconciling physical assumptions with mathematical formalism. The one-dimensional wave equation :

$$\frac{\partial^2 y}{\partial t^2} = a^2 \frac{\partial^2 y}{\partial x^2} \tag{i}$$

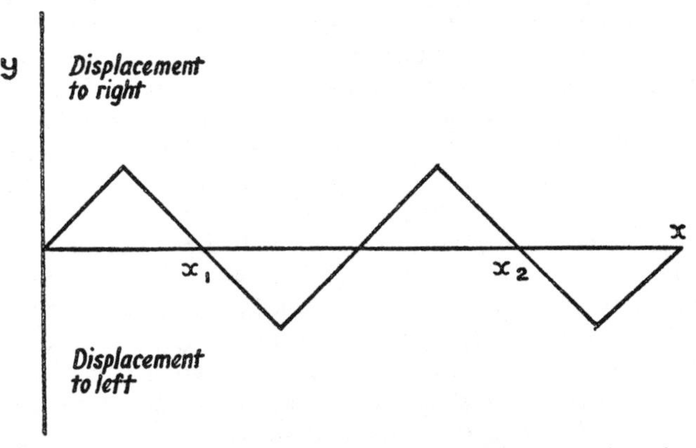

FIG. 5 A discontinuous periodic function. x is the distance along the elastic string, and y is the displacement along the string of the point which is at x in equilibrium. This periodic solution of equation (i) indicates a discontinuous change of density in the string at the points x_1, x_2, where the particles surrounding x_1, x_2 'pile up'.

had been derived in connection with the vibrating-string problem, by considering the string as made up of n particles, and then letting n become infinite, but although D'Alembert suggested that this could be applied to the propagation of sound, he also argued that if sound has the form of discontinuous pulses, as was generally believed, then a function representing such pulses cannot be a solution of the wave equation. Daniel Bernoulli again showed his preference for extending existing physical hypotheses rather than inventing abstract mathematical formalisms, by attempting to solve the problem in the discontinuous case by means of superpositions of simple harmonic oscillations, a procedure which only found its formal justification in the much later work of Fourier.

Euler, however, aided by Lagrange's independent justification of the use of discontinuous functions, published in 1759 an application of the wave equation to sound propagation, and maintained, in contradiction to Leibniz's principle of continuity, that since *any* periodic disturbance of the functional form

$$y = \Phi(x+at)+\Psi(x-at) \tag{ii}$$

will satisfy the equation, the disturbance may involve discontinuities, as illustrated for instance by Fig. 5. In this case, however, the wave equation is not satisfied at discontinuities of the functions in (ii), although it is satisfied everywhere else, and in his later writings Euler takes the bold, and formally correct, step of representing the propagation by (ii) rather than by (i), and obtaining from (ii) the foundations of his theory of sound with, as usual, the minimum of physical assumptions. The generalisation of the mathematics here is exactly equivalent, methodologically speaking, to generalisations of a theory by natural extensions of a picturable model.[1]

With these mathematical models for hydrodynamics and wave motion, Euler had made it possible to describe the transmission of action through continuous fluid media. He left unsolved the question of propagation through continuous solid media, that is, media which sustain a certain amount of shearing force and therefore have elasticity of shape as well as of volume. This question was raised acutely in the early years of the nineteenth century by the researches of Young and Fresnel on light. Newton's corpuscular theory of light had been generally adopted during the preceding century,[2] but Young reintroduced the wave theory, and suggested that polarisation effects, which had been thought to be a difficulty for the wave theory, might be explained by periodic *transverse* displacements of the aether particles, whereas sound waves are due to longitudinal displacements. Transverse displacements however can be propagated only in a solid medium, and so began the search, which was to last throughout the century, for mechanical models of a solid elastic aether. First a mathematical model for the motion of an elastic solid had to be set up, and here the problem was again to know how far abstraction could be made from particular molecular hypotheses about the constitution of matter, for such

[1] C. A. Truesdell, Editor's Introduction, *Opera Omnia*, 2nd series, XIII, p. xli ; and J. R. Ravetz, ' Vibrating Strings and Arbitrary Functions ', to appear in *Michael Polanyi Festschrift*, 1961, and ' The Representation of Physical Quantities in Eighteenth-Century Mathematical Physics ', to appear in *Isis*, 1961.

[2] Although Euler was a notable adherent of the aether-wave theory.

hypotheses could at that stage be based on little more than guesswork. The theory was first developed by Navier, Poisson, and Cauchy, who each considered matter to be molecular, and postulated certain forces between the molecules. Stokes,[1] however, was able to develop the equations in such a way that no molecular hypotheses were involved, and small elements of the solid were assumed to be homogeneous. He thus provided a mathematical model for continuous solid media as Euler had done for fluid media.

Criteria for Continuous Contact-action

We can now distinguish three ways in which mechanical action might be transmitted from one point to another, as they appeared in physics at the beginning of the nineteenth century. Each exhibited itself to common-sense observation in a familiar type of mechanical process, and each had a corresponding mathematical model in terms of which its most general features could be described.

1. Action by impact, in which elasticity is taken to be an ultimate property of bodies, as in Newton's law of elastic impact.

2. Action at a distance, as in the theories of gravitation and electric and magnetic attraction. Here it was universally assumed that the action is transmitted instantaneously, or at least that the effects of a finite velocity are so small as to be unobservable, for gravitational attractions were always taken to act along the line joining the simultaneous positions of two bodies, and this assumption was justified by correspondence with the astronomical facts. Laplace had calculated [2] that to ensure this correspondence the velocity of transmission, if finite, must be at least a hundred million times that of light.

3. Action in a continuous medium, described in terms of a field theory. In the fluid and elastic solid theories, matter is regarded as continuous, and no account is taken of small-scale atomicity,

[1] 'On the theories of the internal friction of fluids in motion, and of the equilibrium and motion of elastic solids' (1845), *G. G. Stokes's Mathematical and Physical Papers*, I, pp. 113ff.

[2] *Mécanique Celeste*, 1799, Book x, Chap. vii, 22; 'Sur le Principe de la Gravitation Universelle' (1773), *Œuvres Complètes*, VIII, pp. 219ff. Although Laplace speaks of an 'impulsion' of a fluid which causes motion towards the attracting body, and identifies the velocity of this fluid with the velocity of transmission of gravity, he does not suggest, like Euler, that gravity must be reducible to contact mechanism. In fact, in his *Exposition du Système du Monde* (1796) he says 'There is no need at all to posit vague causes, impossible to submit to analysis, and which the imagination modifies to its liking, in order to explain these phenomena' (Introduction to Book IV).

application to molecular matter being justified by some such rubric as 'Consider an element which is small compared to macroscopic dimensions, but large compared to single molecules', so that the effect of molecularity is smoothed out. But this particular device should not be allowed to obscure the main point of interest here, namely, that there exist mathematical models which can deal with action in strictly continuous media. The propagation of such action, unlike that of action at a distance, generally takes time, its velocity depending on the mechanical properties of the medium. Also, action in a continuous medium certainly satisfies the principle of contiguity, and it therefore offers an alternative mode of contact-action to the impact and pressure of traditional atomism.

During the nineteenth century, however, the question of action at a distance came to be profoundly modified by a new awareness of the relation between physical and mathematical models, and a consequent ambiguity about the status of field theories. The point can be illustrated simply from the theory of gravitation. The attraction at any point due to a given distribution of mass can be expressed as the force which would act on a unit point mass placed there, or, after developments due to Lagrange, Laplace, and Poisson, in terms of the gravitational potential V at the point, where V satisfies the Laplace equation

$$\frac{\partial^2 V}{\partial x^2} + \frac{\partial^2 V}{\partial y^2} + \frac{\partial^2 V}{\partial z^2} = 0$$

at points in empty space, or the Poisson equation

$$\frac{\partial^2 V}{\partial x^2} + \frac{\partial^2 V}{\partial y^2} + \frac{\partial^2 V}{\partial z^2} = -4\pi\rho$$

at points where the mass density is ρ. Thus both the attractive force and the potential are field functions. But there is a physical difference between a gravitational field of this kind and the velocity field of a fluid. In the latter case the field function is an actual property of material at every point of the field, but in the gravitational case the potential function V is 'potential' in the sense that it does not necessarily describe a material property of the field, for it may have a value in empty space; it describes a potential property, namely, the force that *would* be exerted *if* a small mass were introduced into the field at that point.

But the difference between the two types of field should be

looked at a little more carefully. The fluid field is said to be a continuous material medium because it has properties (other than velocity) which render it observable in ways familiar for ordinary liquids and gases ; there is no question of the fluid having been introduced as a hypothetical substantive for the verb ' to flow ' in the theory of hydrodynamics. Properties of the fluid other than its velocity are detectable and there is therefore no difficulty about saying that the fluid transmits action continuously from place to place. The gravitational field on the other hand apparently has no properties except the potentiality for exerting attractive force on masses introduced into it. It has no other detectable properties which would lead us to describe it as material, and our first reaction is to look upon the introduction of the function V as an *ad hoc* mathematical device describing action at a distance, but in no sense exhibiting it as continuous action through a material field. It might however be the case that gravitation is transmitted through the medium of an aether which, while not having the properties of ordinary matter, might nevertheless have some properties other than the potentiality of exerting attractive force, and in this case the potential V would be a property of this aether in exactly the same sense as velocity is a property of the material fluid. This suggests that an important consideration in deciding whether or not a field is to be regarded as a physically continuous medium rather than a mere mathematical device, lies in its possession of detectable properties other than the one property for which it was introduced. A condition of this kind is often suggested as a criterion of the physical ' reality ' of a theoretical entity, and it led Faraday to express his dissatisfaction with Newtonian gravitation, as we shall see later.[1]

But independent detection was not the only consideration which weighed with the nineteenth-century physicists. They were prepared to regard a field as a physically continuous medium on other and less stringent terms, for example, if propagation was affected by material changes in the intervening space, if it took time, if a mechanical model could be imagined for the action of a medium in producing the observed effect,[2] or if energy could be located in the space between interacting bodies. Any one of these three conditions might be regarded as sufficient and no one of them was individually necessary. Thus, gravitation remained an action

[1] cf. P. W. Bridgman, *The Logic of Modern Physics*, p. 56, and *infra*, p. 222.

[2] This condition was later seen to be vacuous, since *any* interaction satisfying the principle of least action can be represented by a mechanism in infinitely many different ways (*supra*, p. 5).

at a distance throughout the nineteenth century, in spite of its description by a potential theory, because it did not satisfy any of these criteria, whereas the electromagnetic field theory began to take on the characteristics of continuous action because it satisfied all of them. It is sometimes suggested in modern (post-relativity) works [1] that a finite velocity of propagation is a necessary as well as a sufficient condition for continuous action, and that this is why instantaneous gravitation could not be regarded as such an action, but in classical physics there is instantaneous transmission of pressure and of longitudinal waves in an incompressible medium, and yet this would certainly be regarded as continuous action.

Faraday : the Physical Nature of Lines of Force

It was Faraday who first suggested that action at a distance was inadequate as an account of electric and magnetic forces. Poisson had shown that a potential could be introduced in electrostatics in a way similar to the introduction of the gravitational potential, and although Faraday admitted that he was not mathematically competent to judge this treatment, he became convinced, on experimental rather than mathematical grounds, that the intervening medium must be regarded as the carrier of electric and magnetic action in a more than formal sense. He wished to picture concretely the events going on in the medium, and although he did not himself contribute to the mathematical field theory, his pictures of lines of force emanating from charged conductors and from magnets became the basis of extensions of that theory in the hands of Kelvin and Maxwell. His use of pictorial representations led him to discussion and reinterpretation of the meaning of action by contact and action at a distance, and this is worth some detailed study.

It was in regard to electric induction that Faraday first became convinced that action was propagated through a medium and not at a distance. His experimental researches led to three conclusions which seemed to point to the existence of an active medium :

[1] cf. M. Born (*Einstein's Theory of Relativity*, trans. Brose, London, 1924, p. 132) : 'in this treatment of electrostatics [by Laplace, Poisson, et al.] . . . we are not dealing with a true theory of contiguous action. . . . For the differential equations refer to the change in the intensity of field from place to place, but they contain no member that expresses a change in time. Hence they entail no transmission of electric force with finite velocity but, in spite of their differential form, they represent an instantaneous action at a distance.'

The view that the condition of finite velocity is sufficient for the *reality* of the field is expressed by Landau and Lifshitz, who remark (*The Classical Theory of Fields*, Cambridge, Mass., 1951, p. 41) that in relativity mechanics a particle creates a field in which action is propagated in time : 'This means that the field itself acquires physical reality', and 'Interactions can occur at any one moment only between neighbouring points in space (contact interaction).'

1. The induction of electric charge between conductors across an insulating medium depends quantitatively on the nature of the insulator.

2. If the insulator is cut and the parts separated, opposite charges appear on the two separated surfaces.

3. The lines of induction are curved, as illustrated by the spark of a discharge, and by experiments showing how the force on a charged ball due to a charged insulator is affected by the presence and shape of intervening conductors, which may make induction ' turn a corner '.[1]

Faraday concluded that the insulating medium propagates the electric induction by means of its own particles, each of which is itself a conductor and becomes polarised, one side having a negative charge and the other a positive charge. This action takes place between contiguous particles along curved lines, and Faraday thinks that the fact that the lines are curved is '*strong proof*' that induction is ' an action of continuous particles affecting each other in turn, and not an action at a distance.'[2] In the *Experimental Researches* of 1837 he speaks of induction as an action between ' contiguous particles ' and adds a note a year later to explain more carefully what he means by ' contiguous ' :

> 'The word *contiguous* is perhaps not the best that might have been used here and elsewhere ; for as particles do not touch each other it is not strictly correct. . . . By contiguous particles I mean those which are next.'[3]

So that ultimately it seems that even this action is at a distance if regarded on the atomic scale. Faraday is explicit about this in a correspondence with Dr Hare of Pennsylvania beginning in 1839.[4] Dr Hare had objected that rarefication of the air between conductors does not affect the transmission of electric induction, and that therefore the material medium cannot be essential. He suggested that an imponderable matter should be postulated, so that the polarisation of this matter between conductors would ' connect the otherwise imperfect chain of causes '. Faraday replies that his use of ' contiguous ' includes a vacuum in which the particles of air may be separated by distances of the order of half an inch, but he will not commit himself here on the hypothesis of an imponderable

[1] *Diary*, London, 1933, III, pp. 72, 87 : *Experimental Researches in Electricity and Magnetism*, London, I, 1839, §1224

[2] *Diary*, III, p. 88 [3] *Exp. Res.*, I, §1164*n* [4] ibid., II, 1844, pp. 251ff.

aether. Two years earlier, however, he had made the same suggestion: 'May inductive action be transmitted by other particles than those of ponderable matter, as by the particles of the supposed ether?'[1]

Physically, action at a distance on the atomic scale is here retained by Faraday, because he regards each particle of the intervening medium as being itself a conductor, and the properties of insulators as arising from the assumption that the conducting particles are *not* in contact with one another—if they were, there would be no insulators. But five years later he has realised that if this argument is applied to conductors, it implies some highly paradoxical conclusions, and he is led to make a fundamental attack upon the current view of the atomic constitution of matter which gives rise to them. In a letter of 1844 published in the *Philosophical Magazine* entitled 'A speculation touching Electric Conduction and the Nature of Matter',[2] he characterises this current view as follows: atoms have a certain volume and are endowed with powers which hold them together in groups, but they do not touch. Thus only space is continuous throughout matter considered as an aggregate of atoms. It follows, as we have seen above, that in insulators, space must be an insulator, but in conductors, if their atoms do not touch, space must be a conductor. Such a contradictory conclusion is absurd and shows that the current theory must be false. Faraday now goes on to close the escape-route suggested by the possibility that in conductors atoms *do* touch. He shows that, in the pure metal potassium, 45 atoms take up the same volume as is occupied by 70 atoms of potassium plus 210 of oxygen and hydrogen when they are combined in potassium hydrate. Thus even if the atoms of the hydrate are very closely packed, those of the metal must be some distance apart, and yet the metal potassium is a good conductor.

Still, the phenomena of physics and chemistry seem to require atoms as centres of force, and this being so, as little as possible should be assumed about them. This principle of economy, together with the contradiction involved in assuming that an atom is an extended 'little unchangeable, impenetrable piece of matter' leads Faraday to the view, which he ascribes to Boscovich, that an atom is a *point* with 'an atmosphere of force grouped around it'.[3] The properties of a body, such as conduction, relation to light or magnetism, solidity, hardness, specific gravity, must then belong, not to a 'nucleus' abstracted from its powers (for this is in any case inconceivable), but to the forces themselves.

[1] *Diary*, III, p. 213 [2] *Exp. Res.*, II, p. 284 [3] ibid., p. 290

> 'But then surely the *m* [the atmosphere of force] is the *matter* of the potassium, for where is there the least ground (except in a gratuitous assumption) for imagining a difference in kind between the nature of that space midway between the centres of two contiguous atoms and any other spot between these centres? A difference in degree, or even in the nature of the power consistent with the law of continuity, I can admit, but the difference between a supposed little hard particle and the powers around it I cannot imagine.'

It follows, in contrast to the orthodox view, that 'matter' is everywhere continuous and that 'atoms' are highly elastic and deformable, mutually penetrable, and that

> 'matter fills all space, or, at least, all space to which gravitation extends . . . for gravitation is a property of matter dependent on a certain force, and it is this force which constitutes the matter. . . . This, at first sight, seems to fall in very harmoniously with . . . the old adage, "matter cannot act where it is not".' [1]

This paper seems to mark a decisive transition from continuous action understood mechanically to continuous action understood in terms of forces filling space.[2] The transition was foreshadowed in Boscovich and Kant, but Boscovich does not equate his matter with its force, and still regards his theory in terms of action at a distance, while Kant, though equating matter with *repulsive* force, distinguishes between this, which acts by contact, and *attractive* force which acts at a distance. Maxwell interprets Faraday's work as a replacement of concepts of action at a distance by continuous action in this sense:

> 'Faraday, in his mind's eye, saw lines of force traversing all space where the mathematicians saw centres of force attracting at a distance: Faraday saw a medium where they saw nothing but distance: Faraday sought the seat of the phenomena in real actions going on in the medium, they were satisfied that they had found it in a power of action at a distance impressed on the electric fluids.' [3]

[1] ibid., p. 293

[2] I am indebted to Dr J. Agassi for pointing out to me that in a previous article 'Action at a Distance in Classical Physics', *Isis,* 46, 1955, p. 337, I did not sufficiently emphasise that in this and some of his other papers, Faraday himself initiated the view of 'physically real' force pervading space, even in the case of gravity (see also *infra*, p. 222). This is the view that I called 'mathematical' in my article, in contrast to the 'mechanical' view that the parts of matter are extended and distinct from empty space. Here I have consistently referred to the 'field' view in contrast to the mechanical.

[3] *Treatise on Electricity and Magnetism*, Oxford, 1873, I, p. x

And again :

> 'This [conception of lines of force] is quite a new conception of action at a distance, reducing it to a phenomenon of the same kind as that action at a distance which is exerted by means of the tension of ropes and the pressure of rods.' [1]

Faraday's attitude towards speculative hypotheses has an ambivalence similar to that of Newton. He follows up his speculations regarding atoms, in an address to the Royal Institution, 'Thoughts on Ray Vibrations' in 1846, where he suggests that, instead of regarding a quasi-material aether as the carrier of light vibrations, these should be seen as vibrations in the lines of force which, in his view, constitute matter and pervade all space. But he apologises for these speculations :

> 'I do not think I should have allowed these notions to have escaped from me, had I not been led unawares, and without previous consideration, by the circumstances of the evening on which I had to appear suddenly and occupy the place of another.' [2]

Elsewhere he is at pains to point out that use of the term 'lines of force' does not commit him to any particular theory of their nature :

> 'The term *line of magnetic force* is intended to express simply the direction of the force in any given place, and not any physical idea or notion of the manner in which the force may be there exerted ; as by actions at a distance, or pulsations, or waves or a current, or what not.' [3]

The lines of force here are simply what would be represented by iron filings scattered in the field, and Faraday uses their patterns as what we have called a model, to suggest further developments :

> 'It would be a voluntary and unnecessary abandonment of a most valuable aid, if an experimentalist, who chooses to consider magnetic power as represented by lines of magnetic force, were to deny himself the use of iron filings. By their employment he may make many conditions of the power, even in complicated cases, visible to the eye at once, . . . By their use probable results may be seen at once, and many a valuable suggestion gained for future leading experiments.' [4]

[1] 'Action at a Distance', *Scientific Papers of James Clerk Maxwell*, II, p. 320
[2] *Exp. Res.*, III, 1855, p. 452
[3] ibid., p. 402, cf. pp. 328ff., 368ff.
[4] ibid., p. 397

Although Faraday wishes fundamentally to replace the dualism of matter and force by a conception of all-pervading but continuously differentiated force, the distinctions marked in common speech by 'matter' and 'space' must remain, though, in this view, on a less fundamental level. Hence the question arises as to how the forces acting in apparently empty space are to be described. Faraday uses a terminology derived from the idea of a bundle of elastic strings stretched under tension, and also phrases like 'conducting power' which express an analogy between magnetic force and electric current. But here again all that is intended is a convenient way of stating the fact that iron concentrates the lines of force by saying that iron is a better conductor of magnetic force than is air. The analogy cannot necessarily involve the further statement that something is travelling along the lines of force as charges travel along lines of current flow, although it may suggest that this is a profitable hypothesis. In modern terminology the word 'tension' has been retained in speaking of lines of force, while the phrase 'conducting power' has not, but both are 'dead metaphors' in the sense that they do not involve any important *physical* analogy between elastic strings and lines of force, or between electric currents and lines of force. There are important *mathematical* analogies which were worked out later on the basis of Faraday's suggestions, and which we shall consider in connection with the work of Kelvin and Maxwell.

Faraday: Criteria for Action at a Distance

The various forces propagated through space have different characteristics, and Faraday compares these in the orthodox terminology of 'action at a distance' and 'continuous action', without implying anything about the ultimate nature of matter and force, in papers of 1851 and 1852. Here he gives important experimental criteria for the distinction between action at a distance and continuous action, arising out of what is known of the electric and magnetic forces. We have seen how the properties of electric induction led him to the view that this is not an action at sensible distances, but requires the intervention of the material medium. He is concerned to discover whether the same can be said of magnetic action:

> 'How the magnetic force is transferred through bodies or through space we know not:—whether the result is merely action at a distance, as in the case of gravity; or by some intermediate

agency, as in the cases of light, heat, the electric current, and (as I believe) static electric action.' [1]

In a paper entitled 'On the physical character of the lines of magnetic force' (1852), in which Faraday remarks that he is leaving 'the strict line of reasoning' and entering upon 'a few speculations',[2] he suggests some criteria by which different kinds of action may be recognised :

(i) Can transmission of action be affected by changes in the intervening medium, as regards, for instance, a bending of the lines, or polarity effects ?
(ii) Does the transmission take time ?
(iii) Does it depend upon the 'receiving' end ?

These questions are answered with respect to gravity, radiation, and electric and magnetic force. First, with respect to gravity :

(i) Nothing in the intervening medium affects a line of gravitational force between two particles. The line is straight, no matter what other particles may be in the field, and the action between any pair of particles is independent of that between any other pair.[3]
(ii) It seems impossible to prove whether or not gravity takes time. 'If it did, it would show undeniably that a physical agency existed in the course of the line of force.'
(iii) The action of gravity is dependent upon the mass of both reacting particles and their distance apart. 'So gravity presents us with the simplest case of attraction ; and appearing to have no relation to any physical process by which the power of the particles is carried on between them, seems to be a pure case of attraction or action at a distance, and offers therefore the simplest type of other cases which may be like it in that respect.'

Second, with regard to radiation :

(i) Lines of radiation are affected by the properties of the intervening medium both in curvature and in transverse orientation about their axis (polarity).
(ii) They require time for their propagation.
(iii) They are not dependent upon a second reacting particle. Here 'we obtain the highest proof, that though nothing ponderable

[1] ibid., p. 330 [2] ibid., p. 407
[3] Faraday has not understood that the resultant gravitational attraction at various points of a number of mass particles may be represented by a curved line of force, just as in electrostatics.

passes, yet the lines of force have a physical existence independent, in a manner, of the body radiating, or of the body receiving the rays.'

Thirdly, with regard to electric induction :

(i) Lines of electric induction are affected by the material medium, but it is not certain whether, in a vacuum, they would be straight like those of gravity, or curved. No condition of polarity has been observed.
(ii) No time has been shown to be required for their propagation.
(iii) A second reacting particle is required.

Fourthly, with regard to electric current :

(i) Current is affected by the medium as regards direction and quantity, and it is essentially related to a *material* medium.
(ii) Time is required for propagation even in good conductors.
(iii) The lines of flow are either limited as in a discharge, or endless and continuous. In both cases the current depends upon two extremities, as for instance two charged conductors, or the plates of a voltaic cell.

There are thus three types of forces exerted over a distance :

Gravity, ' where propagation of the force by physical lines through intermediate space is not supposed to exist ' ;
Radiation, ' where the propagation does exist, and where the propagating line or ray, once produced, has existence independent either of its source, or termination ' ; and
Electricity, ' where the propagating process has intermediate existence, like a ray, but at the same time depends upon both extremities of the line of force.'

Is magnetic action like any of these ? Have the lines of magnetic force a physical existence, and if so, is it static like electric induction, or dynamic like an electric current ? Faraday answers his three questions in the case of magnetic lines of force as follows :

(i) They have not been shown to be affected in any way by any medium other than iron. On the other hand it seems that the external lines must be curved in the case of a rectangular bar magnet in vacuum, since they begin at one pole and end at the other, and Faraday ' cannot conceive curved lines of force without the conditions of a physical existence in that intermediate space.

If they exist, it is not by a succession of particles, as in the case of static electric induction, but by the condition of space free from such material particles.'

(ii) No time has been shown to be required for propagation of magnetic action.

(iii) The lines are dependent on opposite poles at their extremities.

Magnetic lines of force have many properties in common with those of electric induction and current, and can probably be said to be ' real ' in the same sense. The chief evidence for this Faraday takes to be their curvature, and the fact that current is induced in a circuit by mere motion in a magnetic field. He remains agnostic as to the precise state of matter or aether which accounts for them, whether a current or a stress or any other modification of the medium.

In 1854, after Kelvin had shown the mathematical equivalence of various ways of representing magnetic action and the mathematical analogy between heat flow, current flow, and electric and magnetic lines of force, Faraday declared himself strengthened in his view that the lines of force represent something physically real[1]; in other words he was prepared to take mathematical analogy with other physical processes as evidence for physical reality.

Maxwell : Mechanical and Field Theories of Continuous Action

Beginning with the work of Faraday and his mathematical successors, one must distinguish clearly between two ways of considering the problem of action at a distance, ways which may be called respectively the mechanical view, and the field view.

The mechanical view is inherited from the physics of the seventeenth century, and considers the question of action in terms of a theory of substance or matter having mechanical properties : extension, duration, motion, mass, force. In the nineteenth century mechanical models for theories are still sought, but it is clear that in terms of these no answer to the mechanical problem of action is possible. The reason for this is that the mechanical models are no longer thought of as literal descriptions of entities existing in nature, but only as interpretations, in terms of mechanical devices, of phenomena that are described mathematically but whose ultimate nature cannot be regarded as crudely mechanical.

Take, for example, Maxwell's mechanical model of the electromagnetic aether. He accounts for the propagation of electro-

[1] *Exp. Res.*, III, p. 529

magnetic effects by a quasi-material elastic medium in which tubes of magnetic force are vortex filaments causing tension in the medium along their length and pressure laterally. The vortex motion is made possible by 'idle' particles between one vortex and another; the flux of these particles in a conductor represents electric current; their displacement in an insulating medium produces dielectric effects. This model applies to so-called free space, or aether, as well as to the interior of matter, and Maxwell showed that his equations of the electromagnetic field can be derived from it, giving the propagation of electromagnetic disturbances with the velocity of light.

If this model were intended as a description of the ultimate particles of nature, it would be difficult to say whether it involves action at a distance or not. The equations of the model are those of a continuous elastic fluid medium, but the question of whether this medium is ultimately continuous or discrete is left undecided, because it depends on more detailed knowledge of the molecular constitution of matter and aether. If matter and aether ultimately consist of discrete simple atoms, the chances are that some action at a distance, over atomic distances at least, will have to be postulated, otherwise it becomes impossible to account for the cohesion of solids. But Maxwell is concerned only to explain 'action between distant bodies without assuming the existence of forces capable of acting directly at *sensible* distances.'[1] On the question of whether aether is discrete or continuous, Maxwell remarks:

> 'It is often asserted that the mere fact that a medium is elastic or compressible is a proof that the medium is not continuous, but is composed of separate parts having void spaces between them. But there is nothing inconsistent with experience in supposing elasticity or compressibility to be properties of every portion, however small, into which the medium can be conceived to be divided, in which case the medium would be strictly continuous. A medium, however, though homogeneous and continuous as regards its density, may be rendered heterogeneous by its motion, as in Sir W. Thomson's hypothesis of vortex-molecules in a perfect liquid. . . . The aether . . . is probably molecular, at least in this sense.[2]

[1] 'A Dynamical Theory of the Electromagnetic Field' (1864), *Scientific Papers*, I, p. 527 (my italics—M. B. H.)

[2] 'Ether', *Scientific Papers*, II, p. 774. Ultimate elasticity of a continuous medium was exactly the suggestion made by Leibniz (*supra*, p. 161), but neglected owing to lack of mathematical expression.

This illustrates the sort of difficulty that arises when one attempts to take the mechanical models as literal descriptions of nature, and to wring from them an answer to the question whether action at a distance occurs in nature or not. One generally becomes involved in an infinite regress : action between bodies at finite distances is explained by continuous stress in the intervening medium ; this stress is explained by the molecular constitution of the medium, which may itself involve action at a distance, and so on. There are other examples of the same sort : Helmholtz introduced terms representing viscosity into his equations of aether motion, but, as Kelvin pointed out,[1] viscosity produces thermal motion of the particles themselves and presumably viscous forces between aether particles would, if the idea is to be taken literally, involve thermal motion of particles within the particles. Again, in Lorentz's theory of the electron, the force on a macroscopic charged particle is replaced by forces on its constituent electrons, and then forces on charges *within electrons* are spoken of.[2]

Such language is legitimate if it is merely a way of speaking about the appropriate mathematical equations, but no solution to the problem of action at a distance is to be found by taking such mechanical models literally. Maxwell himself is careful to explain the status of his molecular vortex model :

> ' I propose now to examine magnetic phenomena from a mechanical point of view, and to determine what tensions in, or motions of, a medium are capable of producing the mechanical phenomena observed.'
>
> ' The conception of a particle having its motion connected with that of a vortex by perfect rolling contact may appear somewhat awkward. I do not bring it forward as a mode of connexion existing in nature.[3]

And in a later paper :

> ' I have on a former occasion attempted to describe a particular kind of motion and a particular kind of strain, so arranged as to account for the phenomena. In the present paper I avoid any hypothesis of this kind ; and in using such words as electric momentum and electric elasticity in reference to the known phenomena . . . I wish merely to direct the mind of the reader

[1] *Baltimore Lectures*, Cambridge, 1904, pp. 91ff.
[2] H. A. Lorentz, *Theory of Electrons*, Leipzig, 1909, pp. 13, 14
[3] ' On Physical Lines of Force ' (1861), *Scientific Papers*, I, pp. 452, 486

to mechanical phenomena which will assist him in understanding the electrical ones. All such phrases in the present paper are to be considered as illustrative, not as explanatory.

In speaking of the Energy of the field, however, I wish to be understood literally.' [1]

On the other hand, the field aspect of the problem of action at a distance became increasingly important as the nature of mechanical models came to be better understood. The problem may be said to have been that of reinterpreting action at a distance and action by contact so that the concepts remained relevant to a physics whose fundamentals were becoming more abstract and less mechanical, and whose structure was more easily understood in terms of mathematical than of mechanical models.

Kelvin showed in a series of mathematical papers beginning in 1842 [2] that the same mathematical formalism could be used to express the laws of fluid flow, of heat flow, of electric and magnetic phenomena, and of elasticity. Thus, a source of fluid or of heat is the analogue of an electric charge, magnetic pole, or source of electric current ; lines of flow are analogues of lines of force ; temperature is an analogue of potential, and so on. Kelvin also showed that Faraday's representations in terms of lines of force were consistent with the inverse-square law, and that this followed without assuming any physical hypothesis about the nature of the lines of force. These lines are mathematically defined when the distribution of centres of force is known.

Kelvin remarks that no physical hypothesis follows from the fact of these analogies. Fourier did not deduce that heat is a material fluid from the laws of heat flow, and Coulomb did not deduce ultimate attractions and repulsions at a distance from the inverse-square law, but the analogies are bound to suggest that, if heat is propagated from particle to particle in a continuous medium then electric and other actions may be propagated in a similar manner. Maxwell is equally cautious in his use of the analogies ; he proposes to treat lines of force as if they were lines of flow of an incompressible fluid, but he adds :

'The substance here treated of must not be assumed to possess any of the properties of ordinary fluids except those of freedom of motion and resistance to compression. It is not even a

[1] 'A Dynamical Theory of the Electromagnetic Field', ibid., p. 563

[2] Sir W. Thomson (Lord Kelvin), *Papers on Electrostatics and Magnetism*, London, 1872, pp. 1, 15, 42, 52, 340 ; *Mathematical and Physical Papers*, Cambridge, 1882, I, p. 76

hypothetical fluid which is introduced to explain actual phenomena. It is merely a collection of imaginary properties which may be employed for establishing certain theorems in pure mathematics in a way more intelligible to many minds and more applicable to physical problems than that in which algebraic symbols alone are used.' [1]

Even Faraday's conceptions are essentially mathematical :

'As I proceeded with the study of Faraday, I perceived that his method of conceiving the phenomena was also a mathematical one, though not exhibited in the conventional form of mathematical symbols. . . . Faraday's methods resembled those in which we begin with the whole and arrive at the parts by analysis, while the ordinary mathematical methods were founded on the principle of beginning with the parts and building up the whole by synthesis.[2]

Thus, the notion of potential developed by Laplace, Poisson, Green, and Gauss finds its proper interpretation in terms of Faraday's theory, whereas in the conventional mathematics of the inverse-square law and action at a distance, potential has to be regarded as a summation of the effects of individual particles.

In general, opinion in physics has, following Maxwell, regarded field theories derived from the analogies with fluid flow and with elastic media as the new type of continuous-action theory. Continuous action has now come to mean that each point of space can be characterised by certain mathematical quantities representing the energy present there without implying that any mechanical events are happening :

'. . . we may regard Faraday's conception of a state of stress in the electro-magnetic field as a method of explaining action at a distance by means of the continuous transmission of force, even though we do not know how the state of stress is produced.' [3]

And before long physicists ceased to ask how the state of stress is produced in a mechanical sense, or even to allow that the question has any meaning. The stress-field throughout space became fundamental : the field was not to be explained in terms of matter, matter was rather a particular modification of the field. Larmor, for example, writing in 1900, remarks :

[1] 'On Faraday's Lines of Force' (1856), *Scientific Papers*, I, p. 160
[2] *Electricity and Magnetism*, I, pp. x, xi
[3] 'Action at a distance', *Scientific Papers*, II, p. 321

'It is not superfluous to repeat here that the object of a gyrostatic model of the rotational aether is not to represent its actual structure, but to help us to realise that the scheme of mathematical relations which define its activity is a legitimate conception. Matter may be and likely is a structure in the aether but certainly aether is not a structure made of matter.' [1]

Faraday's criteria for continuous action began to be accepted as sufficient, namely, that the action should be affected by material changes in the intervening space, that its propagation should take time, and that, as became explicit in Maxwell, electromagnetic energy could be located in the field and shown to be transformable into other kinds of energy according to the conservation law [2]: 'In speaking of the Energy of the field . . . I wish to be understood literally.'

Poynting later developed this conception of a substantial field-energy.[3] He remarks that in the theories of Faraday and Maxwell, the energy of the field is not simply carried along by the currents, but resides in the intervening medium. This must follow, if continuity of motion is asserted, because, if a particle be placed in the field at a point previously empty of matter, it may acquire a kinetic energy, and this energy must come through the surrounding space. 'The alternative that it appeared in the body without passing through the space immediately surrounding the body need not be discussed.' The alternative would violate the conservation of energy, because if electromagnetic action is propagated between bodies in a finite time, and if its energy is not present in the intervening free space in the interval, it is not conserved during that interval. Poynting's concept of energy-flux enabled descriptions to be given of the flow of energy between the electromagnetic field and other interacting physical systems, including its transformations into heat, chemical, kinetic, and other forms of energy. Thus energy could be seen to be in some respects similar to a fluid in satisfying equations of conservation and continuity, although the energy-distribution cannot in fact be defined uniquely over the field, and there are various difficulties in describing the flow in some physical situations. Maxwell was able also to associate momentum with the electromagnetic field, so that momentum as well as energy

[1] *Aether and Matter*, Cambridge, 1900, p. vi *n*

[2] The distribution of energy cannot in general be defined uniquely for any given field, but the total energy associated with the field has a definite value.

[3] 'On the transfer of energy in the electromagnetic field', *Phil. Trans. Roy. Soc.*, CLXXV, 1884, p. 343; 'On the connection between electric current and the electric and magnetic inductions in the surrounding field', ibid., CLXXVI, 1885, p. 277

is conserved in the passage of action from one point to another, and this was later confirmed experimentally in the phenomenon of radiation pressure.

Hertz : Interpretations of Maxwell's Equations

To admit with Maxwell that the 'reality' of the field consists in the presence of energy within regions devoid of matter, and that it is the mathematical equations describing the field which are important, is not to abandon all concrete pictures of what is going on in the field. Maxwell's theory, for example, involved the conception of a 'displacement current' in free space which completes the current circuit in any case where, as in the charging or discharging of a condenser, charge is moving in material dielectrics in an apparently open circuit.[1] Since the aether is here being regarded as a polarised dielectric, by analogy with material dielectrics, in the sense that similar equations describe the behaviour of both, it is possible to have alternative pictures of the distribution of energy both in material dielectrics and, analogously, in the aether. In his Introduction to the collection of papers *Untersuchungen über die Ausbreitung der Elektrischen Kraft*,[2] Heinrich Hertz gave a careful analysis of the various possible interpretations of action across free space, ranging from direct action at a distance to action carried entirely by the aether. He distinguishes four standpoints :

1. 'There is a kind of spiritual affinity'[3] between the reacting bodies in the sense that force is present only when there are two or more bodies. This is the conception of Coulomb's inverse-square law, and is almost abandoned in electricity, although it is still used for gravitation.

2. It is assumed that one body by itself 'continually strives to excite at all surrounding points attractions of definite magnitude and direction' with which 'we fill the surrounding space', but no change is assumed in the space itself, and hence the distance forces are still supposed to be independent of any medium. This is the standpoint of the potential theory.

3(*a*). Force is transmitted both by action at a distance and by polarisation of small parts of the medium, which then in turn act at a distance. This, says Hertz, is Helmholtz's interpretation, and here polarisation depends on a medium, but distance forces do not.

[1] *Electricity and Magnetism*, I, §62

[2] German ed., 1892 ; 1st English ed., *Electric Waves* (trans. D. E. Jones), London, 1893

[3] *Electric Waves*, p. 22

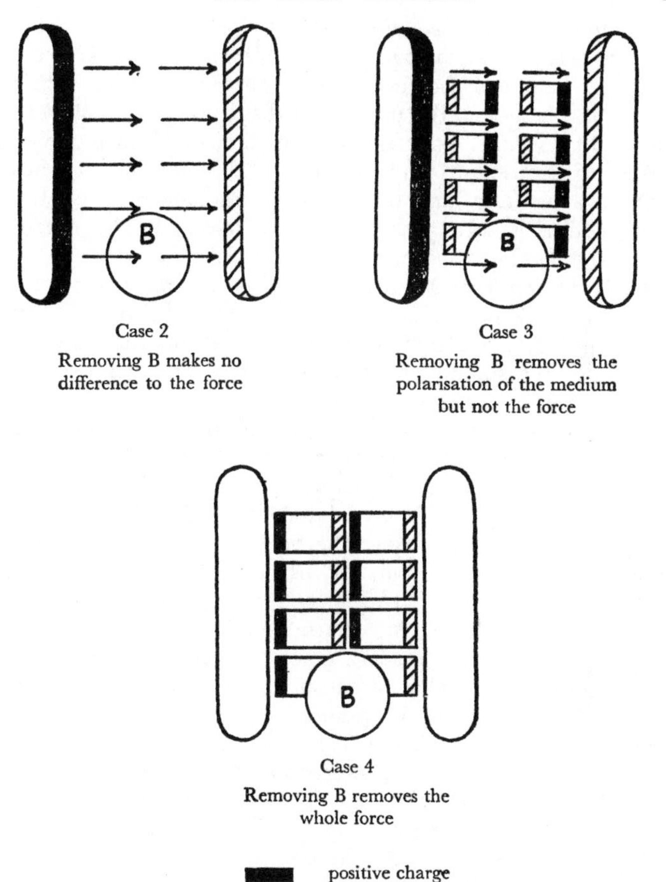

Fig. 6 Hertz's interpretations of electric induction

(*b*). There is, however, a limiting case of this interpretation in which all the energy of interaction is assumed to be present in the medium and none in the charges acting at a distance. The mathematical expression of this case leads to Maxwell's equations, but it does not coincide with Maxwell's physical interpretation.

4. There are no distance forces; only polarisations are really

present, and these may be describable by some mechanical hypothesis. The mathematical expression of this point of view is identical with that of 3*b*, but physically it is quite different, for here electricity (assumed as fundamental) is not regarded as the cause of polarisation of the medium, rather polarisation is fundamental and regarded as the cause of electricity. Thus nothing at all would happen in the absence of a medium, and the ordinary terminology of electricity and magnetism is merely shorthand for describing the effects of action in the medium whatever these may be.[1]

Hertz considers that Maxwell himself uses both conceptions 3*b* and 4 in ways which sometimes appear contradictory, but that his developed thought should be understood in the sense of 4, which is also the standpoint Hertz adopts in his own theoretical account of Maxwell's theory.

It should be recognised that none of these four viewpoints imply that any mechanical account of the structure of the aether can be given. In fact in 4 the field in 'free space' has come to be regarded as more fundamental than the bodies whose behaviour led empirically to the theory of the field. But more fundamental in what sense? Hertz admits in his comments on 4 that there is a physical difference between the pictures 3*b* and 4, and it would seem natural to extend the notion of 'model' to include, not only the mechanical models which had been discredited, but also such electrical analogies between the aether and material dielectrics as are suggested by these pictures. Hertz, however, returns to the positivist view of the early nineteenth century, and regards the pictures as genuine alternatives only in so far as they are distinguishable by empirical test. Thus the three interpretations, of Helmholtz (the limiting case 3*b*), of Maxwell (a mixture of 3*b* and 4), and his own (4), are the *same* theory, since they are all expressed by Maxwell's equations :

> 'This, and not Maxwell's peculiar conceptions or methods, would I designate as 'Maxwell's Theory'. To the question, 'What is Maxwell's theory?' I know of no shorter or more

[1] There appears to be a contradiction between the directions of polarisation of particles of the aether in 3*b* and 4. But the diagram for 4 is misleading, since in this case there are *no* particles, but only a state of the aether giving rise to positive charge on the left-hand plate and negative charge on the right-hand plate. In the case of a material dielectric, where there *are* discrete particles, the pattern of diagram 4 will repeat itself in the aether between each particle, so it is beside the point to claim as did Duhem (*Les Théories Électriques de J. C. Maxwell*, Paris, 1902, pp. 124ff.) that 4 is falsified by experimental results on dielectrics. What has happened is that the aether is no longer pictured as analogous in this respect to actual dielectrics.

definite answer than the following:—Maxwell's theory is Maxwell's system of equations. Every theory which leads to the same system of equations, and therefore comprises the same possible phenomena, I would consider as being a form or special case of Maxwell's theory; every theory which leads to different equations, and therefore to different possible phenomena, is a different theory.' [1]

Later he apologises for the somewhat abstract and colourless appearance of his own rendering of the theory, in which he has tried not to introduce any conceptions which are experimentally untestable, but

'If we wish to lend more colour to the theory, there is nothing to prevent us from supplementing all this and aiding our powers of imagination by concrete representations of the various conceptions as to the nature of electric polarisation, the electric current, etc. But scientific accuracy requires of us that we should in no wise confuse the simple and homely figure, as it is presented to us by nature, with the gay garment which we use to clothe it. Of our own free will we can make no change whatever in the form of the one, but the cut and colour of the other we can choose as we please.' [2]

This is as good a statement as any to be found in the writings of a physicist of the doctrine that the essential meaning of a scientific theory is exhausted by its testable content. Hertz, however, overlooks the impossibility of distinguishing *a priori* between the 'simple and homely figure', and its clothes. It can never be certain that there are not testable consequences of one of his three forms of Maxwell's theory which differ from those of the others, or that there is not another interpretation of the equations which leads to different consequences from any of them. The latter in fact proved to be the case when Maxwell's equations were incorporated into the theory of relativity. But Hertz need have looked no further than the history of the conceptions of electric action he describes under 1, 2, and 3(*a*) for examples of the essential difference between theories whose testable consequences were at a certain stage identical. This can be illustrated by considering two problems of the mid-nineteenth-century electrical theory, namely, the force between current-elements, and the time of propagation of electromagnetic effects.

[1] *Electric Waves*, p. 21 [2] ibid., p. 28

The Continental Action-at-a-Distance School

During the years in which Faraday and Maxwell were developing the electromagnetic field theory, the notion of direct action at a distance according to the inverse-square law still remained fundamental on the Continent. In contrasting the Continental action-at-a-distance theories with those of Faraday and himself in which there is ' action through a medium from one portion to the contiguous portion ', Maxwell suggests that

> ' The comparison, from a philosophical point of view, of the results of two methods so completely opposed in their first principles must lead to valuable data for the study of the conditions of scientific speculation.' [1]

The Continental methods go back to experimental and theoretical developments connected with the mutual action of magnets and electric charges. In 1820 Oersted published his discovery of the effect of a current-bearing wire on a pivoted magnet in its vicinity. It appears that many years had elapsed before he recognised this effect, although he had long been searching for some evidence of connection between the powers of electricity and magnetism, but always on the assumption that any force between them must act along the line joining pole to charge. When the effect was discovered, however, it showed that an electric current exerts a force *at right angles* to its own direction, causing the magnet to take up a position perpendicular to the plane of the current circuit. This was the first apparent experimental proof that forces at a distance may be other than direct attractions or repulsions, and Oersted himself was inclined to explain it by postulating vortices of electric matter surrounding the current and exerting force on the magnet.

Shortly after this discovery, Ampère showed experimentally that not only does a current circuit act upon a magnet, but that two current circuits also act upon each other, and that the actions of a small plane circuit and a magnet directed at right angles to it are identical. Then in a brilliant theoretical investigation,[2] Ampère derived an expression for the force acting between two small current elements, regarding this as the fundamental formula of electrodynamics, derived directly from experiments and thus analogous

[1] *Electricity and Magnetism,* II, §502

[2] ' Théorie Mathématique des Phénomènes Électro-dynamiques ', *Mém. de l'Institute,* VI, 1823, p. 175

to the law of gravitation in mechanics. His intention was to show that the forces acting between current elements are directed, as in gravitation, solely along the line joining them, and for this purpose he assumes no hypotheses about their 'physical causes', but by arguments concerning symmetry between the elements, guided by experimental results, he arrives at a formula which permits him to express the mutual force as a direct attraction or repulsion, together with an oblique force whose value is indeterminate because the experimental conditions involve the use of at least one closed current instead of two current elements. Ampère held, however, that this oblique force can be considered to be zero.

Ampère's expression for the force contained merely the geometrical relations of the conducting elements and the currents flowing in them. Attempts were now made to interpret the terms in the formula by means of physical hypotheses concerning the transport of charge in the conductors, but in which the intervening medium was ignored, action being supposed to take place instantaneously at a distance. The most usual hypothesis about the current was that of Fechner, according to which currents always flow in conductors in pairs: a current of positive electricity in one direction, and an equal current of negative electricity having the same velocity in the opposite direction. Making this assumption, Gauss and Weber derived from Ampère's formula two different expressions for the force involving the velocity of the currents in each element. It was not clear, however, that such expressions would satisfy the principle of conservation of energy, for in the proof of conservation for distance forces given by Helmholtz in his paper 'On the Conservation of Force' (1847),[1] he had assumed that forces between parts of a system are attractions and repulsions (central forces) depending only on the relative positions of the centres and not on their velocities, and had stated that non-central forces and forces depending on velocity must violate the conservation principle. Gauss's formula was immediately shown to violate it, and in the case of Weber's, although in general it was consistent with conservation, Helmholtz was able to describe particular circumstances under which perpetual increases in the energy of a system would be obtained.[2] Hence neither of these formulae nor others derived from similar procedures could be regarded as satisfactory, and it was later shown that, although Ampère's theory

[1] Translated in *Scientific Memoirs*, ed. Tyndall and Francis, London, 1853, p. 114. Helmholtz assumes in this proof that propagation of distance forces is instantaneous.

[2] 'On the Theory of Electrodynamics', *Phil. Mag.*, XLIV, 1872, p. 530

gave adequate results for closed currents, nevertheless, if current *elements* exist, then the force between them must contain components other than those along the line joining them.[1]

The Faraday-Maxwell method, on the other hand, depended on seeing the forces between currents in terms of the magnetic field produced by the currents in the intervening medium. Since this involved current circuits regarded as equivalent to magnetic shells with their boundaries coincident with the circuit, it did not give results for current elements, and therefore did not provide a means of deciding between the various expressions for the mutual force, which were in any case all equivalent for closed currents. According to Maxwell's theory, however, there are only closed currents, since any apparently discontinuous current is completed by the displacement current within the dielectric of the condenser, even if this is free space. This assumption of displacement currents in free space appeared very dubious to Maxwell's contemporaries, and its eventual acceptance was largely due to the discovery that Maxwell's equations, containing the displacement term, are relativistically invariant, and that they therefore provide a satisfactory electromagnetic theory within the framework of the special theory of relativity. Thus eventually not only electric and magnetic energy, but also current, in the generalised sense of Maxwell's equations, are shown to exist in the absence of matter or charge.

In this case Maxwell's leap beyond the immediate induction from experiments characteristic of Ampère's method turned out in the end to be fully justified, but it did not at first provide an answer to Ampère's original problem. Helmholtz in fact expressed his satisfaction that Maxwell's theory

> 'proves that there is nothing in electrodynamic phenomena to compel us to attribute them to an entirely anomalous sort of natural forces, to forces depending not merely on the situation of the masses in question, but also on their motion.'[2]

But the problem soon assumed a new importance, and at the same time became experimentally tractable, when the motion of electric bodies in high vacua began to be studied. J. J. Thomson then calculated[3] from Maxwell's theory the mutual forces of two moving charged particles, on the assumption that they are spherical, and

[1] For an assessment of these theories, see J. J. Thomson, 'Report on Electrical Theories', *Report of the British Association*, 1885, p. 97.

[2] *Phil. Mag.*, XLIV, 1872, p. 532

[3] 'On the Electric and Magnetic Effects produced by the Motion of Electrified Bodies', *Phil. Mag.*, XI, 1881, p. 229

in 1889 Heaviside gave [1] a corrected formula based on the same methods, namely,

$$\mathbf{F} = e\mathbf{E} + \frac{e}{c}\mathbf{v}_\wedge \mathbf{H},$$

for the force on a moving charge e with moderate velocity $\mathbf{v}$ in an electromagnetic field **E, H.** The same expression was obtained in 1892 by H. A. Lorentz following a suggested formulation of Ampère's expression by Clausius,[2] in which the force depended on the absolute velocities of the moving charges, not only on their relative velocities as in the formula of Weber. This force formula was added by Lorentz to the four equations of Maxwell to provide a theory in which electromagnetic effects were understood as being due to discrete moving charges.

To complete the solution of Ampère's problem, all that was required was the derivation from Maxwell's theory of the potentials due to a moving charge, and this was published by A. Liénard in 1898 and by E. Wiechert in 1900. The Liénard-Wiechert potentials in combination with the Lorentz law of force show finally that the resultant force between two moving charges is *not* in general directed along the line joining them.

Thus Ampère's problem was solved in terms of both the Faraday-Maxwell and Continental traditions, but at the expense of arbitrary assumptions which were different in the two cases. From the point of view of Maxwell's theory, it was somewhat unnatural to have to discuss discrete electric charges,[3] and the significance of regarding elementary charges as small charged conducting spheres was far from clear. On the other hand, the procedure by which the Clausius and Lorentz formulae were derived from Ampère's theory had an arbitrariness of its own, even though discussion of discrete charges was more in keeping with Ampère's point of view than that of Maxwell.

[1] 'On the Electromagnetic Effects due to the Motion of Electrification through a Dielectric', *Phil. Mag.*, XXVII, 1889, p. 324

[2] See Whittaker, *Aether and Electricity, The Classical Theories*, pp. 392–6

[3] In the chapter on Electrolysis in *Electricity and Magnetism*, I, §260, Maxwell finds himself driven to postulate a natural unit of electricity, the same for all chemical substances. But this is how he describes the notion of 'one molecule of electricity':

'This phrase, gross as it is, and *out of harmony with the rest of this treatise*, will enable us at least to state clearly what is known about electrolysis, and to appreciate the outstanding difficulties. . . . It is extremely improbable that when we come to understand the true nature of electrolysis we shall retain in any form the theory of molecular charges, for then we shall have obtained a secure basis on which to form a true theory of electric currents, and so become independent of these provisional theories' (my italics—M. B. H.). Faraday was also reluctant to have to consider atoms even as point-singularities in his continuous force-field; cf. *Exp. Res.*, II, p. 289.

These later developments presupposed the propagation of electric action with the velocity of light, and the growth of this conception provides another example of the contrasting methods of the medium and the action-at-a-distance schools. In his great papers of 1861 and 1862 [1] on the mechanical conception of the electromagnetic field, in which Maxwell developed his field equations, he demonstrated that disturbances are propagated in the field with velocity *c*, where *c* is a constant which appears in the equations as the ratio of the current in electrostatic units, to its equivalent magnetic moment per=unit=area in electromagnetic units. In 1856 Kohlrausch and Weber had measured this constant,[2] and found it to be close to the value of the velocity of light, a 'coincidence' remarked in the following year by Kirchhoff, who discovered that disturbances are propagated along telegraph wires with this velocity.[3]

On finding that electromagnetic waves are propagated through free space with a velocity close to that of light, Maxwell made his momentous identification of light and electromagnetic radiation. Maxwell was not, however, the first to suggest that the propagation of electric action in a finite time is of cardinal importance for electromagnetic theory. Faraday had suspected it, although in 1852 he considered that it had not been shown experimentally. As early as 1845, Gauss had written to Weber that he regarded as the corner-stone of electrodynamics the demonstration that actions are propagated between electric particles in time, like those of light. He had not himself succeeded in demonstrating this, and expresses his 'subjective conviction that it will be necesssary in the first place to form a consistent representation of *how* the propagation takes place'.[4]

Gauss's Continental successors were not able to provide this consistent representation, for they still preferred to speak in terms of action at a distance, although there were attempts by Riemann, C. Neumann and Betti to develop somewhat *ad hoc* mathematical expressions for the propagation of potential in time, without postulating any medium. These theories were criticised on mathematical grounds by Clausius,[5] and on methodological grounds by Maxwell, who remarks that their persistent adherence to action

[1] 'On Physical Lines of Force', *Scientific Papers*, I, p. 451

[2] *Ann. der Phys.*, XCIX, 1856, p. 10

[3] *Ann. der Phys.*, C, 1857, p. 210; *Phil. Mag.*, XIII, 1857, p. 406

[4] *Werke*, V, p. 629

[5] 'Upon the new Conception of Electrodynamic Phenomena suggested by Gauss', *Ann. der Phys.*, CXXXV, 1868, p. 606; *Phil. Mag.*, XXXVII, 1869, p. 445

at a distance must be due to an *a priori* objection to an intervening medium :

> ' Now we are unable to conceive of propagation in time, except either as the flight of a material substance through space, or as the propagation of a condition of motion or stress in a medium already existing in space. In the theory of Neumann, the mathematical conception called Potential, which we are unable to conceive as a material substance, is supposed to be projected from one particle to another, in a manner quite independent of a medium, and which, as Neumann has himself pointed out, is extremely different from that of the propagation of light. . . .
>
> But in all of these theories the question naturally occurs :— If something is transmitted from one particle to another at a distance, what is its condition after it has left the one particle and before it has reached the other ? ' [1]

By the time Maxwell wrote these words, however, his notion of a medium had been widely accepted on the Continent, and workers there were advancing on the basis of his own theory. L. Lorenz had arrived independently at field equations practically equivalent to Maxwell's,[2] although without his conception of the part played by the medium, and had derived solutions representing the so-called *retarded potentials*, that is, scalar and vector potentials propagated in a vacuum with speed c, so that their arrival at a distance r from their source takes place at a time r/c after their emission. In two papers of 1875 and 1876,[3] giving another formulation of the force between moving charges, Clausius accepts the propagation of electric action through a medium, and remarks that in this case absolute velocities of the charges may be involved as well as their velocities relative to each other (the medium being supposed to be absolutely at rest), also actions other than those along the line joining the charges may be assumed, and energy need not be conserved by the motions of the charges alone, for the effect of the medium has to be taken into account. Helmholtz had also come to recognise the importance of the Faraday-Maxwell method, and in 1879 he encouraged Hertz, then his pupil, to begin the series of researches which led to experimental verification of Maxwell's theory of electromagnetic waves. But in regard to the mechanical models of Kelvin and Maxwell, Helmholtz says :

[1] *Electricity and Magnetism*, §866
[2] *Ann. der Phys.*, CXXXI, 1867, p. 243 ; *Phil. Mag.*, XXXIV, 1867, p. 287
[3] *Phil. Mag.*, I, 1876, pp. 69, 218

' English physicists . . . have evidently derived a fuller satisfaction from such explanations than from the simple representation of physical facts and laws in the most general form, as given in systems of differential equations. For my own part, I must admit that I have adhered to the latter mode of representation and have felt safer in so doing ; yet I have no essential objections to raise against a method which has been adopted by three physicists of such eminence.' [1]

In these two examples of the law of force and the velocity of propagation, the action-at-a-distance and field theories were originally expressed in equivalent mathematical formalisms, in the sense that nothing could be said in one without being immediately translatable into the other, and that both led to the same experimental results. But their natural development was very different : in the case of the force-law, the action-at-a-distance school provided the framework of discussion within which the Lorentz law of force was eventually accepted, but it was a somewhat arbitrary addition to Maxwell's theory ; while on the other hand the finite velocity of propagation emerged naturally from Maxwell's theory, but could only be injected *ad hoc* into the Continental theories. It is not a very profitable exercise to discuss which school contributed more to electromagnetic theory, but it is historically and philosophically important to demonstrate that two theories may be different in essential respects in spite of equivalence of form and identity of experimental content. To deny this is to forget that science grows, and that it is dangerous to assume that the last word has been said about any theory.

Gravitation

At the end of the century, gravitation still remained outside the electromagnetic synthesis which seemed in principle to have comprehended the rest of physics. Faraday had argued in his paper of 1852 that gravitation appeared to exhibit pure action at a distance, but in 1855 he expressed his conviction that if this were the case, it violated what he called the principle of ' conservation of force ' or of ' power '. Newton himself did not accept action at a distance as sufficient, and, Faraday goes on, ' I cannot help believing that the time is near at hand, when his thought regarding gravity will

[1] Preface to H. Hertz, *Principles of Mechanics* (trans. Jones and Walley), London, 1899 (first German ed., 1894), p. xx. The ' three physicists ' referred to are Kelvin, Maxwell, and Hertz himself.

produce fruit '.[1] According to the action-at-a-distance view, the sun and the earth have no gravitational power when entirely removed from each other, but this power suddenly arises when they are in relation to each other. But Faraday considers ' That a body without force should raise up force in a body at a distance from it, is too hard to imagine ; but it is harder still, if that can be possible, to accept the idea when we consider that it includes the *creation of force* '. There are only three possibilities consistent with the conservation of force :

1. That ' the gravitating force of the sun, when directed upon the earth, must be removed in an equivalent degree from some other bodies, and when taken off from the earth (by the disappearance of the latter) be disposed of on some other bodies ', but no such thing has been observed.

2. That ' it must take up some *new* form of power when it ceases to be gravitation, and consume some other form of power when it is developed as gravitation ', but Faraday himself has undertaken experiments with the object of connecting electricity with bodies moving in a gravitational field, with entirely negative results,[2] and the idea has never been suggested by others.

3. That ' it must be *always* existing around the sun through infinite space '. This is the only possibility remaining : ' This case of a constant necessary condition to action in space, when as respects the sun the earth is *not* in place, and of a certain gravitating action as the result of that previous condition when the earth *is* in place, I can conceive, consistently, as I think, with the conservation of force : and I think the case is that which Newton looked at in gravity ; is, in philosophical respect, the same as that admitted by all in regard to light, heat, and radiant phaenomena ; and . . . is that now driven upon our attention in an especially forcible and instructive manner, by the phaenomena of electricity and magnetism '.[3]

In a paper entitled ' On the Conservation of Force ', published in the same year, Faraday has a quantitative argument. When two bodies approach under their mutual gravitational force, this force increases fourfold when the distance is halved, and the momenta of the bodies are also increased. Where does this power come from ?

[1] ' On some Points of Magnetic Philosophy ', *Exp. Res.*, III, p. 571
[2] ' On the possible relation of Gravity to Electricity ' (1850), ibid., p. 161
[3] ibid., pp. 573, 574

It is very unlikely 'that there should be a power of gravitation existing by itself, having *no relation to the other natural powers, and no respect to the law of conservation of force*'.[1] Faraday is not clear about the distinction between force and energy (he speaks, for example, of the disappearance of 'force' when ice thaws), but in some supplementary considerations added to this paper a year later, he seems to be answering the objection that potential energy is introduced to balance the energy difference in the example he quotes. But potential energy is an energy of the *attracted body* and disappears when that body disappears, whereas Faraday says he intends 'force' to mean, not 'tendency of a body to pass from one place to another', but the *cause* of the gravitational power in space, and so mere satisfaction of the energy equation is not sufficient for him.

But no independent evidence of such a cause could be found; neither was there any satisfactory model of gravitational attraction in terms of aether action, although there were a great many attempts during the nineteenth century to construct *ad hoc* theories of gravitation in terms of mechanical aether motions, some of them based on the theory of Lesage of 1747. In 1864 Maxwell says he has given up the attempt to find a model for gravitation similar to that for the electromagnetic field, because where there is attraction between *like* bodies the energy of the field (potential energy) decreases as the resultant gravitational force increases and the bodies approach with increasing velocity. Thus where there is no resultant gravitational force the medium must possess enormous energy, and

> 'As I am unable to understand in what way a medium can possess such properties, I cannot go any further in this direction in searching for the cause of gravitation.'[2]

In the last published work of his short life, Hertz suggested that apparent mechanical actions at a distance, such as gravitation, should be reducible to the motions of a medium consisting of rigidly connected 'concealed masses', as electric actions had been shown to be reducible to concealed polarisations, but he was unable to pursue the suggestion in detail.[3] In the Preface to the English edition of Hertz's *Electric Waves*, Kelvin expresses his opinion that:

[1] 'On the Conservation of Force', *Experimental Researches in Chemistry and Physics*, London, 1859, p. 443; reprinted in *Correlation and Conservation of Forces*, ed. Youmans, New York, 1885, p. 376

[2] *Scientific Papers*, I, p. 571

[3] *Principles of Mechanics*, pp. 25–41

> 'Absolutely nothing has hitherto been done for gravity either by experiment or observation towards deciding between Newton and [John] Bernoulli, as to the question of its propagation through a medium and up to the present time we have no light, even so much as to point a way for investigation in that direction.' [1]

It was not even clear whether gravity was propagated in time. Larmor, for instance, writing in 1900, asserts that it is known that its speed of transmission 'if finite at all, enormously transcends that of radiation'.[2] The influence of Laplace's calculations to this effect was still strong, but in spite of them, there had been suggestions towards reconciling a speed of propagation of gravity of the order of c with astronomical facts. Laplace had assumed that if gravity were propagated with finite velocity it would show aberrational effects like those of light, but since the development of theories of non-instantaneous action between moving charges, it was possible and more plausible to compare the interaction of gravitating masses with these rather than with sources emitting light. Laws of force like those of Weber and Riemann were investigated in the case of gravitation, and shown to lead to propagation approximately along the line joining the *simultaneous* positions of two moving bodies, that is, with hardly any aberrational effect, so that the success of Newton's assumption of instantaneous propagation could be accounted for as an adequate approximation to the hypothesis of propagation with velocity c.

In general, however, there was no theoretical reason to pursue these speculations until the advent of the theory of relativity. At the end of the nineteenth century gravitation was understood no better than in the seventeenth century, although in the meantime it had served as an essential model for electromagnetism, and had been outgrown by it. Ampère's law of force for charges was often compared to Newton's for masses,[3] and there is a similarity of method as well as of brilliance of achievement. But Ampère's work found its Maxwell within forty years, whereas it was over two hundred before Einstein performed a similar service for Newton.

[1] *Electric Waves*, p. xii [2] *Aether and Matter*, p. 188

[3] cf. Ampère, *Mém. de l'Institute*, VI, 1823, pp. 175ff. ; Weber, *Ann. der Phys.*, LXXIII, 1848, p. 193, and R. Taylor's *Scientific Memoirs*, V, 1852, p. 489 ; Maxwell, *Electricity and Magnetism*, §528

Chapter IX

THE THEORY OF RELATIVITY

Interpretations of the Michelson-Morley Experiment

ALTHOUGH often regarded as part of ' modern physics ', the theory of relativity is from most points of view rather the culmination of classical physics, for it resolves many of the difficulties inherent in the original formulation of electromagnetism, and introduces field concepts into its theory of gravitation, thus continuing the nineteenth-century trend towards field models. The other main ingredient of modern physics, namely quantum theory, on the other hand, has made fundamental changes in the basic physical model, and requires the notions of particle as well as field, and discreteness as well as continuity.

The experimental basis of the special theory of relativity is generally taken to be the Michelson-Morley experiment. This was not the only experiment which found its most convenient explanation in terms of that theory, for there were others concerning the optical and electromagnetic properties of moving bodies which led in the same direction, but since the Michelson-Morley experiment is familiar and comparatively simple, it will be convenient to use it in an analysis of the logical status of the theory.

The experiment is well known and need not be described in great detail.[1] Briefly, its object was to attempt to detect the velocity of the earth through the aether by measuring differences in the velocity of light relative to the earth in different circumstances to a high degree of accuracy. Since the velocity of light relative to the aether (supposed equivalent to absolute ' empty ' space) was shown by Maxwell's theory to be approximately $c = 3 \times 10^{10}$ cm./sec., and since the earth may reasonably be assumed to be in motion with respect to the aether in at least part of its orbit, it was calculated that the velocity of light relative to the earth should differ from c in small quantities of the second order in v/c where v is the velocity of the earth. The Michelson-Morley experiment enables quantities of this order to be detected by an interference method. Light from a source L is partly transmitted and partly reflected at a half-

[1] The first paper of Michelson and Morley published in England is in *Phil. Mag.*, xxiv, 1887, p. 449.

silvered mirror P. Each beam is then reflected at another mirror, one at S_1 and the other at S_2, where $PS_1 = PS_2$. They combine at P and are viewed at Q. Interference fringes will be observed, and their positions can be changed by slightly altering the relative lengths PS_1 and PS_2. If now the apparatus is rotated continuously, and if the light velocity along PS_1 and back is different from that along PS_2 and back for most orientations, due to the velocity of the earth through the aether, then regular changes in the positions of

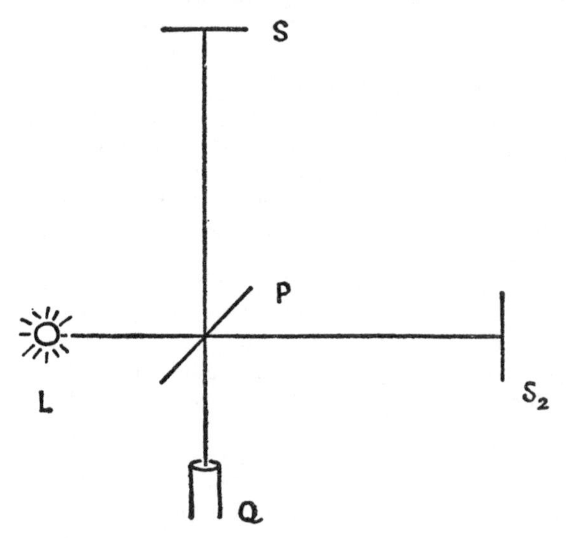

FIG. 7 The Michelson-Morley Experiment

the interference fringes should be observed during the rotation. The result of the experiment, repeated on many occasions, was, however, negative, no shift in the interference fringes being detected. Four possible explanations of this result were examined :

1. *The convection hypothesis*, that the earth in its motion carries the neighbouring aether along with it, at its own velocity.

This had already been investigated by Stokes and Hertz, and had been shown to be inconsistent with the optical behaviour of moving transparent bodies, which required that aether inside matter is not wholly carried along by the motion of the matter.

2. *The ballistic hypothesis*, that the velocity of light is always c relative to its source, so that its velocity in closed paths between the relatively

stationary points P, S_1, S_2 would show smaller variations than those to be expected on the aether-wave theory. This hypothesis implies, however, that the velocity of light would vary with the velocity of its source, and experiments have so far indicated that this is not the case.

3. *The Fitzgerald-Lorentz hypothesis*, that the lengths PS_1, PS_2 are affected by the motion of the apparatus relative to the aether.

This suggestion was made by Fitzgerald in 1892.[1] If it is assumed that all bodies are contracted in the direction of their motion relative to the aether in the ratio $\sqrt{(1-v^2/c^2)} : 1$, then it will not be possible to detect the velocity of the earth in the aether by the Michelson-Morley or any other optical or electromagnetic experiment, for the assumed changes in the relative velocity of light will be exactly compensated by the change of distance over which the light travels. This contraction would not, of course, be measurable independently of the results of these experiments, because measuring rods laid alongside the lines PS_1, PS_2 would suffer the same contraction, and so would appear to indicate a constant length. Lorentz then showed further that if all electromagnetic phenomena in moving systems take place exactly as if the systems were at rest in the aether, not only does the Fitzgerald contraction have to be assumed for moving bodies, but time also has a different measure in systems moving with respect to one another; that is to say a moving clock will appear to run slow compared to one at rest, although the two clocks will agree when there is no relative motion between them. Lorentz derived the equations, known as the Lorentz transformation, which express length and time measures in one system in terms of those in another moving relatively to it with uniform velocity, and according to these a rod moving in a coordinate system S appears to an observer at rest in S to be shorter than it does to an observer moving with the rod, and a clock moving in S appears to the stationary observer to run slow.

It was possible to reconcile this hypothesis with classical electrodynamic theory by ascribing the Fitzgerald contraction to electromagnetic forces intrinsic to a body moving relatively to the aether, but this device was somewhat *ad hoc* because it entailed that motion in the aether is in principle unobservable. Now the impossibility of measuring a quantity postulated in a physical theory does not always mean that the quantity is meaningless, as some operationalists have held, for the unobservable quantity may be an essential

[1] Reported by Lodge at a meeting of the Physical Society, *Nature*, XLVI, 1892, p. 165

ingredient in a theory which is supported by experiment in other ways, but the absence of a means of measurement does suggest that the quantity may be fulfilling no function in the theory, and that therefore it ought to be erased. For a long time the hypothesis of an aether was thought to be necessary because indirectly supported by the evidence for the existence of light waves in empty space, but as the notion grew in the nineteenth century that physical fields could exist in empty space without immediate association with matter, it became more plausible to suggest that electromagnetic waves were merely periodically varying electric and magnetic fields, and that no vibrating material need be postulated to support them. The abandonment of the aether as a useful physical concept was thus a consequence of the failure to find acceptable material models for it and of the realisation that only mathematical models were required, and from this point of view, the impossibility of measuring velocity relative to the aether was merely a confirmation that the aether was fulfilling no function in physical theory. This brings us to the fourth, and more fundamental, method of dealing with the Michelson-Morley experiment ; a method which is equivalent to the Fitzgerald-Lorentz hypothesis in that it leads to the same transformation equations, but is logically more satisfactory in avoiding the *ad hoc* appearance of that hypothesis.

4. *Einstein's hypothesis*, that there is no aether providing an absolute standard of rest, and that the velocity of light in empty space is always c relative to any moving coordinate system.

Since a postulate of the constancy of the velocity of light is a necessary and sufficient condition for the impossibility of velocities greater than c in the special theory of relativity, and hence for the impossibility of instantaneous action at a distance, it is important to understand the status of this conclusion from the Michelson-Morley experiment. Clearly, it immediately explains the negative result of that experiment, in other words it is a sufficient condition, but it is not a necessary condition, for it cannot be deduced from the result of the experiment without further assumptions which we shall now consider.

The theory of the experiment as outlined above rests on the following assumptions and measurements :

(i) With the apparatus stationary, the absence of variation in time of the length difference (PS_1-PS_2) was ensured by the stationary position of the interference fringes. This observation eliminated the possibility of the result being masked by heat effects, air-currents, and so on, all of which had been avoided by careful design of the apparatus. The assumption here is that path-difference causes shift in the fringes, and this could be verified directly by fine adjustment of the position of S_1 on the line PS_1.

(ii) A time-difference in the return of the two beams of light to Q would be indicated by a fringe shift. The expected value of this shift had to be calculated accurately to ensure its observability, and this calculation involved Euclidean geometry and the classical wave-theory of light. Since time-differences of this order of magnitude could never be checked operationally against any mechanical clock, time is here being measured effectively by phase-shift, and hence the classical theory is assumed in taking ' phase-shift time ' to be the same as ' clock-time '.

(iii) The absolute value of the lengths PS_1 and PS_2, which was required in the calculation, was measured to a sufficient accuracy by a meter scale.

(iv) The experiment was not designed to detect an absolute difference in the velocity of light of the order v^2/c^2 at different periods of the year. It could detect only a difference in velocity in any two directions on one and the same occasion. Arago's experiment in 1810 to detect a shift in the focus of a lens after a six-month interval had revealed no first-order variation of the velocity of light over this period, but no second-order experiment of this kind had been devised.

The most conservative hypothesis about the invariance of the velocity of light that can be extracted from these assumptions and the results of the Michelson-Morley experiment is as follows :

H_1. The velocity of light in a closed path containing relatively stationary mirrors is constant in all directions in a local Euclidean reference frame moving with the surface of the earth.

The velocity c is here defined as in classical optics, absolute length being measured by meter scales and small length- and time-differences by phase shifts.

That H_1 is not a ' deduction from phenomena ' even in Newton's sense is sufficiently shown by the other hypotheses that were suggested to explain the Michelson-Morley experiment. In Newton's case, if the relevant experimental generalisations were accepted together with the Rules of Reasoning, the law of gravitation followed, but in this case, if the result of the Michelson-Morley experiment and some general assumptions about economy and space-time homogeneity corresponding to Newton's Rules are accepted, H_1 does not follow deductively, and there is logically a choice between hypotheses, guided but not forced by other experimental results.

Although H_1 is a sufficient condition for the Michelson-Morley result it is not the hypothesis which Einstein in fact adopted as the foundation of the special theory of relativity.[1] He states two postulates, which can be expressed as follows :

1. There is no unique standard of rest. In particular, the velocity of light is invariant, in whichever one of a set of uniformly moving coordinate systems it is measured.

2. Any light-ray moves in a given system of coordinates with velocity c, for any uniform motion of the light-source in that system.

Since it cannot be assumed in a discussion at this fundamental level that the intuitive notion of time-intervals between events at different places is adequate, Einstein gives an 'operational' definition of simultaneity and time-interval at different places as follows : suppose time-intervals at different points of a given coordinate system are measured by clocks of similar construction ; we may then synchronise these clocks by means of light-signals. A emits a light ray at time t_A by A's clock, it is received and reflected by B at time t_B by B's clock, and returns to A at t'_A by A's clock. Then B's time t_B is *defined* to be simultaneous with A's time $\frac{1}{2}(t'_A+t_A)$. This definition makes the velocity of light the same in directions AB and BA in virtue of the measure of time-interval employed, and so, when extended to any pair of observers in the system, it makes H_1 true by definition. The relevance of the Michelson-Morley experiment is then to show that the definition is consistent with measurement of time-difference in the return of two light-rays to one point by phase-shift. It follows, of course, that all other experiments which are more coarse in their time-measurements than this will also have results consistent with the definition.

If we now take from Postulate (1) the special case of invariance of the velocity of light, we may formulate a new fundamental hypothesis for special relativity theory as follows :

H_2. The velocity of light is c with respect to any inertial reference system, for any uniform motion of the light-source in that system.

It is H_2 which is of interest as the necessary and sufficient condition for the impossibility of velocities greater than that of light.

[1] In his paper 'Zur Electrodynamik bewegter Körper', *Ann. der Phys.*, XVII, 1905, p. 891

In order to examine the empirical status of H_2, we may note that various logical gaps between H_1 and H_2 can now be filled by taking account of the results of other experiments, and by making explicit various further assumptions. For example :

1. H_2 refers only to inertial reference frames, that is, the frames which are defined by the Newtonian law of inertia to be those in which that law holds, Euclidean geometry being assumed. Locally to the solar system inertial frames are at rest, at least approximately, with respect to the 'fixed stars', or in uniform motion with respect to them. Now the coordinate frame assumed in describing the Michelson-Morley and other similar experiments is not inertial, since it is carried along on the surface of the earth, but it can plausibly be assumed that the effect of this acceleration is negligible compared to the velocities involved in the experiments.[1] Again, it is assumed that the physical scales and clocks used for measurement in an ideal inertial frame are subject to no forces, whereas these experiments take place in the gravitational field of the earth. Any gravitational effect has to be neglected in interpreting the experiments.

2. The result of the Michelson-Morley experiment refers only to the constancy of c in all directions in one reference frame, not to its constancy in all frames at any time, whatever their velocity. The ease with which Einstein made this generalisation illustrates the importance of theoretical background in the interpretation of experiments. No theoretical considerations led Einstein to expect a second-order variation of c in time or space simply, nor a variation due to one value of the velocity of the earth in its orbit rather than another ; no experiments had been devised at that time to detect such variations, and there was no difficulty in assuming that they did not occur. There were, on the other hand, theoretical reasons for supposing a second-order variation between different directions in one framework, and it was the absence of this that had to be explained.

Some thirty years later this particular gap between H_1 and H_2 was partly bridged by an experiment of Kennedy and Thorndike.[2] This was similar to the Michelson-Morley experiment, but the technique was refined in such a way that variation of the interference fringes over long periods of a year or more could be observed.

[1] J. L. Synge has questioned this assumption : see 'Effects of acceleration in the Michelson-Morley experiment', *Sci. Proc. Roy. Dublin Soc.*, XXVI, 1952–4, p. 45.

[2] *Phys. Rev.*, XLII, 1932, p. 400

All disturbing factors were reduced to negligible proportions, and no variation of the fringes due to the changing velocity of the earth in its orbit was detected.

3. H_2 is more general than H_1 in not being restricted to the passage of light in a closed path in which the source, mirrors, and receiver are relatively stationary. The gap here has however been bridged by other experiments which together show that c does not vary with the velocity of the light-source, or after reflection from moving mirrors, or in passage over an open path from astronomical sources.[1]

H_2 can therefore be said to be a generalisation of equivalent empirical status to H_1, that is to say, both are hypotheses, but neither is further removed from experiment than the other. Let us now see how the hypothesis H_2 leads to the impossibility of causal actions being transmitted faster than light. Consider a coordinate framework set up with physically rigid rods, relative to which coordinates (x, y, z) can be assigned to any point by measuring distances, again, with physically rigid rods. All that is meant at present by ‘ physically rigid rods ’ is those scales which are in practice used for length measurement in physical experiments. Consider also a clock measuring time t such that the reference system K with coordinates (x, y, z, t) is an inertial system, that is, the laws of Newtonian mechanics are locally obeyed in it for slowly moving bodies. This means that the time is measured by one of the mechanical or electromagnetic periodic processes in common use for clocks in physics. It is further assumed that the geometry of the coordinate framework so set up is Euclidean, and that time at different places is synchronised by light-signals as in the definition of simultaneity explained above.

Now consider a reference system K' similar to K, with coordinates (x', y', z', t'), orientated with corresponding axes parallel to those of K, and moving with respect to K with uniform velocity v in the direction of the x-axis. In Newtonian mechanics an event occuring at (x, y, z, t) in K will occur at $(x-vt, y, z, t)$ in K', if the origins of the two systems coincide at $t = 0$. Thus the Newtonian transformation equations are

$$x' = x-vt, \quad y' = y, \quad z' = z, \quad t' = t. \tag{1}$$

[1] See references in Whittaker, *Aether and Electricity, the Modern Theories*, pp. 38, 39. H. P. Robertson (*Rev. Mod. Phys.*, XXI, 1949, p. 378) has shown that the Lorentz transformation (equivalent to H_2) for local frames can be deduced from H_1, together with the results of the Kennedy-Thorndike experiment and the Ives-Stilwell experiment on the Doppler effect due to a moving atomic clock, without any postulates other than the existence of Euclidean frameworks in uniform relative motion in *one* of which light is propagated uniformly and rectilinearly with constant velocity c.

Clearly the velocity of any process relative to one system will differ from that relative to the other by the velocity v of K' relative to K. But we now require the velocity of light to be identical in the two systems, therefore the Newtonian transformation equations must be replaced by others. We can express the hypothesis of the constancy of the velocity of a light-ray as follows :

$$c^2 = \left(\frac{dx}{dt}\right)^2 + \left(\frac{dy}{dt}\right)^2 + \left(\frac{dz}{dt}\right)^2 = \left(\frac{dx'}{dt'}\right)^2 + \left(\frac{dy'}{dt'}\right)^2 + \left(\frac{dz'}{dt'}\right)^2,$$

where (x, y, z, t), (x', y', z', t') are the coordinates of the ray in K, K' respectively. In other words, the equation

$$dx^2 + dy^2 + dz^2 - c^2dt^2 = 0$$

must be invariant to transformations between uniformly moving reference systems. It can be shown that the necessary and sufficient condition for this invariance is that the transformation equations to replace (1) should be

$$x' = \frac{x - vt}{\sqrt{(1 - v^2/c^2)}}, \quad y' = y, \quad z' = z, \quad t' = \frac{t - xv/c^2}{\sqrt{(1 - v^2/c^2)}}. \qquad (2)$$

When v is negligible compared with c, these reduce to the Newtonian transformation (1), and are therefore consistent with the definition of inertial systems as those in which Newton's laws hold for slowly moving bodies.

The equations (2) constitute the Lorentz transformation, and it can be shown that Maxwell's equations, and therefore all electromagnetic phenomena in addition to the velocity of light, are invariant with respect to this transformation.[1] This means that, given a set of uniformly moving inertial reference systems, there are no electromagnetic experiments that could be performed by an observer in any one of them which would indicate that his system was absolutely in motion or at rest, for the same results would be

[1] The Maxwell vector and scalar potentials $(\mathbf{A}, \phi)$ transform like the space-time coordinates according to equations (2). Thus the magnetic part of a field in one reference frame, expressed by $\mathbf{A}$ in that frame, is equivalent to both a magnetic and an electric field in a relatively moving frame, just as the space-coordinate x' in K' is equivalent to an expression containing both x and t measured in K. The distinction between magnetic and electric fields, like that between space and time, depends on the choice of reference frame. This means that results obtained for static situations in one frame can be generalised by considering the equivalent dynamic situations in other frames; for example, the law of force between moving charges discussed above (pp. 218f.) can be derived from Coulomb's law for a stationary charge, together with a Lorentz transformation. Quantities like $(\mathbf{A}, \phi)$, which are Lorentz invariant, are called 4-vectors.

obtained in all systems. The hypotheses H_1 and H_2 are thus supported by various kinds of electromagnetic experiments, and are strengthened thereby, but their status remains that of hypotheses in a hypothetico-deductive system, not Newtonian-type ' deductions from phenomena,' for alternatives to H_1 are conceivable.

Consequences of the Lorentz Transformation

The Fitzgerald-Lorentz contraction and time dilatation are immediate deductions from equations (2), and are now shown to be, not intrinsic effects in bodies moving with respect to the aether, but reciprocal effects appearing to take place in any systems moving with respect to the observer, and in the observer's own system from the point of view of another observer moving with respect to him. There are other immediate consequences of the transformation equations bearing on the propagation of causal action.

1. If t in the simultaneity definition is taken also to be the time of mechanics, then Newton's equations of motion are not invariant to the Lorentz transformation. The conservation of mass and momentum in certain collisions are however invariant if it is assumed that the mass m of a body moving in a given reference frame depends on its velocity v in that frame according to the equation

$$m = \frac{m_0}{\sqrt{(1 - v^2/c^2)}}$$

where m_0 is its mass when at rest in the frame (its *rest-mass*). This relation is supported by electrodynamic and optical experiments, some of which are not directly connected with the experiments which led to H_2, and its status is therefore that of an additional hypothesis, consistent with, but not entailed by H_2. Newton's laws in their original form then become first approximations for slowly moving bodies.

The importance of the mass-equation for this discussion lies in the fact that it shows that no ordinary material body can attain or exceed the velocity of light, for in that case the mass of the body would become infinite or imaginary. This does not, however, show conclusively that no causal action could be propagated with a speed greater than that of light, for such a propagation might not be associated with mass at all, or it might be expressed in such a way that the attribution of imaginary mass was meaningful within the theory and consistent with it, just as negative masses have been postulated in the more recent theory of fundamental particles.

The impossibility of any propagation exceeding the velocity of light is much more firmly based on the next two arguments, which assume only the Lorentz transformation.

2. It follows from the transformation that if causal action were propagated with velocity greater than c in some reference frame, the temporal order of cause and effect would be reversed in some other frames, and this would contradict the fundamental notion of causality, for it would mean that there is no unique temporal direction for two causally connected events. In this case if one makes the convention that for causally connected events effect never precedes cause, and A causes B relative to reference frame K, one could always find another frame K' in which B causes A. But the usual notion of cause and effect is such that ' If cause, then effect ' and not necessarily also ' If effect, then cause ', that is, the causal relation is assymmetrical. But mere change of reference frame cannot affect a causal relation between A and B, therefore causal action, as usually understood, cannot be propagated from A to B at a speed greater than c in any reference frame.

3. Another result of the transformation leading in the same direction is as follows : If any entity had a velocity greater than c in some reference frame, there is some other frame in which it would be in two places at the same time.[1]

These arguments cannot, of course, legislate for the world. Even if the special theory of relativity is accepted as a valid description of the world, there is nothing logically inconsistent in holding that for some processes our usual notions of cause and effect and space and time are inapplicable.[2] What the arguments do show is that we cannot retain those notions and at the same time postulate actions propagated with velocities greater than c. Common-sense ideas about time and space and causality are more fundamental to science than the notion of processes travelling faster than light, and there is nothing in the context of special relativity to induce us to abandon the former rather than the latter, so it may be concluded that if the special theory is accepted, then there is no detectable propagation of action exceeding the velocity of light.

It does not follow, however, that such propagation may not be postulated in a *theory* in such a way that, although the theory

[1] Proofs of this result and that in the preceding paragraph are given in Appendix I on page 304.

[2] It has been suggested by some philosophers that it is *logically* improper to speak of effects preceding their causes. This question will be discussed below (p. 285).

has other observable consequences, it is not in principle possible to observe the resulting causal anomalies. Devices logically similar to this have been resorted to in quantum theory, as we shall see later, in order to avoid such anomalies, but so far it has not been found necessary or convenient in any theory to introduce velocities greater than that of light. The possibility cannot, however, be ruled out, and in that case, in conformity with the general realist view of theories being adopted here, it is also possible that instantaneous action at a distance may take place, although to be consistent with normal notions of causality its anomalous effects must be unobservable.

4. A further deduction from the Lorentz transformations is that there is no absolute time simultaneity, since events at different places at the same time in one framework are not simultaneous in other frameworks. This leads immediately to the conclusion that if there were instantaneous propagation in one frame, it would not be invariant under the transformation. Hence, if it were postulated, even in a theory as just suggested, it would pick out one inertial frame uniquely, and this is contrary to the spirit if not to the letter of the principle of relativity.

5. In Newtonian physics disturbances are propagated instantaneously in rigid bodies and incompressible fluids. Since there can be no instantaneous propagation in relativity physics, it follows that there can be no rigid bodies or incompressible fluids as understood in Newtonian theory. This is an indication that the theories of elasticity and hydrodynamics must be modified to accommodate relativity ideas, as might have been expected, and it would mean no more than that if it were not for the fact that in the way in which the theory has been developed above, physically rigid bodies have been assumed.

The difficulty arises because, in the effort to keep the hypothesis H_2 as close to empirical data as possible, it has been expressed in 'operational' terms. There is, however, no necessity to do this, and no real advantage either, since H_2 has in any case been shown to be a hypothesis and not an immediate inference from observations. The empirical justification for H_2 is not weakened in any way by exhibiting the theory wholly in hypothetico-deductive form, and its logical structure is then seen to be as follows:

We postulate first a set of ideal Euclidean coordinate systems, (x, y, z, t), (x', y', z', t') . . ., in uniform relative motion. The velocity c of a certain process is said to be invariant with respect to these systems, and the development of the theory proceeds as already described. There need be no mention of physical entities until the results of the deductive theory are applied to the results, say, of the Michelson-Morley experiment. The coordinates x, y, z will then be *interpreted into* (not identified with) approximate measurements with ordinary meter scales where such measurements are required by the experiment. In fact, length measurements are never made by meter sticks in precision experiments, except as rough first approximations, which are then corrected by optical methods, as has been described already in connection with the Michelson-Morley experiment. This experiment was designed to ensure that no practical lack of rigidity in the apparatus was sufficient to affect the expected shift of the interference fringes. The fact that the apparatus could not theoretically be perfectly rigid, since no material can transmit disturbances instantaneously, and that in practice it was almost certainly much less rigid than is even theoretically possible, does not affect the result of the experiment at all, for the constancy and equality of the lengths involved was ensured throughout the experiment to the necessary degree of approximation by the stationary position of the fringes. The meaning of the space coordinates x, y, z is given by the theory, not directly by measurements with quasi-rigid rods, and it is the theory as a whole which is given physical meaning by interpretation into sentences descriptive of experimental results.[1] Even relativistic definitions of 'rigidity' do not affect this point, for though they are theoretically interesting,[2] they are still not realised in experimental meter scales, and are therefore not defined by them.

The same remarks apply to the more fundamental objection brought forward by Milne against mention of rigid rods in physics, namely that they are not only theoretically unrealisable as physical entities, but in principle undefinable.[3] Rigid rods cannot be defined as those which retain the same length under all circumstances, for how do we know what is meant by 'the same length' without reference to measurement by rigid rods? This would be a serious objection if we had to define our geometry by means of

[1] Or rather, experimental results are *interpreted as* approximations to the situation described in the theory. See *supra*, Chapter I.

[2] See for example W. H. McCrea, 'The Fitzgerald-Lorentz Contraction—some paradoxes and their resolution', *Sci. Proc. Roy. Dublin Soc.*, XXVI, 1952–4, p. 27. A 'rigid rod' in relativity physics is generally defined as one which transmits disturbances with the velocity of light.

[3] E. A. Milne, *Relativity, Gravitation and World-Structure*, Oxford, 1935, p. 14: 'We must endeavour to build up a physics out of observations without introducing the indefinable concept of the transport of rigid bodies'; cf. his *Kinematic Relativity*, Oxford, 1948, p. 6.

material bodies alone. This in fact cannot be done, because even if we select measuring rods which ' look ' rigid, the assumption that they retain the same length when transported and rotated and throughout periods of time is always a conventional assumption, the simplest that is consistent with the empirical fact that they continue to ' look the same', but not implied by it. Therefore the use of particular material rods can never define a geometry uniquely without further conventional assumptions.

But this only shows that it is necessary to consider the geometry as part of the conceptual model, whose total consequences have to be related to the behaviour of such things as steel rods rather than lengths of elastic. On this view rigid rods do not appear among the fundamental concepts of physics and there is no logical need to go to great trouble to avoid mention of them. The notion that they are fundamental seems to have arisen from the fact that throughout the theory of relativity there has been a residual operationalism resulting from some of the features of the special theory, and to some incautious remarks about the relation between geometry and measuring rods in various presentations of the theory.

From an operational point of view, too, much has been made of the conclusion that there is no absolute time simultaneity, invariant for all reference frames. Einstein showed in his 1905 paper that if simultaneous times within one frame are defined operationally in a simple manner by means of light-signals exchanged between observers, the Lorentz equations can be derived for transformations between reference frames. This approach has been developed by Milne in a far-reaching attempt to base physics on what he calls purely ' epistemological ' principles, that is, on simple operations with light-signals, assuming no particular geometry or means of direct space-measurement. The theory involves a very fundamental reinterpretation of the concepts of physics, and because it appears to assume less, it suggests the possibility that the consequences of the orthodox theory with regard to instantaneous propagation may be evaded ; in fact at an early stage Milne claimed to have evaded them,[1] although later the claim was tacitly dropped.

Milne's Action-at-a-Distance Theory

Milne's primitive conception is that of a set of observers equipped with clocks and means of sending and receiving light signals. A ' clock ' is any device for ordering events at the observer in a time sequence, and the only relation assumed between these events is

[1] *World-Structure,* pp. 96 and 276

that of 'before-and-after'. It should be noticed that here, unlike orthodox relativity, it is not assumed that time and distance as measured by phase-shift have any relation to that measured by ordinary clocks and rods. All that is required of Milne's 'clock' is that it should mark intervals related by 'before-and-after', and no question of the equality of these or their relation to any other process is involved, because this has not yet been defined in the theory. It is further assumed that any observer can read the clock of any other simultaneously with his own, and that they can communicate their observations to each other. Thus the actual observations on which the theory is based are very simple, consisting only of the ordering of emission and reception of light-signals according to the clocks of various observers.

It cannot be said, however, that these processes are operational in the strict sense, since there are no observers with clocks stationed throughout space, and if there were we should be unable in most cases, because of the long time-lag, to communicate with them. This objection is not frivolous, because Milne later identifies his fundamental observers with galactic nuclei, and most of the results of the theory are concerned with large-scale cosmic structure. Hence the set of observers must be regarded as a model, and not as an operational concept, in spite of the 'operational' terminology frequently used by Milne.[1] It might, however, be maintained that this model is an improvement on the somewhat *ad hoc* hypotheses of orthodox relativity (especially general relativity), because its fundamental concepts are fewer and simpler. It is easy therefore to ensure that the model is self-consistent, and this is one of the main advantages of building models in terms of simple 'thought-experiments', even if these experiments cannot actually be performed.

Suppose A and B are two observers and B sees his clock reading t'_2 at the same instant that he sees A's clock reading t_1. He can then plot a graph of the variation of t'_2 with t_1; suppose the functional relation is $t'_2 = \theta\ (t_1)$. Now let A take similar readings of his own clock and B's clock, and suppose the functional relationship to be $t_4 = \psi\ (t'_3)$. A and B are said to have *congruent* clocks if $\theta = \psi$. Milne shows that any two clocks can be made congruent by regraduating one of them, and that a set of congruent observers, or *equivalence*, can be defined which is unique, but which is consistent with various graduations of all the clocks depending on the function θ, which is not completely determined by this definition of the equivalence.

[1] Milne himself admits this in *World-Structure*, p. 269.

The requirement that the clocks of any two observers be congruent is a special case of the so-called ' cosmological principle ', which is assumed in some form in most modern cosmologies, and which asserts that the overall appearance of the universe is the same for any observer, whatever his space-time position.

Milnė now proceeds to show how each equivalent observer can define his own space-time framework by making use only of the clock-readings already described. Suppose A sends out a light-signal at t_1 by his clock ; it is reflected back at B ; and when it returns to A, A sees B's clock reading t'_B and his own reading t_3. A now *defines* the space-time coordinates of B *in A's own reference frame* as follows :

$$\text{Distance of } B \equiv r_B \equiv \tfrac{1}{2}c(t_3 - t_1),$$

$$\text{Time of reflection of signal at } B \equiv t_B \equiv \tfrac{1}{2}(t_3 + t_1).$$

The constant c here is *quite arbitrary* and merely assigns the scale of A's length measurements. The time t_B assigned to the reflection by A is not necessarily the same as the time t'_B read off B's clock at the instant of reflection, and A can derive a *clock running relation* $t'_B = f(t_B)$ between them by means of a series of observations. A can also define an *epoch-distance relation* between r_B and t_B, namely $r_B = c\phi(t_B)$.

B can now perform analogous observations and assign space-time coordinates (r'_A, t'_A) to an event of reflection at A in B's own framework. In a series of observations, B can derive a clock-running relation and an epoch-distance relation, and the condition of congruence requires that these functions should be the same as those found by A, and that the same constant c should be chosen. Thus

$$t_A = f(t'_A) \qquad \text{and} \qquad r'_A = c\phi(t'_A).$$

It now becomes important to consider the significance of the constant c. Since it is the ratio, in all reference frames, of the distance $2r_B$ covered by the signal, to the time of transit $(t_3 - t_1)$, it is clearly the average velocity of the signal on the ordinary definition of velocity. Thus the constancy of the velocity of light for all fundamental observers is a consequence of the definitions included in the model. But this consequence cannot legislate for the facts, and it must now be asked how, for example, the Michelson-Morley experiment is related to the application of the model.

What this experiment and others do for Milne's model is simply to show that the model is applicable to the world. Physical light has, at least approximately, the property of invariant velocity demanded of Milne's signalling process on the ordinary definitions of physical length and time. In one respect, however, the experiment goes further than Milne's definitions require, for whereas these definitions apply only to so-called ' fundamental observers ', which Milne later identifies with galactic nuclei, the Michelson-Morley experiment involves an ' observer ' (the mirror P) which is certainly not fundamental in Milne's sense, since it is carried along on the surface of the earth. Thus if it is assumed in accordance with a very general cosmological principle that the results of the experiment would be the same for all observers whatever their state of motion, Milne's definitions make use of only a limited class of such observers.

There need, then, be nothing surprising in the assertion that the velocity of light in Milne's model is a conventional constant. In a sense, most properties of models are conventional ; the only relevant question is whether they are useful or applicable, and, as Milne himself remarks, it is ' the code of interpretation of these symbols in terms of observations ' [1] which is the important thing. His theory does, however, provide a different intrepretation of the experiments from those discussed in relation to the Michelson-Morley experiment above, for it assumes no length measurements or fixed time-scale, but only that the points P, S_1, S_2 can be regarded as equivalent observers as defined in the theory. Hence it cannot be assumed without further investigation that the Lorentz transformation and its consequences for the propagation of action are necessary consequences of this new interpretation. This must now be studied by describing Milne's two systems of time-keeping.

1. *The t time-scale*

The functional relation between A's reading of B's clock t'_B and of his own clock t_3 is what we have called θ. Thus $t_3 = \theta_{AB}(t'_B)$, where θ_{AB} is characteristic of the observers A and B of the equivalence.

Suppose $\theta_{AB}(t) = a_{AB}t$ where a_{AB} is a positive real number characteristic of A and B. Milne then shows that B is moving with a constant velocity

$$v_{AB} = \frac{(\alpha^2_{AB} - 1)}{(\alpha^2_{AB} + 1)} c$$

relative to A, in other words all members of the equivalence are in uniform motion with respect to one another with all possible velocities

[1] ibid., p. 71

from zero to c. Since Milne has already shown that if two members of an equivalence ever coincide, they all coincide at that instant, this time-scale may be taken to describe a set of observers receding from an initial point at $t=0$ with uniform velocities. It is not *necessary* to assume that the observers coincide at $t=0$, or that they are receding from that point rather than converging upon it, but in view of the obvious analogy with the recession of the galaxies, it is reasonable to consider an equivalence with these properties, since such an equivalence is possible, given the time-scale t as defined above. This is one of the instances where empirical knowledge guides a choice among logical possibilities in the model, and deprives the theory of any claim to be completely *a priori*.[1]

Milne also shows that the transformation equations expressing the space-time coordinates of A's frame in terms of those of B's frame are the Lorentz equations. This important conclusion means that in the t-scale all the consequences deduced from the Lorentz transformation above will apply. In particular, there can be no causal propagation with a velocity greater than c relative to any observer, because if there were, the time order of cause and effect would be reversed for some other observers, and this is contrary to the definition of congruence for observers, according to which the primitive before-and-after relation must hold in the same order for all of them.

Clock graduation on the t-scale is not necessary, however, and we must examine the question of whether a velocity greater than c is possible in one of the other time-scales.

2. *The τ time-scale*

Suppose the clocks are regraduated from the t-scale to a scale τ, where $\tau=t_0 \log(t/t_0)+t_0$, t_0 being the 'date' of the present epoch, measuring the time elapsed since the initial point $t=0$ on the t-scale when all observers were coincident. The clocks still form an equivalence, and it can be shown that in this case they all keep the same time, and that all fundamental observers are relatively stationary. The transformation equations between the reference frames of various observers are

$$\tau_B = \tau_A, \qquad x_B = x_A - ct_0 \log \alpha_{AB}$$

where the constant $ct_0 \log \alpha_{AB}$ measures the distance between A and

[1] It is now recognised that in defining this equivalence, Milne also effectively assumed uniformity of relative velocity between its members by a hidden axiom, and that it is not derivable from the previous considerations alone. See W. H. McCrea, 'Cosmology', *Rep. Prog. Phys.*, XVI, 1953, p. 334.

B. Thus all observers have the same reference frame, except for a change of origin, and there is therefore a public space-time. Also the 'age of the universe' on this scale appears to be infinite.

Milne proceeds to show that according to his theory, the time-scale used in contemporary physics is ambiguous, τ-time corresponding to that used in Newtonian dynamics, and t-time to that used in electromagnetic theory. This is not the place to pursue these consequences of his theory, which begins at this point to depart widely from orthodox relativity, for we wish only to ensure that there is no departure from the consequences of the Lorentz transformation in regard to propagation of action. Two relevant remarks may be made about the τ-scale.

First, it is clear that of the two time-scales, t is more fundamental, because it defines the value of the constant t_0 which appears in the definition of the τ-scale. If fundamental observers were to continue to use the τ-scale over long periods of time, they would either find that mechanical and electromagnetic time-measures begin to diverge widely, which means that we are now (at t_0) living in a privileged epoch, and this is unlikely, or they would find that the measures remain in step, in which case they would have continually to regraduate their τ-clocks, as the age of the universe, t_0, increases on the t-scale. The τ-scale is therefore applicable only for intervals of time in the neighbourhood of t_0.

Secondly, it can be shown (see Appendix II, page 305) that if a velocity is greater than c in τ-time, or in any other time-scale which is consistent with the congruence-relations of the equivalence, it is also greater than c in t-time, and so the causal paradoxes consequent upon the Lorentz transformation could be produced by a simple regraduation of clocks to the t-scale. But the mere regraduation of clocks cannot affect causal processes, therefore there can be no velocity greater than c in any time-scale.

In *Relativity, Gravitation and World-Structure* Milne claims [1] that the velocity of propagation of gravity may be finite or infinite according to the description adopted, as long as the description is relativistic for equivalent observers. This claim is not, however, repeated in *Kinematic Relativity*, where a gravitational and electromagnetic theory is worked out in detail, and where the equations, expressed in t-time, are Lorentz invariant, showing that no velocity greater than c is possible. The only exception that could be made to this would be in a statistical model-universe in which equivalence of description by different observers did not imply that the same

[1] p. 276

events are equivalently described. For example, if an event P_1 causes another P_2 in A's description, and the velocity of propagation of the causal action is greater than c, it is only necessary that there shall be *some* event Q_1 which equivalently causes Q_2 in B's description, and Q_1, Q_2 need not be identical with P_1, P_2. Thus in B's description P_2 may precede P_1, but no causal relation will be asserted between them, and thus the paradoxes would be avoided. But such a model could not be applied to particular local events such as the gravitational action of the sun on the planets. Here gravitational propagation would have to be less than or equal to c.

In spite of its Lorentz-invariance, however, Milne's theory is an action-at-a-distance theory in the sense that no field-energy or momentum are postulated, and conservation of energy and momentum are not taken to be empirical laws but are investigated by means of the fundamental kinematic equations. Whether energy is conserved or not is then partly a matter of how it is defined, for there are alternative general definitions consistent with its observable conservation. The purpose of this somewhat lengthy digression into Milne's theory has been to show that even when the primitive concepts of length- and time-measurement are fundamentally modified, empirical considerations as well as the logic of the theory still lead to the Lorentz transformation. The consequent impossibility of instantaneous propagation of action does not, however, imply that action-at-a-distance theories are forever rejected from physics, for absence of field-energy and momentum now become necessary and sufficient conditions for such theories, irrespective of the finite time of propagation and of the consequences for conservation of energy. What the Lorentz transformation entails is not that action at a distance is impossible, but that if it takes place over space, then it also takes place over time, that is, for the first time in physics it has to be contemplated that causal action 'jumps' over temporal intervals. We shall return to such action-at-a-distance theories later, but meanwhile we shall describe the more orthodox development of general relativity in which finite velocity is taken to indicate the presence of a field, and where gravitation joins electromagnetism in being described by a field theory in which the velocity of propagation is c.

Einstein's Theory of Gravitation

The status of the general theory of relativity is by no means as well established as that of the special theory. The experiments that can be devised to test it are fewer, the interpretation of most of them is more ambiguous, and none of the cosmological theories developed from it has yet proved satisfactory. But the basic idea of a gravitational field theory is firmly established, and orthodox relativity provides a good example of one such theory in terms of which to describe the novel conceptions of action which are thus introduced.

Einstein was led to his general theory mainly by theoretical, and in part aesthetic, considerations. The special theory selects out of all possible coordinate frames the inertial frames, and if these are thought of as somehow the ' natural ' frameworks of space-time, then space-time is still being endowed with some absolute properties, independent of the matter within it. It has the causal property of determining the inertial frame at a given point, and hence of determining how bodies would move there in the absence of all forces, but there is no reciprocal causal action of the distribution of bodies upon space-time. This situation is familiar in Newtonian dynamics, but is foreign to the spirit of relativity, which originates with a denial of the absoluteness of space and time and wishes to replace these two independent concepts by a space-time defined in terms of the *relations* between bodies. Again, inertial frames are defined in terms of test-particles placed outside all fields of force, and it is not possible to determine the set of inertial frames uniquely in the neighbourhood of gravitating matter. In a uniform gravitational field producing in all bodies an acceleration g, one may either take a stationary framework K as inertial, and explain the accelerated motion of apparently free bodies by postulating a gravitational force, or one may eliminate the force by taking a reference frame K' having the same acceleration g, for example one fixed in a lift descending freely from the surface of the earth, in which bodies will move with uniform velocity according to the law of inertia. This elimination of gravitational force by choosing an appropriate reference frame is possible only because gravitation produces the same acceleration in all masses; because, in other words, the ratio of gravitational to inertial mass is the same for all bodies. This is an experimental fact in the sense that inertial mass can be measured independently of gravitation by impact experiments, and its equality (in appropriate units) with gravitational mass has been verified to a high degree of accuracy. Einstein, however, took this apparently arbitrary empirical fact to indicate a deeper theoretical relation between gravitation and inertia, and asserted that both express a single property which shows itself as gravitation or inertia or a combination of both, depending on what coordinate framework is used in the description. Thus Einstein was led to the general principle of relativity, namely that fundamental physical laws should have the same form for all relatively moving reference frames, including those which are accelerated relative to the inertial frames. The study of functions which are covariant for all coordinate transformations, and hence candidates for inclusion in these physical laws, is carried out in the calculus of tensors, and the principle of general relativity together with the assumption that the resulting tensor equations should be simple (in general, of the second order only), severely limits the possible forms of the laws. In some cases the development of the theory is uniquely determined by these two requirements, and so gives a misleading impression of being *a priori*.

Denial of the special status of inertial frames in relation to an absolute space-time means that the existence of local inertial frames in which free bodies move with uniform velocity must be explained in terms of the total distribution of matter in the universe, for if space-time itself does not determine the local motions of bodies, then it must be the spatio-temporal relations of matter that do so. Mach had suggested that this is the explanation of the apparently absolute nature of rotation.[1] The well-known example of the rotating bucket had been used by Newton to support the postulate of the absoluteness of space, for here rotating and non-rotating coordinate axes can be distinguished relative to the water in the bucket by the concavity or flatness of the water surface. When the surface is concave it seems that the water must be said to be rotating absolutely. Mach, however, pointed out that the bucket is not situated in an empty universe, and that we cannot know what would happen if it were. As it is, rotation can be distinguished only relative to the universe of ' fixed stars ', and the principle of relativity of rotation can be saved by supposing that the distribution of matter in distant parts of the universe determines local, non-rotating, inertial frames. This however seemed to presuppose a curious kind of action at a distance between masses, and one which would have to depend predominantly on masses at great distances, since masses at short distances do not have the effect of changing the local inertial frame, and any theory must explain the apparent local irrelevance of the distribution of mass in the universe. If Mach's principle is to be reconciled with the requirement of continuous action in a field, then inertia as well as gravitation must be expressed as a field property, and this would be expected also from the

[1] *Science of Mechanics*, 2nd English ed., p. 229

equivalence of inertia and gravitation to which the general principle of relativity leads.

The problem of general relativity, then, is to develop a gravitational field theory satisfying the principle of the equivalence of all reference frames. The novel feature which it introduces into field theory in attempting to carry out this programme is that it regards space-time itself as the field, and associates gravitational potentials with the geometical character of space-time, instead of describing the field in terms of potential functions which depend on coordinates in a previously given Euclidean space and time as in classical field theory. The theory as it stands at present is open to many theoretical and empirical objections into which we need not enter here. But certain *a priori* epistemological objections have also been alleged against the attempt to identify geometry with a physical field, and these do require examination from our present point of view. We shall first outline the fundamentals of the theory in a very elementary way.

We have seen that a uniform gravitational field can be made to disappear by taking appropriately accelerated coordinate axes. This already indicates that gravitational force is not something external imposed upon space-time, but that it is intimately related to the reference frame in which we choose to describe the motion of bodies. Now the principle of relativity states that all reference frames are equivalent, so the motions must be described in such a way that the difference between the two mutually accelerated reference frames is not mentioned. We know that relative to one frame (that in which no gravitational field appears) the path of a ray of light is described in special relativity by

$$ds^2 = c^2dt^2 - dx^2 - dy^2 - dz^2 = 0. \tag{3}$$

This expresses the passage of light over a space interval dr calculated in terms of Euclidean geometry, so that

$$dr^2 = dx^2 + dy^2 + dz^2$$

in a time dt at velocity c. Now the quantity ds^2, as we have seen, is invariant under the Lorentz transformation between uniformly moving reference frames, but it cannot (logically) be made invariant for relatively accelerated frames, for example for those rotating relative to the inertial frames. If therefore we wish to retain the general principle of relativity with regard to the velocity of light, the only possibility is to drop the requirement that the interval *ds* be calculated according to Euclidean geometry.

It can be shown that, if there are inertial frames in any region of space-time, that is, if there are some frames relative to which free particles move uniformly in straight lines, as would be the case in a uniform gravitational field, then relative to these and to any other accelerated frames, the interval *ds* between two point-events is invariant if

$$(ds)^2 = \sum_{p,q=0}^{3} g_{pq}dx^p dx^q, \qquad (4)$$

where cdt, idx, idy, idz are now replaced respectively by dx^0, dx^1, dx^2, dx^3 (the affixes numbering the coordinates and not denoting powers) and where the g_{pq}'s are all constant. In this case a transformation of coordinates from a non-inertial to an inertial frame will bring the g_{pq}'s back to the values

$$\begin{matrix} 1 & 0 & 0 & 0 \\ 0 & 1 & 0 & 0 \\ 0 & 0 & 1 & 0 \\ 0 & 0 & 0 & 1 \end{matrix}$$

corresponding to (3). The equation (4) indicates, however, that in the general case the geometry is non-Euclidean, since ds^2 is not calculated according to the theorem of Pythagoras. This equation, expressing the value of ds^2 in terms of the coordinates, determines the *metric* of the geometry, and is the simplest generalisation of Euclidean geometry implied by (3).

It is not, however, always the case that inertial frames can be found in any region of space-time. In a non-uniform gravitational field such as that of the sun, there are no reference frames relative to which ' free ' particles (that is, those free from visible constraints) move uniformly, and a ' gravitational force ' has to be introduced to explain their deviation from uniform motion. Now the metric (4) suggests a way in which a varying gravitational field can be described in terms of the geometry of the region. If the g_{pq}'s are allowed to be functions of (x^0, x^1, x^2, x^3), then it is not (logically) possible to arrive at an inertial framework by any transformation of coordinates, but under certain conditions (described by saying that g_{pq} is a *covariant tensor of the second order*) ds^2 remains invariant. The variable g_{pq}'s then play the role of gravitational potentials, and the momentum and energy densities of the field are defined in terms of them and are shown to satisfy a law of conservation. These functions need no longer be thought of as independent field functions imposed upon a Euclidean space-time, for they are now intrinsic characteristics of the geometry of the region, and the gravitational field theory can be assimilated to the non-Euclidean geometry of curved spaces. For example, in two dimensions instead of four, such a geometry applies to curved surfaces, whereas Euclidean geometry applies to flat surfaces. For this reason, regions of space-time where there are inertial frames are called ' flat ', and regions with non-uniform gravitational fields are ' curved '. Just as free particles on a curved surface move along the shortest (or longest) path between two points (a *geodesic*), so free particles in a gravitational field move along the geodesics determined by the metric of the field.

A field theory developed in this way seems conceptually farther from the notion of action at a distance even than the electromagnetic field theory. For here it is not that the presence of the field-sources (matter, electric charge, etc.) causes a field of force to appear in previously empty space, but the properties of space itself are determined by the presence of matter. For the geometrician accustomed to thinking in terms of abstract spaces the difference between the two models is very great ; in one case geometry itself is the model, whereas in the electromagnetic case the model is still a refined version of fluid-flow. The mathematical advantages of the geometrical model have led to various attempts, upon which Einstein was engaged until the end of his life, to incorporate electromagnetism and quantum theory with gravitation in a unified field theory of the geometrical type, but so far no fully satisfactory theory of this kind has been produced.

Einstein himself did not consider that he had shown his theory to be consistent with Mach's principle, but some progress has recently been made in this direction. In 1953 D. W. Sciama [1] developed a simple world-model in which inertia is assumed to be dependent on the presence of the rest of the matter in the universe. By postulating a simple form of gravitational potential due to this matter, he derived an equation of motion for a test-particle in the presence of the rest of the universe together with a mass M stationary with respect to the universe, and this equation is a combination of Newton's law of motion with the usual gravitational force due to

[1] ' On the Origin of Inertia ', *Monthly Notices of the Royal Astronomical Society*, 113, 1953, p. 34

M, showing that the inertial frame of the test-particle is that which is at rest with respect to the universe. The centrifugal and Coriolis forces for rotation were also derived. Sciama claimed that such a theory differs from general relativity in that (i) it enables the amount of matter in the universe to be estimated from a knowledge of the gravitational constant ; (ii) the principle of equivalence of reference frames is a consequence of the theory, not an initial axiom ; (iii) it implies that gravitation must be attractive. Recently W. Davidson [1] has developed this theory and appears to have shown that, contrary to Einstein's view, general relativity is consistent with and may incorporate Mach's principle, and that Sciama's three results are also derivable from it. Hence Mach's principle probably does not involve an action at a distance of a new and peculiar sort, but only a generalisation of relativistic gravitation.

There have been some fundamental objections to the programme of identifying field quantities with the geometry of space-time, and these have usually arisen where an operational philosophy of science has been tacitly presupposed. Several writers have seen a logical circle in the method used to determine the geometry of a given region of space-time : the intrinsic geometry is dependent on the distribution of matter in the region, but in order to describe the distribution of matter measurements must be made and therefore a geometry assumed. Thus Whitehead wrote that

> '... measurement on [Einstein's] theory lacks systematic uniformity and requires a knowledge of the actual contingent physical field before it is possible '.[2]

This was also one of the considerations which led Milne to formulate his alternative theory of gravitation, for he says that in Einstein's theory

> 'A conceptual scheme [the geometry] is first formulated, in which it is posited that the matter-in-motion can be described. The rules by which it can be verified observationally that the matter is in fact present to such and such an extent, and moving

[1] 'General Relativity and Mach's Principle', *Monthly Notices of the Royal Astronomical Society*, 117, 1957, p. 212

[2] *The Principle of Relativity*, Cambridge, 1922, p. 83. Whitehead's own theory developed in this book is an action-at-a-distance theory in the sense that, like Milne's, it does not postulate field-energy and momentum, and uses flat space-time. It has been shown to violate conservation of momentum and to predict a secular acceleration of the centre of mass in two-body systems, which might be experimentally detectable in the case of double stars (A. Schild, 'Gravitational Theories of Whitehead's type', *Proc. Roy. Soc.*(A), 235, 1956, p. 202).

in such and such a way, are formulated as a means of interpretation of the symbols used. The symbols, the conceptual element, come first, the interpretation in observation second. . . . In our treatment, on the other hand, we have abolished any conceptual element by *beginning* with observations (temporal experiences), instead of beginning with symbols and later fixing their observational interpretations.' [1]

The description Milne gives here of Einstein's theory is that of a hypothetico-deductive system, and this it has in common with most physical theories, including Milne's own, for as we have seen, his kinematic relativity is not without hypothetical elements. But his objection calls attention to a difficulty, which we have already mentioned, about hypothetico-deductive theories which are expressed in *abstract* terms, namely that if the symbols in the theory are initially given no interpretation, the theory ceases to be testable because observations can be arbitrarily interpreted to fit into it. This seems to be what Milne is afraid of here, and intends to overcome by using a hypothesis interpreted from the beginning in terms of equivalent observers and their measurements. The difficulty arises increasingly with the development of mathematical models in modern physics, and appears serious in the general theory of relativity, because there are at various stages apparently arbitrary identifications of the symbols in the theory into physical terms, for example, the interpretations of the g_{pq}'s as gravitational potentials, and various elements of a certain second-order tensor as momentum components and energy.[2] Three points may be made in connection with this difficulty. First, in a complex formal theory like relativity, what we have called *formal tests* may be possible, so that identifications which first appeared arbitrary are supported by correct predictions deduced from the formalism. Second, no mathematical theory is purely formal, and there is a 'naturalness' about, for example, the argument leading to the gravitational potentials outlined above, which means that even extensions of the theory requiring new interpretations are not wholly arbitrary and untestable.

[1] *World-Structure*, pp. 182, 183

[2] Milne himself does not escape criticism on the grounds of arbitrariness of interpretations (W. H. McCrea, 'Cosmology', *Rep. Prog. Phys.*, XVI, 1953, p. 334). Similar criticisms of Eddington's theory have been made, namely that when his theoretical calculations of physical constants have not coincided with their experimental values he has 'found reasons' for reinterpreting the significance of the experiments. His reasons were no doubt sometimes good ones, but the procedure is clearly open to abuse. Eddington himself apparently thought that 'definitions', that is 'experimental interpretations', in physics must be circular and hence arbitrary (cf. his 'cyclic method of physics', *The Nature of the Physical World*, 1928, p. 260).

Third, the charge of untestability is not always a capital one, particularly not in the case of new theories which are establishing a new fundamental model.

Conventional and Factual Aspects of Geometry

There are some other logical questions raised by the theory of relativity, not because it is hypothetico-deductive in form, but because there are a number of alternative theories which all appear observationally equivalent. The fundamental *logical* proposition to which all such theories must conform is that one cannot assert *both* that the velocity of light is invariant for all possible reference frames, *and* that the geometry of light rays is Euclidean, but it appears that one might decide to postulate one or other of these, and that the resulting theories would then be observationally equivalent.[1] It seems therefore that the geometry of physical space becomes wholly conventional, and that the geometrisation of gravitational fields as in relativity is merely an arbitrary choice among many possible interpretations, none of which can be said to be factual.

This, however, is to simplify the question unduly. It is clear that the suggestion that the velocity of light is invariant everywhere and in all frames of reference is derived, empirically speaking, from a wide extrapolation of the Michelson-Morley experiment. It is not possible to perform this experiment everywhere nor in frames which differ greatly from inertial frames (for example, those rotating very rapidly with respect to the earth) and so we do not know what the results of such experiments would be, but it is possible to build a world-model on the assumption either that there would be no shift of interference fringes, or that there would be a shift of a determinate amount. In order to consider the Michelson-Morley experiment in a non-inertial frame, say on a rapidly rotating platform, we may describe it in a way somewhat different from before, and say that it is essentially a method of comparing the geometry of rigid rods (the base of the apparatus) with that of light rays. The result of the experiment in an inertial frame according to Einstein's hypothesis is then described by saying that the geometry of light rays is inconsistent with a Euclidean geometry of *space* as measured by rigid rods, but consistent with a Euclidean *space-time* measured by rigid rods and physical clocks in which different reference frames are related by the Lorentz transformation. The

[1] The subsequent discussion is indebted to H. Törnebohm, *A Logical Analysis of the Theory of Relativity*, Göteborg, 1952, pp. 48ff.

Fitzgerald-Lorentz hypothesis on the other hand would then interpret the experiment as showing that the geometry of light rays is consistent with Euclidean geometry, but that of rigid rods is not, because lengths change as they are rotated.

From an operational point of view, a geometry must be that to which either rigid rods or light rays or both conform, at any rate approximately, so that to say that the geometry of space is such-and-such is to say that rigid rods and/or light rays will exhibit such-and-such relations. From the hypothetico-deductive point of view, there is a third possibility, namely that a geometry may be postulated in the hypothesis to which neither light nor rigid rods conform, but which is convenient for other reasons. Let us examine the conventional or factual nature of the various possibilities by considering possible results and interpretations of an ideal Michelson-Morley experiment performed on a rapidly rotating platform. There are two possible results :

1. There is no shift of interference fringes.
2. There is a shift of fringes of a determinate amount.

There are alternative interpretations of each result :

1*a*. The generalised Fitzgerald-Lorentz hypothesis, namely that light conforms to a Euclidean space-time and its velocity is therefore (logically) not invariant, and the apparent invariance shown by non-shift of the fringes is due to systematic changes in the rigid base, that is, rigid rods do not conform to Euclidean geometry.

1*b*. The Einstein hypothesis, namely that the velocity of light is invariant and hence light conforms to a non-Euclidean geometry, and so also do rigid rods, since the experiment with this interpretation shows that distance measured by light is the same as that measured by the base.

In hypothesis (*b*) both light and rigid rods conform to the same geometry, which may now be called the geometry of space-time. In hypothesis (*a*) Euclidean geometry may be called the geometry of space-time, but rigid rods do not conform to it, so the initial simplicity of the geometry is counteracted by the necessity of invoking *ad hoc* forces to explain the non-conformity of rigid rods to the geometry. Such distorting forces will have to be what Reichenbach calls ' universal forces ',[1] that is, they act on all material bodies

[1] *The Philosophy of Space and Time* (trans. M. Reichenbach and J. Freund), New York, 1958, p. 13 (first German ed., Berlin, 1928)

alike, as distinct from 'differential forces', such as heat, whose effects can be measured directly because they act differently on different materials. This procedure has an air of arbitrariness, but is not necessarily to be ruled out if other aspects of the Euclidean model are satisfactory. Theories in which Euclidean space-time is assumed have been developed by Milne and McCrea,[1] and these are not concerned with questions about general invariance for all relatively accelerated frames or in all gravitational fields, and do not need to assume anything about the behaviour of rigid rods in such frames.

There may be other difficulties about postulating Euclidean geometry for light rays. It might be concluded from very general considerations, without any detailed assumptions about geometry, that the stability or apparent expansion of the universe require the space of the universe to be closed, that is to be in some sense finite. In that case Reichenbach has pointed out that the assumption of Euclidean space-time would lead to what he calls 'causal anomalies'.[2] For example, when the closed space of the surface of the earth is projected on to a flat Euclidean plane, the north pole is represented by the point at infinity. If a light signal were sent across the pole, its description in terms of Euclidean geometry would involve its passage over an infinite distance in a finite time :

> 'If the principle of normal causality, i.e. a continuous spreading from cause to effect in a finite time, or *action by contact*, is set up as a necessary prerequisite of the description of nature, certain worlds cannot be interpreted by certain geometries.'

But, of course, this 'principle of normal causality' is not logically necessary or empirically required, and might be abrogated if other considerations made it convenient to do so.

There are also alternative interpretations of the other possible result of the Michelson-Morley experiment, namely that there are fringe shifts :

2*a*. It may be postulated that light conforms to Euclidean geometry and its velocity is therefore not invariant. Then there are two possibilities with regard to rigid rods which can be decided by the actual fringe shift observed : either this is consistent with Euclidean behaviour of the rods, or it is not. If it is, then this hypothesis would appear to be comparatively simple, for both rods and light conform

[1] See W. H. McCrea, 'Cosmology', *Rep. Prog. Phys.*, XVI, 1953, p. 321

[2] 'The Philosophical Significance of the Theory of Relativity', *Albert Einstein : Philosopher-Scientist*, p. 298

to Euclidean geometry, but the general principle of relativity would have to be dropped. If it is not, then universal forces would have to be invoked to account for the non-Euclidean behaviour of the rods.

2*b*. On the other hand, it may be postulated that the velocity of light is, after all, invariant, and hence that it conforms to a non-Euclidean geometry. The fact that there *are* fringe shifts will then entail that the geometry of rigid rods is not the same as that of light—it may be Euclidean or differently non-Euclidean according to what the shifts are, but in any case, universal forces will have to be invoked.

Would it make sense in any of these cases to postulate a geometry that was conformed to by neither light nor rigid rods, to postulate, for example, Euclidean geometry in case (1*b*) ? This would involve *two* kinds of universal forces, but it is possible that simplicity in other directions might compensate this, and if so, then on a hypothetico-deductive view, it must be admitted that to assert the geometry of space-time to be X, where neither the geometry of light nor of rigid rods is X, is nevertheless a possible hypothesis.

Such a conclusion is disputed by operationalists, and its rejection may have far-reaching results. For example, in his *Mathematical Theory of Relativity*,[1] Eddington says that the statement that the radius of curvature of space is of constant length must mean ' constant relative to material standards of length ', and he derives from this some speculations about the apparently arbitrary radius of the electron. This radius, he says, must adjust iteslf at each point of space so that measuring rods, which are composed partly of electrons, always bear a constant ratio to the radius of curvature of the space at that point. Thus even if space were non-homogeneous and anisotropic, it would always appear to be of constant curvature as a result of measurement with material bodies, and hence from an operational point of view its homogeneity becomes analytic. This is an early example of the method which Eddington developed considerably in his later work, according to which the fundamental constants of physics are not contingent facts, but logical consequences of the means of measurement and hence calculable *a priori*. Whatever may be thought of these later claims, and they are as yet imperfectly understood, this early example seems to be based on a logical confusion derived from an operational approach. For the

[1] p. 153

'radius of curvature of space' is not necessarily a physical entity which can be measured by laying material rods alongside it. It is a concept occurring in a geometrical model, and it is not the case that 'the radius of curvature is of constant length' *means* 'constant relative to material standards of length'. Within the model the statement means that a certain well-defined function of the co-ordinate variables is a constant, and this may not be directly interpreted into physical measurements. It is supported indirectly by the success of the whole model in explaining empirical data of various kinds, mostly astronomical, and the question of whether the geometry of material rods leads to a space of constant curvature is then partly an empirical one, and partly conventional, as we have seen above. The fact that other empirical data suggest that a non-Euclidean geometry of a certain kind is an appropriate model cannot entail any conclusions about measuring rods, nor about the electrons associated with them.

We have considered a number of possibilities which are all consistent with existing observational evidence, and of which some are apparently consistent with any possible evidence. Is there in principle any empirical difference between, for example, (1*a*) and (1*b*)? According to our previous discussion of a similar example, namely the alternative Continental and Maxwellian theories of electromagnetism, the answer must be that almost certainly there is an empirical difference. The difference between theories developed respectively in terms of Euclidean and non-Euclidean geometry is so radical, that it is inconceivable that the natural development of each will not eventually lead to testable predictions which will differentiate them. In cases like this it is often possible in principle to persist in alternative descriptions of the same data, but only at the cost of extreme arbitrariness and complexity. If we are prepared to say that Maxwell's field theory was factually different from Weber's action-at-a-distance theory, then we must also say that there are factual differences between alternative geometries and their associated cosmological models, the only difference between the two examples being the greater difficulty of deciding which theory is preferable in the cosmological case. From the discussion of alternative geometries it is possible to summarise some of the criteria of choice for theories of this kind, although no such criteria can ever be conclusive, and all may sometimes have to be discarded :

1. Ease of natural development and fertility in suggesting testable predictions, which when tested are continually confirmed.

2. Comparative lack of *ad hoc* assumptions such as 'universal forces'.

3. Conformity with general principles extrapolated from ordinary experience such as that of 'normal causality'.

Thus it may be said that the assimilation of gravitation and geometry in general relativity is in intention not conventional but factual, and it entails that gravitation is one of the causal processes which consist, in Reichenbach's words, of 'a continuous spreading from cause to effect in a finite time, or *action by contact*'. The criterion for action by contact here is the presence of energy in the field as well as finite velocity of propagation.

It has not yet been shown, however, that gravitational action satisfies a third condition for continuous action which is satisfied by the electromagnetic field. Bridgman has pointed out [1] that while local electrostatic energy in free space can be detected by two independent methods, namely by its action on a test charge and by measuring the magnetic field due to a change of electrostatic energy, in the case of gravitational field energy, only one such test can be made by measuring the action on a test mass. Like Faraday, he suggests, by analogy with the electromagnetic case, that a changing gravitational field may give rise to a new physical effect as yet too small to be detected. If this should prove to be the case, no doubt the intuitive feeling of the physical reality of the field in free space would be strengthened, but meanwhile most physicists are prepared to take the presence of energy in the field as a necessary and sufficient condition for continuous action by contact.

[1] *The Nature of some of our Physical Concepts*, New York, 1952, p. 30. See also his 'Einstein and Operationalism', *Albert Einstein : Philosopher-Scientist*, p. 351.

Chapter X

MODERN PHYSICS

The Empirical Basis of Quantum Mechanics

In the classical and relativistic field theories continuous action is represented by field quantities which have some of the conservation properties of classical matter. Such a representation is always possible in classical physics, and is also necessary if the conservation laws are to be retained for non-instantaneous propagation (apart from the introduction of advanced potentials, to which we come later). The crisis which arose in quantum theory was essentially the discovery that, if the broad principles of the theory are correct, then such a representation is not possible.

In one sense the break with classical physics did not at first seem to be quite as sharp in quantum theory as in relativity theory. Whereas the Lorentz transformation implied at the very outset that the classical picture of space, time, and mass required radical revision, early quantum theory made at every stage as much use as possible of classical theory, and introduced new conceptions only when forced to do so. The correspondence of relativity with Newtonian mechanics when the velocities involved are much less than the velocity of light has to be tested afresh in each formulation of relativity ; in quantum theory on the other hand the correspondence with classical physics under appropriate conditions was ensured from the start, because quantum theory was developed out of classical theory by modifications, which often had the appearance of *ad hoc* assumptions brought in to accommodate particular experiments, and it was guided at every stage by classical analogies. The prevailing mood of relativity theory is *a priori* in the sense that cosmological theories are developed with more regard in the first place for logical convenience than for direct explanation of phenomena, whereas the mood of quantum theory was at first predominantly empirical, building up from classical macroscopic theory and direct observation to theoretical system. Both approaches are legitimate, but there are strong indications that in the end the method adopted in quantum theory may prove inadequate, for microscopic phenomena seem potentially to involve a far sharper break with classical analogies than has so far taken place, and a sharper break even than that involved in the curved-space model of general relativity. The symptoms of deep-seated mathematical difficulty in quantum theory are increasingly evident in the appearance of unavoidable divergent integrals, but the roots of the difficulty can be traced even in classical theory to the introduction there of various kinds of fundamental particle, that is, to the basic notion of atomicity, which, as Maxwell saw, is not a natural companion to the continuous field. Irreducible atomicity has been contemplated before in the history of physics, but *mathematical*, as opposed to qualitative, theories have been developed almost exclusively in terms of field theory, and although the notion of atomicity can be introduced into these by development of the analogy of discrete 'natural frequencies', it may be that this is not enough, and the whole notion of the continuous field and its accompanying mathematical apparatus may have to be abandoned.

It has been suggested, by Eddington and others, that the notion of identical fundamental particles is one which we impose upon nature, and that much of the resulting theoretical structure of physics can be derived *a priori* from the properties of this model. But if we consider the history of the introduction of the model, from the experiments on electrolysis which prompted Maxwell's remark about the gross and discordant nature of the notion of a 'molecule of electricity', to the basic experiments of quantum theory, it becomes clear that, given the then existing structure of physical theory, which did not *seem* to presuppose ultimate explanation in terms of identical particles, this model was not arbitrary, but was, in part at least, forced upon theoreticians by experience.[1] There is no single experiment related to quantum theory in the same decisive way as the Michelson-Morley experiment has been related to relativity theory, nevertheless an account of a few typical experimental results from the early days of quantum theory will suffice to indicate how modification of the classical theories of radiation and of atomic particles was forced, and what is the empirical status of the resulting quantum hypotheses.[2]

[1] Eddington himself maintained that it is the decision to test physical theories by *measurement* which introduces an *a priori* structure, including the fundamental particles, into physics. If this is so, then this structure must have been already implicit in the classical theory which was used to interpret the apparently 'atomic' experiments. How far Eddington's claim to have found such an *a priori* structure is justified is still a matter of debate, but even his *a priori* assumes that a physics based on measurement is possible, and this is an empirical, not a logical, fact. In any case, whether as a result of experiment or of an *a priori* structure in Eddington's sense, the commitment of modern physics to atomicity is more than conventional.

[2] A detailed survey of these experiments, with references, will be found in Whittaker, *Aether and Electricity, the Modern Theories*, Chap. iii.

In 1900 Planck showed that in the radiation spectrum of a heated black body the distribution of energy as a function of temperature, which was then in conflict with classical theory, could be explained if it were assumed that radiation energy is not emitted continuously from a hot body, but only in discrete parcels or quanta of amount $h\nu$, where ν is the frequency of the radiation and h is Planck's constant, a small quantity with the dimensions of mechanical action (energy $\times$ time). The assumption was confirmed in 1905 by Einstein's explanation of the energy of electrons emitted from a metal subjected to light-radiation of short wavelength, and in 1907 by his study of the failure of the existing theory of the specific heat of solids at low temperatures. In 1913 Bohr showed that the positions of lines of the hydrogen spectrum, as given experimentally by the Balmer series, resulted from the assumptions that

(i) The hydrogen atom consists of a nucleus and electron revolving round it in one of a number of orbits corresponding to discrete states of energy.

(ii) When in one of these 'stationary' states, the electron does not radiate, although according to classical theory it should do so whenever it is accelerated.

(iii) The electron can 'jump' from one orbit to another and the atom then passes from a state of energy W_1 to another of W_2 and radiates or absorbs energy $W_1 - W_2$, where this difference is equal to $h\nu_{12}$, and ν_{12} is the frequency of the radiation. Since the electron cannot, according to (i), have energies between the stationary values, there is no way of describing its passage between these states by a continuous path, and to ensure its continuous existence in time, the jump must be instantaneous. (This assumption does not, as we shall see later, generate the causal anomalies to be expected on the basis of special relativity.)

These, and other assumptions introduced to account for particular experimental results, led to the empirical generalisation that under some circumstances radiation energy behaves as if it consists of indivisible quanta, or photons, each of energy $h\nu$ and momentum $h\nu/c$. It is important to notice at this point that photons can be called 'particles' only in the sense that they are bearers of this energy, and that angular momentum can be ascribed to them, but that they have none of the other usual mechanical properties of particles. Thus, their rest-mass must be zero, otherwise their mass would always be infinite since they travel at the speed of light;

no meaning can be given to the position of an individual photon except at the moment of emission or absorption by matter, when there may be observable effects such as a scintillation or spot on a photographic plate; and photons are indistinguishable and are created or destroyed whenever radiation interacts with matter. Also the photon theory has to account for those familiar properties of radiation, namely interference, diffraction, and polarisation, which were originally taken to falsify the corpuscular theory of light. Thus it is clear that photons are not particles in any ordinary sense.

The second innovation of quantum theory was the discovery, predicted by De Broglie, and shortly afterwards experimentally confirmed, that just as radiation sometimes shows particle properties, so the fundamental particles might sometimes show wave properties. Confirmation was obtained by the observation of diffraction patterns produced by the passage of electrons through matter. In 1926 Schrödinger described the wave properties of matter in terms of a wave equation, and at the same time accounted for the stable orbits of an electron in Bohr's atom as stationary wave solutions of the equation, each orbit containing an integral number of waves. The solutions of the wave equation (ψ-functions) were interpreted by Born as a measure of the probability of finding a particle in a given region.[1]

The representation by ψ-functions, having the classical characteristics of spatio-temporal continuity, has led Schrödinger to hope that discontinuities implied by the quantum jumps could be eliminated.[2] But it is difficult to see how such a view can be maintained, for in any but the simplest one- or two-body problems the wave-function requires a multi-dimensional space, and therefore cannot give a picture of the probable distribution of events in three-dimensional physical space; also there is no energy associated with the wave as such, as in the Maxwell field; and in any case discontinuities are bound to arise when transition is made from the theoretical wave representation to observed macroscopic effects such as those of the experiments already described, and all experiments involving detection of individual particles by scintillations, tracks or counters, for these do involve quanta.[3]

[1] To put it precisely: $\psi^*\psi dv$ is the probability of finding the particle in the volume dv, where ψ^* is the complex conjugate of ψ. Thus $\psi^*\psi$ is a probability density.

[2] 'Are there quantum jumps?', *B.J.P.S.*, III, 1952, pp. 109, 233

[3] cf. M. Born, 'The interpretation of quantum mechanics', *B.J.P.S.*, IV, 1953, p. 95; and W. Heisenberg, 'The development of the interpretation of the quantum theory', in *Niels Bohr and the Development of Physics*, ed. Pauli, London, 1955.

The Wave-particle Duality and Uncertainty Principle

Neither Schrödinger's nor any other attempt to interpret quantum theory exclusively in terms of waves or of particles has been succesful, and the practice is now well established, in accordance with Bohr's so-called 'principle of complementarity', of using both interpretations or either, depending on the circumstances. In order to understand the logical significance of this duality it is necessary to be clear first of all that the duality arises directly out of classical physics by natural interpretation of experiments which previously led unambiguously and without any self-contradiction to particle or wave theories. The contradiction arises in quantum theory because such interpretations always imply statements about states of affairs between observations, and here the interpretations of various experiments do contradict each other. The point can be illustrated quite simply by considering a diffraction experiment in which a beam of electrons passes through a slit in a screen and is detected by scintillations on another screen. If the second screen is a photographic plate, the incident radiation will eventually be seen to be distributed in a well-known and predictable diffraction pattern, but if the beam is of low intensity and its arrival at the screen is watched, it is seen to occur in a number of discrete scintillations whose statistical distribution can be predicted from the wave equation, but whose individual positions at any given time cannot be predicted.

So far the experiment is described entirely in terms of observations (or *phenomena*, to use Reichenbach's term[1]). It is necessary to stress this, because many 'thought-experiments' with ideal particles are described in connection with the foundations of quantum theory, and it is not always clear how much theoretical interpretation is

[1] Much of the subsequent discussion on the wave-particle duality is indebted to H. Reichenbach, *Philosophic Foundations of Quantum Mechanics*, Berkeley and Los Angeles, 1944, pp. 17ff. Reichenbach's definition of 'phenomena' is as follows: 'Using the word "observable" in the strict epistemological sense, we must say that none of the quantum mechanical occurrences is observable; they are all inferred from macrocosmic date. . . . There is, however, a class of occurrences which are so easily inferable from macrocosmic data that they may be considered as observable in a wider sense. We mean all those occurrences which consist in coincidences, such as coincidences between electrons, or electrons and protons, etc. We shall call occurrences of this kind *phenomena*. The phenomena are connected with macrocosmic occurrences by rather short causal chains; we therefore say that they can be "directly" verified by such devices as the Geiger counter, a photographic film, a Wilson cloud chamber, etc.' (ibid., p. 20).

Events which may be postulated to occur between the phenomena, but which are in principle not detectable in this way, Reichenbach calls *interphenomena*. Reichenbach's use of 'phenomena' must be distinguished from that adopted in the first chapter of this book, for his use allows for inferences from classical physics, mine does not. No ambiguity will arise in the present chapter, where Reichenbach's sense is intended throughout.

involved in them and how much could actually be performed or observed. Thought-experiments have the function of rendering intelligible the language of a model, but this is not the same as the function of actual experiments of justifying that particular model, and this is what we require at present. The slit experiment is an entirely practicable one, involving for its immediate interpretation only the knowledge derived from classical physics about what apparatus is a source of radiation producing scintillations on a screen.

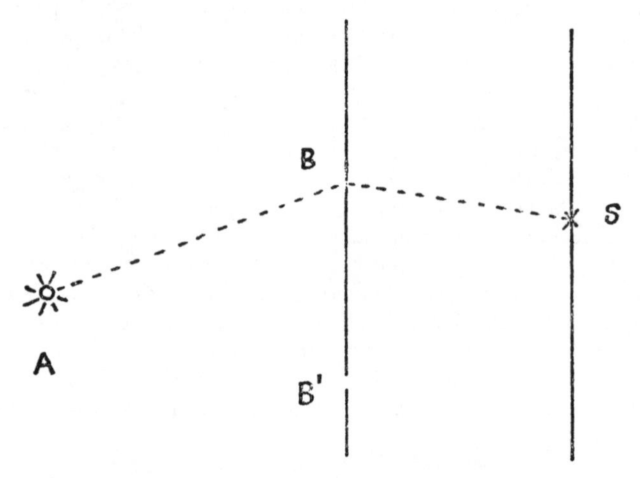

FIG. 8 The Slit-Experiment

Now if attention is concentrated on an individual flash at *S*, classical physics would naturally lead us to describe the *inter-phenomena* between the screens by saying that a particle emitted at *A* travels in a straight line to *B*, is there deflected by collision at the slit and travels in a straight line to *S*. The fact that the next flash is probably not at *S* is naturally explained by saying that the collisions at the slit are not identical for each particle, but have a statistical distribution which is shown up by the eventual diffraction pattern on the screen. If now a second slit is made in the first screen at *B′* and an individual flash is observed at *S*, the classical particle interpretation is unable to say which of the paths *ABS* or *AB′S* was taken by the particle. But assuming, as can be tested in other ways, that the direction of emission at *A* is quite random, the classical interpretation will assert that the final distribution on the screen will be obtained by superposing the intensity distributions due to B and B' alone, since the effect of either slit on the particles cannot depend on the presence or absence of the other slit. It turns out, however, that *this superposition does not correctly describe the distribution observed with both slits open.* At this stage the particle interpretation can be saved only by assuming that the presence of the slit at B' *does* influence the distribution of collisions at B, and vice versa, but this is to introduce a type of influence unknown to classical physics, and one which is not affected by any change in the intervening medium around BB'.

Is it possible to describe the experiment with two slits in terms of the wave interpretation? According to this, diffraction takes place at both slits, and the pattern is the result of interference between the two diffracted waves, giving intensities which are not the sum of the intensities due to B and B' acting alone. The wave interpretation is quite consistent as long as the total diffraction pattern is considered, but entirely fails to account for the appearance of discrete flashes on the screen. The wave in any case cannot, consistently with the quantum postulate, be regarded as the carrier of continuously distributed energy, and as soon as a scintillation screen is interposed, the whole energy is localised in a flash at some unpredictable point of the screen, and the wave as such is destroyed.

We must conclude that, for any given experiment, one or other of the classical descriptions of interphenomena will involve what Reichenbach calls *causal anomalies*, such as apparent action at a distance between slits or between points of a wave-front. If indeed both descriptions are possible, these effects must be instantaneous, for if they were not, it would not be possible to describe the non-zero period of their propagation in the alternative interpretation. For example, the opening of slit B' must be simultaneous with its effect on B in the particle-picture, because in the wave-picture it is the passage of a second wave through B' and its subsequent arrival at the screen which affects the pattern there, and if two waves pass through B and B' respectively at the same instant t, they will arrive together and interact at the screen at a time $t + x/v$, where v is their velocity and x the distance between the slits and the screen. Thus on the particle-picture the deflection at B must also have occurred at t.

But, just because of the possibility of alternative description, these actions can never be used to send instantaneous signals, as can be seen from the following considerations. Suppose that the slit B' is opened at time t. The effect cannot be detected at all at B, but only at the screen S by a change in the diffraction pattern there, and assuming that action cannot be propagated *between* the screens with a velocity greater than c, the change at S cannot take place at a time less than $t + x/c$. Again, the instantaneous collapse of a wave associated with a single particle when that particle appears at S cannot be the result of an experimental operation upon any other part of the wave-front, because by definition there is only one particle associated with it, and this particle can be affected or detected at only one (unpredictable) point of the front.

The unpredictability of individual quantum phenomena, such as flashes on a screen, created the gravest scandal of quantum theory from the point of view of classical physics, and some account of it is necessary before going on to discuss wider philosophical questions raised by the theory. A statement of accepted results without mathematical details will have to suffice. Schrödinger's wave-mechanics, although suffering from the limitations described above, is an adequate formulation for practical purposes, and can be said to be as well founded empirically as most theories in physics. Now it might be hoped that anomalies such as the impossibility of deriving from it predictions about the behaviour of individual particles could eventually be overcome by a more complete theory including wave-mechanics, but supplementing it by detailed descriptions of the individuals composing the statistical aggregates, just as classical statistical mechanics could in principle be supplemented by Newtonian descriptions of the behaviour of individual gas molecules. This, however, can be shown to be an unrealisable ambition. If wave-mechanics is accepted at all, it follows that some events are undetermined, for wave-mechanics is logically equivalent to the alternative formulation by Heisenberg, which includes the latter's uncertainty principle. This principle states that simultaneous measurement of the position and momentum (or certain pairs of functions of these, called *complementary variables*) of a particle is subject to an inherent limitation such that if Δq is the uncertainty of position, and Δp of momentum, then the product $\Delta q \Delta p$ is never less than $h/4\pi$. Thus, if q could be made exact, p would be wholly uncertain and vice versa.

Some writers, including Heisenberg himself, have somewhat confused the logical position of this principle by speaking as if it followed directly from experimental considerations regarding measurement. Thus, in order to measure the position of a particle with high accuracy it must be 'illuminated' by radiation of very short wavelength, but such radiation has high energy and would therefore change the velocity of the particle by a correspondingly large amount. Hence its velocity cannot be measured accurately at the same time as its position. Similarly, if the velocity is measured by illumination of low energy, the position is relatively uncertain. Now clearly such an account is highly theoretical. There are no *operations* which are correctly described as 'illumination' of an electron so that its position can be 'seen' and so measured: the position of an electron at any except isolated points of its career can be determined only by inference from macroscopic events. The empirical situation is therefore more complicated than these thought-experiments suggest. Moreover, the experiments do not make it clear why the disturbance of the observed particle by the observation cannot be allowed for, as the effect of inserting a cold thermometer into a hot liquid can be allowed for in measuring the temperature of the liquid. The reason why such allowance cannot be made in quantum theory is that the uncertainty principle *entails* that the results of the thought-experiment are relatively uncertain, but it is not entailed by them. The empirical basis of the principle has to be sought, not in such ideal situations, but in the actual experiments which lead to wave-mechanics. Then in every case of actual unpredictability which is examined it is found that Heisenberg's principle intervenes to circumvent any attempt to obtain sufficient information to enable a prediction to be made. Thus in the diffraction experiment described above, the direction of deflection of a single particle at the slit could be predicted only if detailed information were available about the particles of the screen BB', but it is just this information, involving positions and velocities of the colliding particles that, according to Heisenberg's principle, is impossible to obtain.

Sub-quantum Theories

It is not possible either in current quantum theory to save causality in an ontological sense by suggesting that particles *really have* exact position and momentum at all times and that the results of collisions are *really* determined, even though we can never obtain the information to enable us to predict them. Such an assumption was shown by Von Neumann to lead to self-contradictions in the theory.[1] But even if Neumann's proof leaves no loopholes, a point on which there is some doubt, still any conclusion regarding the ultimate

[1] *Mathematische Grundlagen der Quantenmechanick*, Berlin, 1932, Chap. iv. For a critical account of Von Neumann's proof, see P. K. Feyerabend's review of the English translation, *B.J.P.S.*, VIII, 1958, p. 343.

uncertainty of atomic events is based on the assumption that the state of a system is always at least partly defined by ' observables ' (position, momentum, etc.) satisfying the rules of current quantum mechanics and defined by analogy with classical physics. But we are not compelled to make this assumption, for, just as causal laws such as those of Newtonian mechanics may be regarded as approximations to the behaviour of the physical world when all disturbing features are small enough to be neglected, so the statistical laws of quantum mechanics may be approximations to causal laws at a sub-quantum level, where the specification of systems has to be in terms quite other than ' observables ' derived from the particle and wave analogies.

In accordance with this line of reasoning it has been suggested by D. Bohm [1] that there is no decisive reason why future progress may not yield a theory penetrating deeper into the microcosm than the present quantum theory, and possibly having the features of continuity and causality which quantum theory lacks, and Bohm and his collaborators have developed such theories, although so far without decisive success. Heisenberg makes two objections in principle to suggestions of this kind. First,[2] if any such new theory is formally equivalent to quantum mechanics within the limits of possible observation, then no existing kinds of experiments could decide between the theories, and Bohm's sub-quantum theory would be unfalsifiable. Bohm's reply to this is that there are regions where existing quantum theory is admitted to be insufficient, namely in dealing with high-energy reactions and short distances comparable to nuclear size (10^{-13} cm.). If, therefore, a new theory were successful in this region, its logical position would be similar to that of relativity *vis-à-vis* Newtonian mechanics : it would be more comprehensive than existing quantum theory, but would approximate to it for low energies and distances greater than 10^{-13} cm. The reply seems justified, for we have seen in general that two theories expressed in wholly different terms can never be equivalent, so long as it is possible that they, or developments of them, will eventually be observationally distinguishable.

Heisenberg's second objection is that Von Neumann's theorem shows that in a sub-quantum theory such as Bohm suggests, particle and wave analogies would have to be abandoned. What is to take their place ? This is incidentally a strange objection to come from Heisenberg, whose name has been associated with attempts to

[1] For a general discussion, see Bohm, *Causality and Chance in Modern Physics*, London, 1957, p. 95. [2] *Niels Bohr*, p. 18

eliminate all models from quantum theory by developing matrix-mechanics entirely in terms of what is experimentally observed.[1] However, both the objection and Bohm's reply are further evidence that models are not dispensable but essential to theories. Bohm replies that, according to the usual procedure in physics, when a model becomes inadequate, it is modified or enriched in various ways, and methods of dealing with the new concepts involved are learnt during the process of modification. He gives some concrete examples taken from various stages of a theory being developed by himself and his collaborators as a theory of the sub-quantum domain, and although it may be that none of these examples will turn out to satisfy the requirements of such a theory, they do illustrate the possibility of using models other than particles and waves, but still taken from macroscopic physics so that they satisfy the criterion of inherent intelligibility.

The model for the sub-quantum domain which Bohm describes is roughly as follows : the basic entities are the fields of various kinds already described in physics, but it is assumed that the appearance of continuity and local homogeneity which they show at the macroscopic level is really the effect of averaging out of a large number of rapid and violent small-scale fluctuations. The analogy here is with small-scale turbulence associated with large-scale continuous flow, and its effect on particles in the field is compared to random Brownian motion which counteracts the general tendency of the particles to fall under gravity. It is not satisfactory, however, to assume that particles and fields are ultimately distinct entities, so it is further postulated that in addition to random fluctuations of the field, there are also small systematic oscillations which appear as the localised properties of atomic particles. Detailed working out of a model of this kind leads to a picture like that of Brownian motion of mist droplets :

> '. . . the particle-like concentrations are always forming and dissolving. Of course, if a particle in a certain place dissolves, it is very likely to re-form nearby. Thus, on the large-scale level, the particle-like manifestation remains in a small region of space, following a fairly well-defined track, etc. On the other hand, at a lower level, the particle does not move as a permanently

[1] cf. Heisenberg, *The Physical Principles of the Quantum Theory*, trans. Eckart and Hoyt, Chicago, 1930, where he holds that *visualisation* must be in terms of the wave or particle pictures, but that ' mathematics is not subject to this limitation ', and the mathematical scheme of the quantum theory ' seems entirely adequate for the treatment of atomic processes ' (p. 11). Why, then, could there not be a mathematical scheme for sub-atomic processes not depending upon these pictures ?

existing entity, but is formed in a random way by suitable concentrations of the field energy.'[1]

Such a theory, if it could be made satisfactory in other ways, would overcome the objection drawn from Von Neumann's theorem, for while the field could be continuous and causally connected on the sub-quantum level, the appearance of particle-concentrations in it on the quantum level need not, and the state of a given system would not, be expressible at all times in terms of observables such as particle position and momentum.

It is not at all certain that any theory of this kind will prove scientifically acceptable, and this Bohm himself freely admits. But it can be said that from a logical point of view such a theory is possible, and that its formulation involves no departure from the usual methods by which science develops. A notion of theories of novel kinds penetrating without limit into the microcosm is far closer to the tradition of physics than the somewhat spectacular utterances of the founders of quantum physics, who have implied that here the limits of human understanding have been reached, for this is to assume that current formulations of quantum theory and current models of physical reality are unalterable. If two models each turn out to be unsatisfactory in isolation, but usable when regarded as complementary to each other, it is curiously conservative to assert that no other models can be conceived, and to elevate the ' principle of complementarity ' to a quasi-metaphysical status, when it should rather be regarded as a consequence of the poverty of our imagination. It may be very difficult to conceive new models, especially when it is remembered that they cannot be entirely abstract formalisms because they must be tied to the observable at some level, but difficulty does not entail logical impossibility. The number of different analogues imagined in the history of science is large and their nature is varied.[2]

Modes of Action in the Quantum Field

The question of modes of action in quantum mechanics is closely

[1] *Causality and Chance*, p. 121

[2] cf. J. A. Wheeler's comment on a new ' relative state ' formulation of quantum mechanics : ' It is difficult to make clear how decisively the " relative state " formulation drops classical concepts. One's initial unhappiness at this step can be matched but few times in history : when Newton described gravity by anything so preposterous as action at a distance ; when Maxwell described anything as natural as action at a distance in terms as unnatural as field theory ; when Einstein denied a privileged character to any coordinate system, and the whole foundations of physical measurement at first sight seemed to collapse' (' Assessment of Everett's " relative state " formulation of quantum theory ', *Rev. Mod. Phys.*, XXIX, 1957, p. 464).

connected with the uncertainty principle, for owing to that principle it is always impossible to detect instantaneous propagation, and so nothing observable compels us to adopt theoretical action at a distance. The apparent actions at a distance described by Reichenbach, for example, have only one detectable terminus, namely the ' phenomenon ', while the other terminus is by definition undetectable because it is the theoretical description of ' interphenomena '. These actions can in fact be made Lorentz-invariant, for all that is required is to describe the interphenomena in each reference frame in such a way that, for example, if a particle is detected at t in a given frame, t is also the time at which the wave-front associated with the particle in that frame collapses. Reichenbach's anomalies are peculiar in another way also, in that no energy passes, either in the ' effect ' of the slit B' on B, or in the collapse of a wave-front when a particle is detected, for if energy did pass, the complementary description of the same event would be impossible. Strictly speaking, then, these effects cannot be described as *actions* at all, since in classical physics both continuous action and action at a distance presuppose that some energy passes from one end-point to the other, the difference between the modes of action depending on whether energy can or cannot be said to be present in the intervening medium during its passage.

We may now distinguish the possible modes of action left open by orthodox quantum mechanics. Firstly, continuous action in the sense of classical field theories is forbidden by the existing formulation which takes the uncertainty principle as absolute, for we find that if we insist on describing interphenomena in terms of either particle or wave propagation alone, we get spurious actions of a discontinuous kind. Thus unless a field theory without such discontinuities proves to be possible at a sub-quantum level, fields can only be introduced into quantum theory in a *quantised* form in which continuous descriptions of action propagated from point to point are impossible. Secondly, if we attempt to escape from the causal anomalies by using particle and wave descriptions according to the complementarity principle, we are in the unsatisfactory situation of having to use both of two contradictory models, and this cannot be regarded as more than a temporary conceptual device unless we are prepared to abandon the whole standpoint that models are potentially true or false descriptions, that is, that relative to some empirical evidence, a decision can be made between contradictory models. In general, however, quantum mechanics has not adopted either of these interpretations, but a third, which Reichenbach calls *restrictive*.

According to this, quantum mechanics makes no statements at all about interphenomena, but only about *states* which are described in terms of observables and which are therefore potential phenomena, that is, a state is the product of an actual or possible measurement of the system. The notion of the complementarity of particle and wave pictures now enters only in the interpretation of these observables into experimental terms, relying on the known classical behaviour of particles and waves and the methods of detecting them, but it does not enter into any representation of the interphenomena. Interphenomena are therefore strictly undetectable, because any attempt to detect them would produce a state defined in terms of observables, that is, a phenomenon. In the restrictive interpretation the theory merely speaks about systems passing from one state into another when measurements are made, the transitions being discontinuous, and describable neither as action at a distance nor action by contact in the classical sense. The ' quantum jump ' of an electron from one energy level to another with emission or absorption of radiation now appears as a successive occupation of energy states, with corresponding creation or annihilation of photons. There is no causal action of one state on the other in the sense that an individual transition can be predicted, and there is no necessary continuity even of the electron itself. Electrons are indistinguishable from one another and therefore have no individuality, and one can envisage the process, not as a transition of one existing entity, but as annihilation of one electron in the first energy state and creation of another in the second state. A transition between states is detected in a single experimental operation, that is, there is no separate description of a particle leaving one state and then entering another, so that it is meaningless, within the restrictive interpretation, to speak of the duration or any other property of the transition. This is neither a continuous action nor an action at a distance—it is not an action at all.[1]

It does not follow, however, that continuous-action descriptions may not be possible on the macroscopic scale where the quantum of action h is negligible. Thus quantised field theories approximating to classical theories on the large scale have been developed, and these are able both to describe the interaction of particles with radiation, and to assimilate quantum theory with the special theory of relativity. The first such relativistic wave theory was developed by Dirac in 1928.[2] He then suggested a Lorentz-invariant form of

[1] cf. E. H. Hutten, ' On the principle of action by contact ', *B.J.P.S.*, II, 1951, p. 45
[2] *Proc. Roy. Soc.*(A), CXVII, 1928, p. 610 ; CXVIII, 1928, p. 351

Schrödinger's equation whose solutions represent a relativistic density, that is a 4-vector whose temporal component is the Schrödinger particle probability density $\psi^*\psi$. He showed that if $\psi^*\psi$ is to be positive, as it must be to represent a probability density, the particles involved must have angular momentum $\pm\frac{1}{2}h/2\pi$ (abbreviated 'spin $\frac{1}{2}$'), as had already been shown to be the case for electrons. This relativistic wave equation leads, however, to solutions representing particles in negative as well as positive energy states. Such solutions do exist in classical mechanics, but there they are not serious, because energy changes take place continuously, and if a positive solution is once taken to represent a state, it must always be taken, thus negative energy states can never be reached. In quantum mechanics, however, transitions take place discontinuously: an electron may 'jump' from one energy state into any other, and negative states are as accessible as positive.

Dirac overcame this difficulty by his 'hole' theory, which makes use of the exclusion principle for electrons. In the simple case where their interaction can be neglected, this principle states that there cannot be more than two electrons, each having spin of different sign, in a given energy state.[1] Now Dirac postulated that in the absence of an external field, all the negative energy states arising as solutions of his equation are occupied, so that there can be no transitions into them, and that electrons in these negative states produce no Coulomb field. If there is an external field, however, an electron may pass out of a negative into a positive state, and the effect of such a transition will be the appearance of *two* particles, namely the ordinary electron of positive energy, and a 'hole' in the negative state behaving like a particle of *positive* energy and *positive* charge (because total charge must be conserved). This second particle is, except for its charge, in all respects like an electron, and is called a *positron*. Thus arose the first suggestion of 'anti-particles', and the notion that particle-pairs may be created or annihilated, provided that mass-energy and charge are conserved.

[1] The principle follows from the fact that electrons obey Fermi-Dirac statistics, that is, a total state containing many electrons is not only not distinguishable from other states produced by mere interchange of the electrons (since electrons are indistinguishable from each other), but these states are not different individual states *but the same state*, and weighted as such for statistical purposes. Thus electrons are not indistinguishable but separate individuals, *they have no self-identity*. This has been used to support the 'operational' interpretation of quantum mechanics according to which what *we* cannot distinguish has *therefore* no separate identity. This is a sufficient interpretation, but it is not a necessary one, for the facts may be read rather as indicating the inappropriateness of any talk about particles, and hence as reinforcing the demand for new models in quantum mechanics. Pounds, shillings, and pence in a bank balance also obey Fermi-Dirac statistics, indistinguishable billiard-balls do not.

Such pair-productions have now been experimentally confirmed in considerable detail, not only for electrons, but also for the particles of the nuclear domain.[1]

The possibility of creations and annihilations of electrons and positrons with disappearance or appearance of photons, leads to important modifications in the meaning of *vacuum*, that is, space initially empty of particles. If there is sufficient energy in an electromagnetic field in vacuum, electron pairs can always be created, and hence the vacuum can be polarised, incidentally giving an unexpected justification of Maxwell's analogy between the aether and dielectrics. But this means that the notion of an electromagnetic field in vacuum in the Maxwell sense is strictly inapplicable, for, unless it is weak, the original field cannot be distinguished from the field of the resulting particles. In particular, since the number of particles in a field may be continually fluctuating, the analogue in the new formalism of Schrödinger's function $\psi^*\psi$ can no longer be interpreted as probable particle-density, but it can represent probable charge-density, for this is conserved in pair-production. Now charge-density, unlike probability-density, may have negative values, and the restriction in Dirac's relativistic wave theory to particles of spin $\frac{1}{2}$ can be dropped, and other relativistic wave equations developed which describe particles of integral spin, of which important applications are the quantised Maxwell field consisting of photons of spin 1, and also the meson fields of nuclear theory which we shall consider presently.

Meanwhile it should be noticed at this point that the ' vacuum ' in the sense of quantum field theory has introduced into physics an active field which is more clearly independent of ' observable ' matter even than the classical electromagnetic field. There is nothing startling about this for those field-theorists who, since Maxwell, have interpreted matter as itself a manifestation of a physically prior field, but it is an important counter-instance for those who wish, in spite of classical field theory, to maintain the priority of matter to field. Their argument, on the basis of classical theory, would go something like this : all the empirical data of physics are associated with ponderable matter which is found to exert physical force at a distance in a limited number of ways, namely gravitational, electric, and magnetic. No field is ever postulated in physics which cannot be traced to bodies acting in one of these ways, therefore we are not compelled to regard matter as a mere manifestation of the field, on the contrary, the field is a

[1] See Heitler, *Quantum Theory of Radiation*, 2nd ed., Oxford, 1944, pp. 187, 201–17

construction for the purpose of economically describing interactions between bodies, and has no necessary concomitant in reality. Causal activity may be effective at a distance, but is never found dissociated from bodies. The only apparent exception to this rule in classical physics was the absolute space of Newtonian mechanics, for here space itself, independently of bodies, appeared to have effects, namely the determination of inertial frames for the motion of bodies. But even this exception was in principle removed by relativity theory, since there the geometry and causal properties of space are determined by the relative positions of bodies, and so it appears that there is now no case which can undermine the positivist interpretation according to which fields are denied a reality status which is accorded to bodies. I have already argued against this positivist view by maintaining that fields must be understood as models, and that, in a carefully defined sense, models satisfying scientific criteria are intended as real, or factual, descriptions, and that therefore the usual physical interpretation of fields as more fundamental than matter is philosophically well grounded. This argument is independent of the empirical question whether physics does or does not demand the introduction of causal activity which does not presuppose the presence of matter. But the existence of the vacuum field in quantum field theory provides another argument in favour of the ' real field ' interpretation. Here a field is introduced in terms of which not only the motions of particles are described, but also their creations and annihilations, and if the positivist wishes to pursue his interpretation consistently, he has to say that bodies not only act upon one another at a distance, but also create and annihilate particles at a distance, and although it is possible to maintain this (positivism is always irrefutable), it is perhaps an additional piece of implausibility of the kind which may wear away the positivist position.

The Meson and Maxwell Fields

The most fundamental recent developments in quantum field theory have been concerned with the atomic nucleus. All nuclear phenomena make it immediately apparent that the forces here involved are not ordinary electromagnetic forces, and must be regarded as a third type of physical interaction in addition to those of gravitation and electromagnetism. Their range is very short, of the order of nuclear distances, and the binding energy of neutron and proton is very great, of the order of 100 times what would be produced by Coulomb forces acting at the same distance.

Also the energy depends on the number of particles present, not on the square of the number as with the Coulomb energy e^2/r, and this means that there is saturation at a limited number of particles when the addition of another would not increase the total force sufficiently to retain it. In this respect nuclear energies are analogous to the binding energies between atoms in a fluid, or between two H-atoms. In the latter case, if the electrons are of opposite spin they produce a saturated H_2 molecule, with energy slightly less than the sum of the energies of the separated atoms, so that the energy of the molecule is approximately proportional to the number of atoms present.

Analogies of this kind between nuclear and electromagnetic fields indicate that, in spite of their superficial dissimilarities, they are all special cases of the same fundamental theory. The general quantum field theory can in fact be applied either to the Maxwell or the nuclear field by suitable choice of field operator, which in the Maxwell case is the 4-vector $(\mathbf{A}, \phi)$, and all forces can then be represented in terms of exchanges, creations, and annihilations of the particles introduced by quantising the field. This is already clear in the case of charged particles in the quantised Maxwell field, for these emit, absorb, and exchange photons whenever there are transitions between energy states, and in a similar way, chemical valency forces can be described in terms of exchanges of electrons between atoms. Nuclear forces between protons and neutrons are represented by interchange of *mesons*, and the existence of these new particles has been amply confirmed in high-energy experiments and in cosmic radiation. Mesons have masses intermediate between those of the electron and proton, they may be neutral, or positively or negatively charged, and are unstable, that is, they have a short mean-lifetime. Although the existence of particles of this kind was first predicted by the field theory of Yukawa, the correspondence between theoretical meson fields derived by considering the simpler kinds of field operator, and meson fields as experimentally detected, is not yet very close.

An interesting result of the general field theory is that it enables not only radiation to be described in terms of particle emission and absorption, but also the interactions of source particles, for example the Coulomb, Lorentz, and Yukawa forces for Maxwell and nuclear fields respectively. Such a description introduces the notion of *virtual* exchanges of field particles, in which there is apparent short-term violation of conservation of energy by spontaneous emission and subsequent absorption of, for example, a photon, by the same or another source-particle, thus restoring the energy-balance. If this exchange takes place in a short enough time, it will be impossible to detect a small energy unbalance, since energy and time are complementary variables in the sense of the uncertainty principle. Hence these exchanges are called *virtual* as opposed to *real* emissions of radiation, in conformity with the quantum mechanical convention that 'reality' is ascribed only to what is detectable in a classical sense, that is, to radiation which can produce observable effects on photographic plates, and so on. Also, no doubt, the word 'virtual' is meant to entail that these exchanges cannot 'really' take place because they violate conservation of energy, and this is 'impossible'. But it does not seem necessary to insist on this. If the uncertainty principle is correct and ultimate, we cannot know that energy is conserved in detail because in order to measure it accurately we need a correspondingly long time, and in a dynamic process this may not be available. Its apparent conservation may be connected with the value of the constant h which sets a lower limit to observability. If, on the other hand, a theory of the sub-quantum level proves to be possible, the missing energy may be found to be connected with motion at that level, and there would then be no need for an ontological distinction between virtual and real emissions.

This is admittedly speculative, and it should be pointed out that in the present formulation of field theory the description in terms of virtual interactions is merely a pictorial representation of the first-order terms of the expression for interactions of the source particles. But since the formalism is in any case suspect because of the divergencies which are found within it, it may be the case that the first-order approximation will turn out to be more reliable than the whole expansion.[1] At any rate, let us assume for the moment that this is so, and that the ontological distinction implied by the language of 'real' and 'virtual' can be disregarded, then the picture presented of the origin of the Coulomb and radiation fields would go far towards satisfying those who since the seventeenth century have tried to express inverse-square attractions and repulsions in terms of the motions of subtle particles. The phenomena of pair-production suggest that there are continual real and virtual creations and annihilations of photons in the field, the distinction between real and virtual being essentially a distinction between photons that are *free* and those that are *bound*, for since real radiation has definite wavelength and hence momentum, there is a finite

[1] This situation arises for the classical radiation field of an accelerating charge, see *infra*, p. 280.

probability of finding it anywhere in space, whereas virtual photons are bound to a limited region of space surrounding their point of creation, since if they had time to get far, observable energy conservation would be violated. Consequently the momenta of bound photons is to the corresponding degree uncertain. It is found that if the lowest energy level of a field with static source-particles (positive or negative charges) is considered, this corresponds to a field containing only bound photons (the *vacuum field*), and that the actual energy of the virtual photon-exchanges is the Coulomb energy of the source-particles. This exchange process cannot be understood to be virtual in the sense of being merely *potential*, for clearly if A is in possession of a ball without *actually* throwing it to B, no force acts between them. 'Virtual' has always to be understood as non-energy-conserving, and therefore short-time.[1]

In a similar way, nuclear forces can be represented as virtual exchange of mesons, and here the force between like sources may be attractive, so that the theory overcomes the difficulty which Maxwell found for the case of gravitation, in accounting for gravitational *attraction* rather than *repulsion* by means of a field similar to the electromagnetic field. The reserves of energy in the vacuum field, which Maxwell found necessary but implausible, are available in the meson field. These potentialities of the general field theory naturally lead to the hope that the gravitational field can be shown to be a particular case of interaction by means of virtual particles, either particles peculiar to gravitation, or particles already recognised in electromagnetic and nuclear theory. Such a theory in flat space-time might be alternative to the replacement of gravitational force by a Riemannian metric as in general relativity.[2] Modern physics thus provides two modes of macroscopic continuous action in terms of which the attempt to unify gravitational, electromagnetic, and nuclear fields may be made, namely a generalised space comprehending quantum and nuclear as well as classical phenomena, and an account of interactions via virtual particle-productions in vacuum in a flat space; or these might be combined

[1] cf. R. E. Peierls, 'A survey of field theory, I', *Rep. Prog. Phys.*, XVIII, 1955, p. 428. The general picture of virtual interactions can also be used to describe transitions of an electron 'over' a potential barrier in terms of virtual capture of photons sufficient to give the necessary short-term energy, and in this description the radiative damping of accelerating charges and infinite self-energy of point charges appear as continual spontaneous emissions and reabsorptions of photons by the same charge.

[2] See, for example, R. H. Dicke, 'Gravitation without a principle of equivalence', *Rev. Mod. Phys.*, XXIX, 1957, p. 363. As early as 1921, H. A. Wilson suggested that if the dielectric constant (i.e. polarisability) of the aether could be regarded as variable near to matter, gravitational force, and the deflection of light in a gravitational field, might be accounted for ('An electromagnetic theory of gravitation', *Phys. Rev.*, XVII, 1921, p. 54).

in a theory containing virtual particle-productions and also satisfying the principle of general relativity, and hence requiring curved space.[1]

The Action-at-a-Distance Theory of Wheeler and Feynman

The development of quantum field theory is seriously hampered by divergence difficulties, and since quantum theory has been found only to increase the number of infinite self-energy terms which already made their appearance in the classical theory of point charges, it is clear that in its present form quantum electrodynamics is not wholly correct. It can, however, be rendered usable by various devices which consist essentially in finding a unique and self-consistent way of neglecting the infinite terms whenever they appear. But whatever modifications in the theory are found to be necessary in order to eliminate the divergencies and the unsatisfactory procedures of dealing with them which occur at present, the phenomena of pair-production have been amply confirmed, and any future theory must account for these.

Divergence difficulties have been one of the motives for some recent attempts by Wheeler and Feynman to reintroduce the conception of action at a distance. Since it can no longer be maintained that this takes place instantaneously, action at a distance is now understood to mean direct particle interaction involving no independent field with its own energy and momentum. It has been assumed by most physicists that such action would violate the conservation principles, but Wheeler and Feynman have been able to show for classical electrodynamics, following an early attempt along the same lines by Schwarzschild,[2] that this is not so if the advanced as well as retarded solutions of Maxwell's equations are taken into account. This leads to the apparently paradoxical result

[1] The use of multiply-connected spaces is suggested by Wheeler (' Geons ', *Phys. Rev.*, xcvii, 1955, p. 534). A charge would be represented by one mouth of a ' wormhole ' in the space, and lines of force would form continuous closed loops through the hole and outside it in physical space, so that the opposite end of the hole would appear to be oppositely charged. Events at one end would have retarded effects at the other, but the action would not be propagated through physical space.

[2] K. Schwarzschild, *Göttinger Nachrichten*, cxxviii, 1903, p. 132. It is a commentary on the impossibility of writing a comprehensive history of science of the kind attempted by Whittaker that, in his *History of the Theories of Aether and Electricity—The Modern Theories*, this paper is mentioned in a footnote (p. 246), together with twelve others which gave ' formulae for the electric and magnetic vectors of the field ', but nothing is said about advanced potentials. In fact Schwarzschild derived a time-symmetric expression for the force between point charges from the Liénard-Wiechert potentials, without using the field vectors. The moral for history of science would appear to be that if one's principle of selection is not made explicit it will in fact be that of interest and fashion contemporary with (or a little preceding) the historian. Wheeler and Feynman's preliminary account of their theory appeared in 1941, Whittaker's second volume in 1953.

that one event affects another at a distance r, at a time r/c seconds *before* it occurs, as well as r/c seconds afterwards, but before discussing this difficulty, let us see how the introduction of advanced potentials eliminates some of the divergencies due to self-energy.

In an Introductory Note to the paper they call 'III',[1] Wheeler describes their motive as being 'to clear the present quantum theory of those of its difficulties which have a purely classical origin'. These arise from the fact that field theory was developed before the notion of elementary point charges, and it is the notion of point charges, when imported into a field theory initially alien to it, which has led to the infinite self-energy difficulties. The phenomenon of radiation self-damping, for example, leads to problems when the velocity of a particle is not small compared with c, and Wheeler and Feynman take this as the chief application of their theory in Paper III.

In classical electrodynamics an accelerated charge emits radiation and simultaneously loses energy. The resulting damped motion is described by superimposing on the forces causing the original acceleration, a force of radiative reaction whose value is well defined and experimentally confirmed when the charge is moving slowly, but for which no satisfactory general expression is available. Dirac has suggested that this reaction, which he calls the radiation field, may be regarded quantitatively as the result of taking half the difference between the advanced and retarded fields of the accelerating particle, the advanced field representing a spherical wave converging from infinity, and the retarded field a wave diverging to infinity, and evaluating this quantity at the particle. The result is finite everywhere, it agrees with the known expression for the radiative reaction at low speeds, and it treats the particle as a point charge. Wheeler and Feynman accept this as a reasonable suggestion, but as it stands it is clearly somewhat *ad hoc*. In terms of classical field theory no account can be given of the origin of the field converging on the particle from infinity, and if on the other hand we take the standpoint of action at a distance as in Schwarzschild's theory of advanced potentials, it is still necessary to account for the origin of the field, and the account must be in terms of particles other than the accelerating particle, for on this view fields themselves have no energy. Thus Dirac's radiation field cannot immediately be translated 'into the language of action at a distance',

[1] 'Interaction with the absorber as the mechanism of radiation', *Rev. Mod. Phys.*, XVII, 1945, p. 157 (referred to below as 'III'). This paper appeared before 'II', which is 'Classical electrodynamics in terms of direct interparticle action', *Rev. Mod. Phys.*, XXI, 1949, p. 425, and 'I' has not appeared.

and it is Wheeler and Feynman's belief that field and action-at-a-distance representations should be intertranslatable, and should be regarded as 'complementary tools in the description of nature', for they suspect that both lead in different ways to divergencies, and should therefore be used alternately, as are the particle and wave pictures of ordinary quantum theory. Wheeler concludes in a note that 'the union of action at a distance and field theory constitutes the natural and self-consistent generalisation of Newtonian mechanics and the four-dimensional space of Lorentz and Einstein'.[1] Talk of complementarity here does not seem to be any more satisfactory from a logical point of view than in the particle-wave context, and the action-at-a-distance theories, like those of the immediate post-Maxwell era which we have already described, have an air of being parasitic upon the progress of field theories. However, the development of an action-at-a-distance representation is in any case of interest, whether or not it is intertranslatable with field theory, or leads necessarily to unwanted divergencies.[2]

Wheeler and Feynman supplement the radiation field of Dirac by a suggestion first made by Tetrode in 1922, in which one of Faraday's criteria for action at a distance is revived, namely that if radiation is an action at a distance it must depend on a receiver as well as a source. Tetrode wrote:

> 'The sun would not radiate if it were alone in space and no other bodies could absorb its radiation. . . . If for example I observed through my telescope yesterday evening that star which let us say is 100 light years away . . . the star or individual atoms of it knew already 100 years ago that I, who then did not even exist, would view it yesterday evening at such and such a time.'[3]

Wheeler and Feynman go on to discuss models in which all or almost all radiation in a space is absorbed by particles near the boundary of that space, and which are such that no awkward paradoxical effects arising from advanced potentials are observable. Their theory makes the following assumptions:

[1] III, pp. 157*n*, 159 and II, p. 426. The suggestion that the antithesis between action at a distance and action by contact may be regarded rather as a complementarity is made by W. H. McCrea, 'Action at a distance', *Philosophy*, XXVII, 1952, pp. 70ff.

[2] In an attempt to see what a quantum theory of action at a distance would look like, H. J. Groenewold suggests that developments which are not translatable into quantum field theory may be possible, in which case the action-at-a-distance theory would become of fundamental importance, but he concludes his paper by expressing the fear that such a theory would in general be intractable ('Unitary Quantum Electron Dynamics', *Koninklijke Nederlandsche Akademie van Wetenschappen*, LII, 1949, p. 3).

[3] *Zeits. für Phys.*, x, 1922, pp. 325, 326

1. An accelerated point charge in otherwise charge-free space does not radiate.

2. The fields which act on a given particle arise only from other particles.

3. These fields are represented by one-half the retarded plus one-half the advanced Liénard-Wiechert solutions of Maxwell's equations.

4. Sufficiently many particles are present to absorb completely the radiation given off by the source.

Then if we imagine a charge in a completely absorbing system, its acceleration generates a disturbance, the retarded part of which sets the absorber particles in motion so that their *advanced* fields converge upon the source-particle at the moment of its acceleration. The total value of this field is equal to Dirac's radiation field, thus providing an explanation of the radiative reaction on the source, and at other points the radiation field combines with the half-retarded, half-advanced field of the source to give the full retarded field of experience. Thus there is translatability in this case between the action-at-a-distance and orthodox retarded field theories, the difference being that, once the postulates of the former are accepted, it gives a much simpler description.

It seems, however, that there must be some features present in a theory of advanced fields which are not only not present in the orthodox theory, but which are so paradoxical as to constitute a decisive objection to any theory. First, it seems as though we must become involved in an infinite regress, for, to take the example described above, the *advanced* field of the source-particle must cause the absorbers to emit an *advanced* field which arrives at the source *before* the source starts to accelerate, and hence affects its motion before the initial impulse is given to it. This objection is answered by Wheeler and Feynman's analysis of a completely absorbing system, in which they show that the advanced field of the source is exactly cancelled by the advanced field of the absorber. If the absorber is not complete, however, there are some advance effects, but it is plausible to suppose that these are not serious, since they will in practice become attenuated as we go backwards in time, just as the ordinary retarded effects are attenuated as we go forwards. Hence we have to picture an event as surrounded symmetrically in time by a succession of advanced effects increasing in magnitude

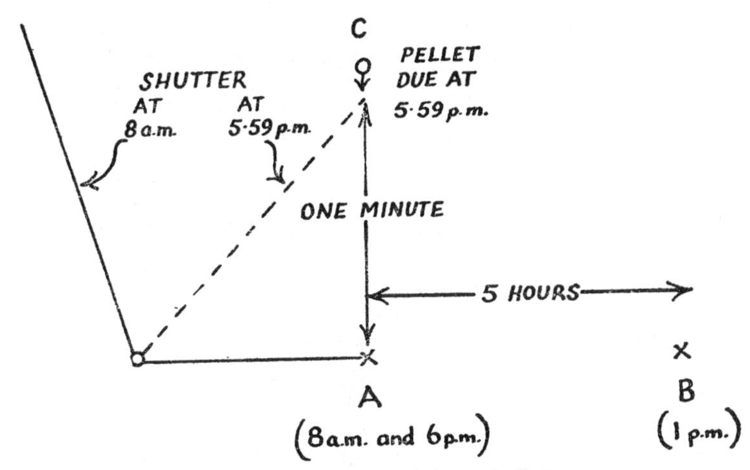

FIG. 9 The paradox of advanced effects

up to the event, and then of retarded effects decreasing and finally disappearing.[1]

The next difficulty seems to be that if there are any back effects which are detectable in principle, we shall be in the paradoxical position of being able to prevent the cause occurring as soon as we have observed its prior effect, although if the cause does not in fact occur, the effect should not have occurred previously. Wheeler and Feynman attempt to resolve a paradox of this kind [2] by postulating sufficient attenuation of back effects, and also that physical effects are continuous, and involve, according to their theory, continuous reciprocal interaction of past and future. Consider charged particles A and B in otherwise charge-free space and at a distance of 5 light-hours apart. Suppose a mechanism is set to hit A with a pellet at 6 p.m. Then A starts to move at 6 p.m. and by advanced effect accelerates B at 1 p.m., which in turn accelerates A, although with diminishing effect, at 8 a.m. Observing the motion of A at 8 p.m. can we not prevent the arrival of the pellet at 6 p.m.? But then why did A accelerate at 8 a.m.? The suggested consistent solution of the paradox is that since discontinuous forces are not found in nature, the intervention must itself be continuous with the rest of the process, and human agencies must be eliminated. Suppose a mechanism is arranged so that the motion of A at 8 a.m. starts a shutter moving continuously to reach the path of the pellet near

[1] III, pp. 165, 175 [2] II, p. 427

A just before 6 p.m. Then at 6 p.m. the pellet will strike the edge of the shutter a glancing blow and then strike *A* with sufficient force to cause the advanced effect to start the shutter moving at 8 a.m.

Wheeler and Feynman claim that this solution resolves the paradox, but this is not so. All that has been done is to show that there are self-consistent states of affairs involving advanced effects, but if a situation were arranged deliberately to create the paradox it could not be resolved in this way. Suppose for example that the shutter moved in such a way as to block the path of the pellet from soon after 8 a.m. until after 6 p.m. This need not involve any discontinuities, but would effectively generate the paradox, for suppose the pellet to have started moving along the path *CA* at 5.30 p.m., then either it hits *A* at 6 p.m. or it does not. If it does, its advanced effect will have brought the shutter right across its path before 5.30, and therefore it cannot hit *A*. If it does not, there will have been no advanced effect to start the shutter moving, the way to *A* is clear, and it does hit *A*. If Wheeler and Feynman's mechanism can be set up, so can this one, and there is no solution in terms of classical physics. We shall consider in a moment whether there may be a solution in terms of quantum physics.

There is one effect in the theory of Wheeler and Feynman which must be expected to show up the advanced potentials even in a completely absorbing universe if the theory is correct. This effect was first pointed out by Dirac, and follows from his formal device of using advanced potentials to calculate the radiation field on an accelerating particle. It is found that if a pulse of radiation is sent towards a particle initially at rest, the particle experiences a *pre-acceleration*, that is, it starts to move before radiation travelling at speed c would have arrived, the time-interval being of the order of time required for radiation to cross a region e^2/mc^2 around the charge. Dirac suggested that this effect might be due to the transmission exceeding the velocity of light within this region, thus indicating a failure ' of some of the elementary properties of space-time ',[1] as if the electron had after all to be regarded as occupying a finite region where it transmits disturbances instantaneously like a classical rigid body. Wheeler and Feynman consider that it is more satisfactory to ascribe pre-acceleration to the explicit effect

[1] ' Classical theory of radiating electrons ', *Proc. Roy. Soc.*(A), CLXVII, 1938, p. 160. Dirac has since developed a new type of theory in which self-interaction problems do not arise for classical electrons and hence no pre-acceleration is introduced (' A new classical theory of electrons ', *Proc. Roy. Soc.* A, CCIX, 1951, p. 291 ; CCXII, 1952, p. 330 ; CCXXIII, 1954, p. 438).

of the advanced potential of the radiation reaching the particle before the arrival of the impulse itself, so that the cleanness of the cut between past and future 'is limited to times of the order of e^2/mc^3 or greater'. They go on:

> 'Those phenomena which take place in times shorter than this figure require us to recognise the complete interdependence of past and future in nature, an interdependence due to an elementary law of interaction between particles which is perfectly symmetrical between advanced and retarded fields.' [1]

Reversibility of Cause and Effect

Novelties of this kind in the view of causation which is implied or suggested by modern physics raise the important question as to how far science can modify common sense in this fundamental respect and still remain consistent and intelligible. It must first be remarked that scientific theory in general does not presuppose any particular mode of causal connection between events, but only that it is possible to find laws and hypotheses, expressed in terms of some model, which satisfy the criteria of intelligibility, confirmation and falsifiability. The mode of causal connection in each case is shown by the model, and changes with fundamental change of model. The requirements for a mode of causality are thus roughly those of coherence and correspondence: the causal behaviour of the model must not be self-contradictory, and it must correspond in observable situations to those common-sense notions of causality which are empirically demonstrable. Thus causality acting backwards in time cannot be ruled out without investigation.

Mr A. E. Dummett has maintained [2] that quasi-causal sequences may be said to go from a later to an earlier event, if they satisfy four conditions, which may be expressed as follows:

1. There is repeated concomitance of the two events.

2. There is no discoverable causal explanation of the earlier event in terms of simultaneous or preceding events.

3. There is no discoverable causal connection between the events; for example they must not be effects of the same remote and earlier cause.

[1] III, p. 181
[2] 'Can an effect precede its cause?', *Aris. Soc. Supp. Vol.* XXVIII, 1954, p. 27

4. There is a satisfactory causal account of the later event which does not mention the earlier.

But Dummett suggests that such quasi-causality is never explanatory, because explanations must be in terms of mechanisms whose causality is always forwards in time, and for this reason he thinks that there could be no science of any paranormal phenomena, such as those of precognition, which seem to reveal a backwards causality. This, however, is to hold that only one kind of causal model, namely that of classical physics, can be scientific, and in view of the necessary modifications of classical causality which have already taken place in modern physics, it is not possible to maintain this. The question remains, however, whether backwards causality and other non-classical concepts can be made self-consistent.

Four types of violation of classical physical laws and principles of causality have been mentioned in connection with quantum theory and the theory of Wheeler and Feynman :

(*a*) Non-conservation of energy in virtual transitions.

(*b*) Impossibility of a causal account of quantum transitions.

(*c*) Reversal of time-order of cause and effect implied by advanced potentials.

(*d*) Apparent breakdown of causal order if speeds of propagation of action exceed the velocity of light.

It is necessary to inquire in each case, firstly, whether there are inherent self-contradictions in these notions, and secondly, how far the freedom given by the existence of uncertainty limits is required in order to reconcile them with experience.

The first type (*a*) is an example of violation of macroscopic physical laws within the limits of the uncertainty principle, and if experimental evidence of an indirect kind suggests it, then it raises no necessary difficulties either of internal consistency or of correspondence. The remaining points involve violations of classical causality which appear more radical than violation of particular physical laws, but even so, in the case of (*b*) mere unpredictability is not in itself paradoxical, nor, in the situations where it is asserted, contrary to experience.

The case is different with (*c*) and (*d*). Let us assume that advanced potential theory is not wholly translatable into retarded field theory, that there are therefore some explicit advance effects, as for example pre-acceleration, and that these are detectable.

Then the only escape from the possibility of paradoxical consequences would be to assume that the advanced effect occurs within the uncertainty limits in such a way that no macro-mechanism (including a human agent) could be made to detect it in time to prevent the later occurrence of the 'cause', and although this might be expected to be the case for phenomena involving individual micro-particles, the effect of pre-acceleration has not yet been studied in the context of quantum theory and no definite answer can be given. All that can be said is that the absoluteness of the uncertainty principle seems to be necessary in order that advanced effects shall not lead to paradoxes, for if a paradox can even be described within a model, without necessarily being practicably realisable, that model is internally inconsistent. Thus classical physics cannot contain explicit advanced effects, but quantum physics may do so, because the paradoxes involved require detection and intervention on a micro-scale, and it is to these that the uncertainty principle sets absolute limits.

A consequence of the absoluteness of the uncertainty principle would be, however, that no description in terms of pre-acceleration could be wholly realistic, for as we have seen, we cannot give *any* realistic description of events within uncertainty limits, and if a description is attempted, we fall into a complementarity dualism which shows that the models being used are alternative conceptual devices only. The only circumstances under which pre-acceleration could be a realistic description would be that, first, an acceptable sub-quantum description involving it were developed, and also there were still absolute limits to intervention. Such a situation would be similar to that in respect of causality if it were shown that after all causal sub-quantum systems could be postulated and described, but that we could in principle, because of an analogue of quantum uncertainty, never obtain sufficient information to make actual predictions.

Similar considerations regarding the reversal of cause and effect apply to point (*d*). If in one reference frame K, an event A occurs before B and is described as the cause of B in the sense of classical causality although the speed of transmission of action from A to B exceeds c, while in another frame K', B occurs before A, then an agent in K' could intervene (event C) to prevent the occurrence of A in both frames. In K the order of events would be A (which both occurs and does not occur), C, B (which both must have occurred if A occurred and cannot have occurred if A did not occur). We have to assume here that signals reaching the agent and his own actions are also transmitted at speeds exceeding c, but if there is any macro-effect where this is possible, it is reasonable to assume that it could be utilised in such a mechanism for intervention. Thus this supposition leads to a paradox unresolvable within classical physics.

Here again, however, the supposed causal action can be postulated to occur within the uncertainty limits in any given frame, for in that case it would be part of a description of interphenomena, and at most one of the pair of events A, B would by definition be undetectable (not a phenomenon). The uncertainty limits are Lorentz-invariant, and hence events undetectable in one frame are undetectable in all frames. If B is undetectable and A is detectable, the causal paradox cannot arise in any case, because the agent in K' will not know that B has occurred until A occurs and hence will not know when to intervene. B will be regarded as an advanced effect in K', but this need not be inconsistent with relativistic invariance if advanced effects are admitted at all. If, however, A, which is the so-called ' cause ' in K, is undetectable, and B, the ' effect ', is detectable, then it seems that the class of such causes that can be postulated is limited by the requirement that it must not be possible to do anything to prevent their occurrence. Consider Dirac's suggestion about pre-acceleration. This envisages an impulse travelling in the neighbourhood of an electron with a speed greater than c. Suppose the cause, namely the arrival of the impulse in this neighbourhood, say at time t_1, is interphenomenal, and that the effect (now supposed to occur later than t_1, say at t_2, in K) is phenomenal (detected by emission of a photon for example). In a frame K' moving with suitable velocity with respect to K, the photon will be detected at t_2', *before* the instant t_1' which is the transform of t_1. But as in the case of the advanced-potential explanation of pre-acceleration, it is likely that an account could be given such that no pre-arranged mechanism nor human action could be triggered off at t_2' to block the arrival of the impulse at t_1'. Again, however, unless the uncertainty principle is merely a limitation on intervention and not an absolute limitation on interphenomenal description, such an account would be conceptual only.

An unusual light is thrown on the nature of the uncertainty principle by this discussion. Just as the principle of operational meaning can be seen to be, not a restriction on theoretical description, but a charter of theoretical freedom, because anything may be said theoretically which is not operationally impossible; so the uncertainty principle, if in some sense absolute, can be seen to confer considerable freedom upon theoretical speculation. Neither explicit advanced effects nor transmissions exceeding the velocity of light are possible unless they can be described inside the uncertainty limits, and there may be other such unusual but useful conceptions to which the same freedom is given : for example, the idea of minimal lengths and time-intervals has been suggested in quantum theory, but these must never be introduced in such a way as to fall into Zeno's paradoxes about indivisibles. Here again paradoxical effects could be rendered unrealisable by the uncertainty limits. These considerations are strong arguments for the convenience of the uncertainty principle, or, if we are interested in reality rather than convenience, they are grounds for the fear that if it is shown to be possible both to describe and to manipulate the sub-quantum level, then, without the freedom of theoretical conception conferred by uncertainty, such description may prove to be intractable.

Chapter XI

THE METAPHYSICAL FRAMEWORK OF PHYSICS

Some Heuristic and Metaphysical Considerations

VARIOUS strands are woven into the history of the concept of action at a distance: psychological, heuristic, metaphysical, and scientific; and in the use of the concept in physics they will no doubt always remain confused. There has undoubtedly been a general preference for continuous-action theories of some kind, and this may be put down partly to psychological or heuristic considerations. Metaphysical arguments have also been used on both sides, but in so far as scientific opinion since the seventeenth century has tended to come to agreement on the mode of action to be assumed, and to persist in that agreement over periods of time long compared to the usual lifetime of particular theories, it seems to have done so primarily on scientific grounds. But this, as our historical survey has shown, is not a straightforward scientific question admitting of direct experimental test.

We have seen that in the early stages of science, fundamental models are derived from familiar processes, and these exhibit both action by contact and action at a distance. In his investigation of notions of causality among young children, Piaget finds both ideas present: it is thought that the sun pursues the clouds, which are therefore acting upon the sun at a distance, but on other occasions the clouds are thought to drive the sun before them by their breath. Piaget remarks:

> 'The antinomy is analogous to that which we find in primitive peoples. Everything acts upon everything else, but everything is omnipresent. In point of fact the question does not arise for the child, or at any rate, it takes quite a different form. For in one way the action of the motor on the moving object is of a psychological order: the external force commands or arouses a desire or a fear. In this sense, it acts at a distance. On the other hand, the actions are for the most part accompanied by material transferences, and in this sense, there is contact. Thus to the extent that there is action at a distance the explanation is psychological, and to the extent that there is contact, the explanation is physical. But originally—and this is why one is apt to be

misled—the two activities are not differentiated, because of the lack of any precise boundaries between thought and things, between the ego and the external world.' [1]

The preference for action-by-contact theories in physics was historically connected with the objectification and depersonalisation of nature and the desire to eliminate from explanations of it the 'psychological' analogies of organism, command, and attraction, in favour of the analogy of mechanism, and it was a fact that most familiar mechanical devices acted by contact.

Such considerations may help to explain the deep-seated intuition of what is possible which forms a climate of thought favourable to continuous action. But fashionable climates of thought are apt to be rationalised in metaphysical systems, and these raise the further question as to whether there is anything necessary *a priori* about the dictum that 'matter cannot act where it is not'. Our discussions of those metaphysical systems in which the assertion is made, namely those of Aristotle, Descartes, and Leibniz, suggest that the dictum is either empirically falsifiable, if its terms are defined sufficiently precisely, or it functions as a directive to look for, or perhaps merely to postulate, a medium which is material *in some sense*, wherever action appears to make jumps. This brings us to the heuristic aspect of the question.

There is no doubt that apparent assertions about action at a distance and continuous action are used as regulative principles from time to time in physics, in the form 'Do not postulate unobservable intermediate entities' by Ockham, Newton (in the *Principia*), and most subsequent positivists down to Bohr, and on the other hand 'Always look for continuously acting causes', by the atomists, Descartes, Newton (not in the *Principia*), Faraday, and Bohm. At first sight it looks as though the second directive will encourage the construction of more fruitful models, because models conforming to it will have to contain descriptions of 'interphenomena' as well as 'phenomena', and in general (that is, in theories not governed by the uncertainty principle), it is eventually possible to devise further experiments to detect the interphenomena predicted. Continuous action therefore appears to be more powerful as a predictive model, and to make more claims upon the facts, and this would seem to explain why action at a distance has usually been associated with positivist views of theory, where it is desired to assert no more than is already justified by experiment. But further

[1] J. Piaget, *The Child's Conception of Causality*, trans. M. Gabain, London, 1930, p. 119

consideration of historical examples suggests that this is too simple a view of the matter. It does not explain why the model of attractions and repulsions at a distance was useful throughout the eighteenth century, nor why the recent theory of Wheeler and Feynman is fertile in revolutionary suggestions, even though these may turn out to be unacceptable. There is no simple equation between action at a distance and positivism, We can in fact distinguish two kinds of action-at-a-distance theory, depending on whether the motive for its development is suspicion of the hypothetical, or simply a desire to find alternatives to continuous-action models.

Into the first category would come the suggestions of Ockham, the Newton of the *Principia*, Ampère, as well as many of the Continental school down to Schwarzschild. Into the other category, and often drawing inspiration and details of development from these, come the Newton of the *Optics* and his eighteenth-century successors including Kant, and Weber, Tetrode, Wheeler, and Feynman. The eighteenth-century model of attracting and repelling particles was as much a speculative construction as its Cartesian or Democritan predecessors. There is confirmation of this in the 'positivist' protests against it made by Lavoisier, Fourier, and others, and in its replacement by mathematical theories merely descriptive of phenomena at the beginning of the nineteenth century, and also in the apparently curious preference of the elastic fluid theorists for action which jumps only across *short* distances. Gravitational action without a medium at solar distances was rarely acceptable, and on the other hand Faraday and Maxwell were at times prepared to accept action at insensible distances within a general framework of continuity. Thus action at a distance within the framework of a general positivist position may be regarded with almost metaphysical fervour, but as a speculative model, alternative to continuous action, it may not be entirely incompatible with the latter, or rather, the ultimate decision between them may be left open, without affecting the heuristic usefulness of the model.

Action-at-a-distance theories are therefore not necessarily lacking in the content which makes good models. So they may, like continuous-action theories, become candidates for reality in so far as it is accepted that theories are in intention factual descriptions and not merely heuristic devices. But theories which, when taken as a whole, are factual in the narrower sense of having falsifiable consequences, may still contain elements which are individually unfalsifiable, and this appears to be the case with their assumptions about modes of action.

Action at a distance and contact action form one of those pairs of apparently contradictory principles which continually reappear in the history of science : atomicity and continuity, mechanism and vitalism, determinism and freedom, causation and teleology, body and mind. Unlike some of these, however, the dispute about action at a distance does not now raise extra-physical problems bearing on the nature of life and mind, although, as we have seen, this has not always been the case, for prior to the eighteenth century the problem was closely associated with some of these other antitheses, and was itself a dispute about the place of animistic explanations in physics. And it may be that its obvious connection with the phenomena of extra-sensory perception will in the near future raise all these problems again. But in the meanwhile, just because the relevance of the dispute is almost wholly in physics, it is perhaps possible to see more clearly what are the essential scientific features of these antitheses without becoming involved in other questions of greater complexity.

There is clearly a sense in which the dispute can be said to be a metaphysical one, for it shows all the signs of progressive withdrawal from tentative empirical positions as soon as they are falsified, which characterise the metaphysical assertion. Matter acts only by contact, so, faced with matter attracting at a distance with no apparent material medium, subtle matter of dubious status and properties has to be postulated. All attempts to describe these precisely are refuted. So the medium is described in terms of stresses and tensions, in such a way that energy is the only material property which is located in it, and this is said to show that action is after all continuous. Again, although it is shown as conclusively as is possible in science that action can only be propagated with a finite velocity, this is not taken as a final refutation of action at a distance, for the antithesis turns up again within the framework of finite velocities, and presumably it will continue indefinitely to take new shapes as new theories are developed. Clearly no single or simple assertion of any kind, metaphysical or empirical, is being made when it is said that matter can, or cannot, act at a distance.

In their most general form, statements asserting ultimate action at a distance or continuous action are unconfirmable and unfalsifiable.[1] The thesis that ' Every action between bodies is made up of spatially indivisible intervals ' cannot be confirmed, because, given two bodies apparently acting across empty space, there may

[1] Statements of this kind are discussed by J. W. N. Watkins, ' Confirmable and Influential Metaphysics ', *Mind*, LXVII, 1958, p. 344.

either be practically unobservable activity going on within it, or the meaning of continuous action may be changed in such a way that the intervals become divisible, yielding parts in which the newly defined action is still found. On the other hand the thesis cannot be falsified because every apparently continuous action may turn out to be composed of smaller indivisible parts. It follows that the antithesis that 'Every action between bodies takes place by a continuous action through the intervening space' can be neither falsified nor confirmed.

These conclusions depend essentially upon the vagueness of the principles when stated in this form. We are not told how to recognise *true* action at a distance or *true* continuous action, and so any apparent instance of either cannot be taken as a confirmation. The principles may, however, be formulated a little more precisely in order to make one or other of them if not falsifiable at least confirmable. For instance 'In all apparent continuous action there are some hidden discontinuities', or 'In all apparent actions at a distance there are hidden chains of continuous causes'. In this formulation, any discovery of gaps in what are *now* taken to be continuous actions would be a confirmation of the first principle, and any discovery of continuities in what is *now* taken to be action at a distance would be a confirmation of the second, and both might at different times function as an incentive to research, although neither could be directly falsified, because failure to find the gaps or the chains could never disprove their existence. Watkins has argued that since 'all-and-some' statements of this kind are confirmable, and are weak entailments of falsifiable statements such as 'All apparently continuous actions are composed of quantum jumps', or 'All apparent actions at a distance are transmitted by an electromagnetic field', then one may regard them as true-or-false, although they are not falsifiable, and share other features with typically metaphysical statements.

No such formal criteria will serve, however, to distinguish between those metaphysical assertions which are scientifically acceptable and those which are not. Their acceptance or non-acceptance will always depend upon what fundamental model is for the time being regarded as explanatory, and we have seen that the criteria of choice of fundamental model are very various, ranging from conformity with received metaphysical systems, to more directly empirical considerations such as correspondence with theories already confirmed over a limited domain, or the comparative simplicity of theories based on different fundamental models. The progress of science in the direction of greater universality, consistency, and simplicity of its theories, has the effect of undermining hitherto unquestioned assumptions about fundamental models, and also of removing prejudices which stand in the way of development of new and unconventional speculative systems. The problem of theory-construction is not that of excluding metaphysics, but of finding a sufficiently comprehensive framework of concepts within which specific theories can be proposed and tested.

Theoretical Aspects of Extrasensory Perception

It is impossible to close a study of action at a distance without a brief consideration of a field of investigation where the causal and action axioms of physics, as so far discussed, appear to be most seriously challenged. I refer to the field of extrasensory perception or parapsychology, or, to use the more modern and theory-neutral terminology, ψ-phenomena. This is not the place for an inquiry into the empirical credentials of these phenomena, for in any case such inquiries can be found in abundance elsewhere,[1] but it is relevant to ask, if ψ-phenomena are accepted as facts, what is their bearing on this discussion of modes of action? There is, however, one point about their empirical significance that ought first to be mentioned.

Research into ψ-phenomena is concerned both with spontaneous individual occurrences of an ' occult ' nature, and with deviations from chance expectation in card-guessing experiments and the like, and it is the latter type of research which has gained for parapsychology such scientific recognition as it is now accorded, for obviously it is only here that the possibility of controlled experimentation exists. Now a ground of criticism of the ψ-interpretation of these experiments has been their statistical nature. Leaving aside radical criticisms which, if justified, would also invalidate use of statistics in fields other than parapsychology, there is also the objection that since *any* series of random guesses can be made to show significant deviations from chance expectation if treated in *some* way, the deviations of apparently astronomical magnitude obtained in card-guessing experiments can have no significance, however scrupulously they are carried out. This objection seems to be based on a view of the use of statistics which would imply that *no* statistical theory in science is falsifiable, since the way in which the series of guesses is examined for significant deviation is controlled by expectations implied in existing scientific theory and

[1] See particularly *Proceedings* and *Journal of the Society for Psychical Research*, *Journal of Parapsychology*, and works by J. B. Rhine and S. G. Soal.

practice. The guessing series is correlated, not with any arbitrary sequence, but with the actual order of the cards, with at most an allowed time-displacement of two or three cards, that is, the guess as card N is turned up is sometimes compared with the actual card $N \pm 2$. Even allowing for this, and for the fact that large numbers of tested subjects are not found to show significant deviations from chance guessing and are discarded from the experiments, the odds against the runs of correct guesses which are obtained are still found to be astronomical. The method of calculating odds is determined by expectations based on existing physical theory, which presupposes that uncontrollable events (shuffled cards) are distributed randomly, and that odds of millions to one against chance do not happen by chance. If the latter assumption were not made, no statistical hypothesis could ever be refuted, for since such a hypothesis cannot *forbid* the occurrence of rare events, refutation can only be said to take place when odds of an arbitrarily fixed amount are exceeded. There is a finite probability on the dynamical theory of gases, for example, that all the molecules of oxygen in the atmosphere will congregate in one region of space, leaving the rest of us gasping in an atmosphere of nitrogen, but if this were to occur it would certainly be taken to be due to the operation of external causes, or, if that appeared impossible, it would be taken to refute the dynamical theory.

If, then, it is accepted that the ψ-experiments have been conducted in such a way as to eliminate every ordinary physical explanation of the deviations from chance obtained, and that the deviations are sufficient to constitute a refutation of existing statistical assumptions derived from physics, it must next be inquired what are the minimum modifications required in the assumptions of physics in order to accommodate these phenomena. Since the feature which distinguishes ψ-experiments from other situations, in which ordinary assumptions of randomness seem to work, is the presence of a human subject guessing at information which is withheld from him by all normal means, it is inevitable that the correlations obtained should be described as some kind of communication of the facts to the subject by abnormal means. Traditionally these abnormal means have been described as telepathy (abnormal communication from mind to mind), clairvoyance (abnormal perception of events or things), and precognition (temporally prior knowledge of events or things). More recently use of these terms, together with the descriptions ' extrasensory perception ' and ' parapsychology ', has come under fire, on the grounds that they imply peculiar sorts of knowing or perceiving, whereas the phenomena indicate that any analogy with normal knowing or perceiving may be misleading. Also it is often not possible to distinguish between telepathy, clairvoyance, and precognition in experimental situations, for there may be alternative interpretations of any given results. For example, if precognition of cards to be turned up at a later time appears to be shown, this may also be interpreted as clairvoyance of the cards in the pack, or as telepathic communication (with or without precognition) from the experimenter, who must at some time be himself aware of the actual card order. If psycho-kinesis (ability to influence the motion of objects by ' willing ') is also accepted as possible, experimental situations become even more difficult to interpret in the traditional terminology. These considerations have led to adoption of the theory-neutral terminology ' ψ ', with the suffix ' γ ' referring to communication phenomena, and ' κ ' referring to the (still problematic) kinetic phenomena.

If the communication hypothesis is accepted, certain of the hitherto assumed axioms of physical science appear to be violated. Professor Broad has listed the following ' Basic Limiting Principles '[1] which he claims are commonly accepted in the Western scientific community, and with which ψ-phenomena seem *prima facie* to conflict :

1.1 No event can begin to have effects before it has happened.

1.2 No event can affect an event at a later time unless the intervening time is occupied by events initiated by the earlier event and causing or contributing to the later event.

1.3 No event can affect another event at a remote place unless a finite time elapses and there is a spatially continuous chain of events between them.

2 It is impossible for an event in a person's mind to produce directly any change in the material world except certain changes in his own brain.

3 A necessary immediate condition of any mental event is an event in the brain of a living body.

4.1 It is impossible for a person to perceive a physical event except by means of sensations produced in his mind via a continuous causal chain from the event to his brain.

[1] ' The Relevance of Psychical Research to Philosophy ', *Philosophy*, XXIV, 1949, p. 291

4.2 *A* can only know what experiences *B* is having by reading, hearing or seeing information given by *B*.

4.3 A person cannot forecast an event except by inference from present data, together with a knowledge of natural law, or by expectations based on habit.

4.4 A person cannot have knowledge of past events except by memory, record or testimony, or by retrodiction based on present data and natural law.

The psycho-physical dualism implied by Broad's statement of Principles (2)–(4) is no doubt a more or less accurate rendering of the views of non-philosophical persons, but, as Broad himself would be the first to admit, they raise some very difficult philosophical issues. It is not necessary to enter into these here, however, because on the level of physical causation it is not clear what it would be to deny that all communication to and from minds takes place via physical changes in brains. These principles reduce in fact to the unfalsifiable 'all-and-some' statement: 'For every mental event there is a corresponding physical event in a brain', and if investigation failed to find one, it would always be open to the investigator to say that it was an event of an unknown kind which he did not yet know how to detect, or that it was technically impossible to detect it with existing instruments. We are here concerned, however, with Principles (1), which assert the priority of cause to effect and the spatio-temporal continuity of causal processes.

First, let us inquire whether Broad is right in claiming that ψ-phenomena appear to conflict with his Principles (1). The answer will depend on one's assessment of the experimental evidence, but, with regard to (1.1), it seems that if the statistical method is accepted at all there is statistical justification for postulating precognition of cards not yet turned up in experiments which are carefully designed to eliminate other normal and paranormal interpretations, and there are also, of course, the spontaneous ostensibly paranormal experiences of 'foreseeing the future' in dreams and waking visions, although the significance of these is at present difficult to assess. In view of the paradoxes implied by time-reversals of causality in physics, it is interesting to note that some kind of test of the occurrence of true precognition (that is, cognition *of* a future event) would be provided by studying the possibility of intervention to prevent an apparently precognised event. Such a study would be impracticable in card-guessing experiments in view of the statistical nature of the results and the fact that the subject does not know which are his ' good ' guesses, but it has been carried out on a large number of reported spontaneous cases by L. E. Rhine.[1] Conclusive evidence of successful intervention is in the nature of the case almost impossible to obtain, because if the ' forecast ' were unfulfilled, it would not be noticed as a ψ-phenomenon, and Mrs Rhine was only able to find a small number of cases in which successful intervention might be said to have occurred, and even in these, alternative explanations were possible. But if a case of successful intervention could be positively established, by showing that without deliberate intervention on the part of someone knowing of the forecast, the forecast would have been fulfilled, this would suggest that there is no true precognition *of* the future, but only some sort of ' warning ' from the future not implying strict causality, or an unconscious inference from present data. If, on the other hand, true precognition did take place, it would seem to imply the impossibility in principle of successful intervention, and hence the existence of far-reaching limitations on human freedom, including, perhaps, the occurrence of self-fulfilling predictions of the Oedipus type.

With regard to Principles (1.2) and (1.3), since they are of the unfalsifiable kind we have already discussed, they cannot be *proved* to be violated, for failure to find continuous causal chains may always mean that we have not looked hard enough or in the right way. No physical trace has for example been found to bridge the time-gap for memory, but we are convinced that such must exist. But various considerations make it *implausible* to suggest that ψ-phenomena should be explained by a theory which satisfies the principles. The main reason for thinking this is that ψ-phenomena appear to be indifferent to time and space. Soal, for example, has carried out experiments over 200 miles which showed no significantly different results from those carried out over 20 feet, and the pre- and post-cognition of cards suggests that whatever process may be at work, it is indifferent to time intervals, forwards and backwards, at least of small amounts. The isolated spontaneous cases of ψ, if veridical, show of course even more spectacular indifference to time and space. Broad holds that it is implausible to postulate a field-type theory for non-precognitive telepathy or clairvoyance involving continuity of action spreading in a finite time, since this would appear to involve attenuation with distance, which is not found in ψ-phenomena, and it would not cover the

[1] ' Precognition and Intervention ', *Journ. Parapsychology*, XIX, 1955, p. 1

otherwise identical cases in which precognition occurs. The other difficulty about physical explanations of this kind is that there is no indication of any mechanism for emitting or receiving such emanations, either in objects or in human brains, and that no energy transmission can be detected. Any such theory is therefore bound to be largely *ad hoc*, but in the present state of the subject this is not necessarily a fatal criticism, for almost any theory which suggests further kinds of experiment would be welcome.

Several theories have been put forward with the object of showing that ψ-phenomena can in principle be accommodated within the framework of the existing physical language, and in particular within the field concepts indicated by Broad's principle. Theories of 'radiation' received by an unknown sense organ were early on the scene, and a suggestion using quantum field theory has recently been made by G. D. Wassermann.[1] Wassermann claims to overcome some of the difficulties of a field interpretation of ψ-phenomena which we have mentioned. For example, he constructs a cascade process of fields in which no attenuation with distance occurs, and he assumes that although energy is required by ψ-fields, their interaction with ordinary matter-fields is absent or extremely small, so that ψ-fields can be propagated over long distances without absorption. The difficulty of physical emitters and receivers is overcome by postulating that animal activity is already 'steered' by 'behaviour fields' associated with the animal body and brain, which also have physiological and neurological functions, and with which ψ-fields interact. Precognition proper is denied, but its appearances are explained by postulating 'copy fields' which reproduce the states of matter and other kinds of fields, but make their transitions more rapidly, so that a subject can be aware of the future state of matter-fields before they occur. Such an explanation would eliminate any precognitive paradoxes, for the 'precognition' would be caused, not by the future event, but by interactions of past events with copy fields, and there might then be intervention to prevent the 'precognised' event without any paradox arising.

This theory is clearly ingenious but inevitably *ad hoc*, and it is not yet clear what testable consequences are implied. In this and in its confident exploitation of current physical ideas it is reminiscent of Mesmer's abortive theory of animal magnetism (1779), in which he attempted an explanation of hypnotic phenomena

[1] 'An Outline of a Field Theory of Organismic Form and Behaviour', *CIBA Foundation Symposium on Extrasensory Perception*, London, 1956, p. 53

in terms of the then current physics of subtle fluids : a universally diffused magnetic virtue was supposed to interact with the heavenly bodies, the earth, and animal bodies according to mechanical laws 'with which we are not as yet acquainted'.[1] But all such theories, in spite of being clothed in the garb of probably transient physical fashions, have generally within them some possibility of providing suggestions for further research.

In so far as quantum field theory retains Principle (1), a theory such as Wassermann's might show that ψ-phenomena need not violate these axioms either. But we have seen that (1.2) and (1.3) are by no means necessary to physics, and that some suggestions have been made which would involve even the abandonment of (1.1). It may therefore seem unnecessary to go to great lengths to preserve them in parapsychology. Is it even perhaps possible that these new suggestions in physics may help to provide an explanatory framework for ψ-phenomena? It has been pointed out[2] that physics offers the following analogies with ψ-phenomena : a 'loosening' of causality with corresponding statistical treatment, time-reversals and action faster than light (as in the suggestions of Feynman[3] and Dirac's pre-accelerations), instantaneous spread and collapse of wave functions according to the probability interpretation, and non-localisation of particles in time and space. But, as Chari has pointed out in his examination of these analogies, many of the suggestions in quantum theory are themselves problematic, for example, the type of causality implied is not agreed, and interpretation of wave functions in terms of probability is non-relativistic and is modified by relativistic field theory. In any case it is not clear how these isolated analogies would help to provide a theory of ψ-phenomena unless there were closer analogies between the

[1] cf. C. L. Hull, *Hypnosis and Suggestibility*, New York, 1933, p. 5

[2] By Jordan, Margenau, and Denbigh. See C. T. K. Chari, 'Quantum Physics and Parapsychology', *Journ. Parapsychology*, XX, 1956, p. 166 ; and K. G. Denbigh, 'Non-localisation as a Model for Telepathy', *Journ. S.P.R.*, XXXVIII, 1956, p. 237.

[3] Feynman ('The Theory of Positrons', *Phys. Rev.*, LXXVI, 1949, p. 749) has suggested an alternative representation of pair production and annihilation in the quantum field involving the motion of a particle backwards in time. Suppose in the normal description an electron A and a positron B are produced at a point P at time t_1, and B then moves to point Q, where at time t_2 ($t_2 > t_1$) there is annihilation of B and an electron C. Then the situation can be alternatively described as the continuous motion of a single electron as follows : C moves towards Q, where at t_2 it proceeds to move *backwards in time* along QP, this motion being equivalent to that of the positron B moving forwards in time along PQ. When C reaches P at the earlier time t_1, it begins to move forwards in time along the final path of A. This description is observationally equivalent to the first one, since the particles A, B, C have no self-identity and can be distinguished in the normal time direction only by charge. This observational equivalence of the two descriptions is only one reason for doubting their relevance to precognition, where reversed time is introduced just because there appears to be *no* description in the normal time direction.

whole structure of quantum field theory and parapsychology, and we have already argued that the analogies are not close enough to permit a simple, non-*ad-hoc* theory of ψ-phenomena in field language. Meanwhile, however, abandonment of some of the principles in physics has undoubtedly loosened the traditional restraints upon theorising elsewhere. This in itself would bring little advantage, for what is needed is not complete freedom, but rather a new discipline based on a satisfactory model, but we have seen that this is what is required in physics also, for difficulties have arisen there which can be traced essentially to the retention of a Newtonian vocabulary and some Newtonian laws. It may be that some fundamental rethinking in physics will prove more helpful to parapsychology than the present piecemeal modifications of the physical language.

Most of those who attempt to theorise in parapsychology have long reached the conclusion that theories of physical type are unhelpful and that some new explanatory concepts are required. But new explanatory concepts are always drawn by analogy from some other conceptual system, and since physics has provided the comprehensive framework for so long, it is difficult to know where else to look. Appeal has been made to various theories of 'mind', generally implying a psycho-physical dualism, as for instance in the notion that there is an analogy between the relation of the perceiving self to the brain, and the clairvoyant to the external event, and between the mind's 'control' of the nervous system and psycho-kinetic influence on dice.[1] A theory of Jungian type involving a 'common unconscious' has been put forward by Professor Price,[2] one based on a multi-dimensional mental and physical space by J. R. Smythies,[3] and the use of Bergsonian concepts is suggested by Broad in the paper already discussed. None of these theories can be said at present to be more than imaginative explorations, but if the time should come when a theory of this type proves satisfactory in parapsychology, it will be time to re-examine the question of action at a distance. We have seen that physics rarely comes to any definite conclusion about this question, and that since spatio-

[1] The 'Shin' theory of R. H. Thouless and B. Wiesner, 'The Psi Processes in Normal and "Paranormal" Psychology', *Proc. S.P.R.*, XLVIII, 1946–9, p. 177. See also Thouless, 'The Present Position of Experimental Research into Telepathy and Related Phenomena', *Proc. S.P.R.*, XLVII, 1942–5, p. 1.

[2] See C. W. K. Mundle, 'Some Philosophical Perspectives for Parapsychology', *Journ. Parapsychology*, XVI, 1952, p. 257. Mr Mundle rightly points out that the fact that a theory of psycho-physical dualism has been discredited in philosophy does not necessarily mean that it is useless in parapsychology. He cites the example of Freud, and one might also remember the philosophical refutations of mechanism by Leibniz and others.

[3] 'The Extension of Mind', *Journ. S.P.R.*, XXXVI, 1951, p. 477

temporal coordinates are the independent variables of physics, action at a distance can always be understood to be an open alternative to continuous action, but if a type of psychic action to which distance and time were *irrelevant* were ever to be accepted in science, it would exhibit action at a distance more clearly than is possible for physics, and although alternative continuous action theories would no doubt still be conceivable, it is difficult to see how they could be anything but highly complex and implausible.

Our discussion has not enabled us to answer the question 'Do bodies act at a distance'—this in any case is not the business of the philosophy of science. It has not enabled us to show that physics itself gives a definite answer, because although we have maintained that physical theories are asserted as facts, the problem of action at a distance is not so much a problem of particular theories as of metaphysical framework. The discussion has, however, enabled us to compose some variations on a theme: namely, the historical, heuristic, and logical importance for physics of ideas and assumptions commonly called metaphysical. A society which is uninterested in metaphysics will have no theoretical science.

APPENDIX I

SUPPOSE the rectangular reference frame K' with coordinates (x', y', z', t') is moving with velocity v in the direction of the x-axis of the frame K, whose coordinates are (x, y, z, t) and whose axes are directed parallel to the corresponding axes of K'. The Lorentz transformation equations are

$$x' = \frac{x - vt}{\sqrt{(1 - v^2/c^2)}}, \qquad y' = y, \qquad z' = z, \qquad t' = \frac{t - xv/c^2}{\sqrt{(1 - v^2/c^2)}}.$$

Suppose a causal action is propagated in K along the x-axis from x_A to x_B with velocity u. Then the time taken from A to B is

$$t_B - t_A = \frac{x_B - x_A}{u}.$$

In K', $\quad t'_B - t'_A = (1 - uv/c^2)\ (t_B - t_A)/\sqrt{(1 - v^2/c^2)}$.
If $u > c$ there are frames K' such that $v < c$ but $uv > c^2$. In these frames, since $t_B - t_A > 0$, $t'_B - t'_A < 0$, that is, the event B which was the effect in K precedes the event A, which was the cause in K.

Hence if causal order is to be invariant, there can be no velocity greater than c in any frame.

If u is infinite, that is if there is instantaneous propagation from x_A to x_B, so that $t_A = t_B$, the causal order in K becomes indeterminate. But in K' the velocity of propagation is

$$\frac{x'_B - x'_A}{t'_B - t'_A} = -\frac{c^2}{v}.$$

This represents a finite velocity greater than c, which has already been shown to be impossible. Hence the velocity cannot be infinite in K.

Again, suppose K' is such that $uv = c^2$. Then $t'_B = t'_A$, but

$$x'_B - x'_A = \frac{(1 - v/u)}{\sqrt{(1 - v^2/c^2)}}\ (x_B - x_A)$$

which must be > 0, i.e. an entity which moved from x_A to x_B in K with a velocity $> c$, would be in two different places at the same time in K', that is, its velocity would be infinite in K'.

APPENDIX II

SUPPOSE a causal action is propagated from P to Q and is described by an observer A situated on the straight line QP produced. A determines the space and time coordinates of the cause-event at P by means of a signal sent out at τ_2 on A's τ-clock, reflected at P simultaneously with the cause-event, and received back at A at τ_3. Suppose the effect-event at Q is similarly determined by A by means of a signal sent out at τ_1 and received back at τ_4.

The coordinates of P and Q are given by

$$\tau_Q = \tfrac{1}{2}(\tau_4 + \tau_1) \qquad X_Q = \tfrac{1}{2}c(\tau_4 - \tau_1)$$
$$\tau_P = \tfrac{1}{2}(\tau_3 + \tau_2) \qquad X_P = \tfrac{1}{2}c(\tau_3 - \tau_2).$$

The velocity of propagation of the causal action in A's τ-time is

$$\frac{X_Q - X_P}{\tau_Q - \tau_P} = c\left(\frac{\tau_4 - \tau_1 - \tau_3 + \tau_2}{\tau_4 + \tau_1 - \tau_3 - \tau_2}\right)$$

which is $\gtreqless c$ according as $\tau_2 - \tau_1 \gtreqless 0$. (1)

If the velocity of propagation is infinite, then we have

$$\tau_4 + \tau_1 = \tau_3 + \tau_2 \qquad \tau_4 - \tau_1 > \tau_3 - \tau_2.$$

Hence $\tau_2 - \tau_1 = \tau_4 - \tau_3 > \tau_1 - \tau_2$, that is, $\tau_2 - \tau_1 > 0$, in agreement with (1).

Now suppose A regraduates his clock to t-time, where

$$t = t_0 \exp\left(\frac{\tau - t_0}{t_0}\right).$$

Then the coordinates of P and Q are

$$t_Q = \tfrac{1}{2}(t_4 + t_1) \qquad x_Q = \tfrac{1}{2}c(t_4 - t_1)$$
$$t_P = \tfrac{1}{2}(t_3 + t_2) \qquad x_P = \tfrac{1}{2}c(t_3 - t_2).$$

The velocity of propagation in A's t-time is

$$\frac{x_Q - x_P}{t_Q - t_P} = c\left\{\frac{e^{(\tau_4 - t_0)/t_0} - e^{(\tau_1 - t_0)/t_0} - e^{(\tau_3 - t_0)/t_0} + e^{(\tau_2 - t_0)/t_0}}{e^{(\tau_4 - t_0)/t_0} + e^{(\tau_1 - t_0)/t_0} - e^{(\tau_3 - t_0)/t_0} - e^{(\tau_2 - t_0)/t_0}}\right\}$$

which is $\gtreqless c$ according as $e^{\tau_2/t_0} - e^{\tau_1/t_0} \gtreqless 0$,

that is, according as $\tau_2 - \tau_1 \gtreqless 0$.

Thus, if a velocity is greater than c in t-time, it is greater than c in τ-time. Similarly it is greater than c in any other time-scale T, since T must be a monotonic increasing function of t in order to preserve the fundamental before-and-after relation between events.

BIBLIOGRAPHY OF MAIN SECONDARY HISTORICAL SOURCES

(This bibliography is not intended to be exhaustive, but contains those sources I have found most useful in writing Chapters II to IX.)

AGASSI, J., *The Function of Interpretations in Science*, London, Ph.D. Thesis, 1956

ARMITAGE, A., ' "Borell's hypothesis" and the rise of celestial mechanics ', *Ann. Sci.*, VI, 1948–50, 268

ARMSTRONG, A. H., *Plotinus*, London, 1953

BAILEY, C., *The Greek Atomists and Epicurus*, Oxford, 1928

BALME, D. M., ' Greek science and mechanism ', *Class. Quart.*, XXXIII, 1939, 1 ; XXXV, 1941, 23

BEARE, J. I., *Greek Theories of Elementary Cognition from Alcmaeon to Aristotle*, Oxford, 1906

BECK, L. J., *The Method of Descartes*, Oxford, 1952

BELL, A. E., *Christian Huygens*, London, 1947

BOAS, M., ' Hero's *Pneumatica*, a study of its transmission and influence ', *Isis*, XL, 1949, 38

— ' Boyle as theoretical scientist ', *Isis*, XLI, 1950, 261

— ' The establishment of the mechanical philosophy ', *Osiris*, X, 1952, 412

— *Robert Boyle and Seventeenth-Century Chemistry*, Cambridge, 1958

BONDI, H., *Cosmology*, Cambridge, 1952

BRANDT, F., *Thomas Hobbes' Mechanical Conception of Nature*, Copenhagen and London, 1928

BRETT, G. S., *The Philosophy of Gassendi*, London, 1908

BURNET, J., *Early Greek Philosophy* (4th ed.), London, 1948

— *Greek Philosophy : Thales to Plato*, London, 1914

BURTT, E. A., *The Metaphysical Foundations of Modern Physical Science*, London, 1932

CARR, M. E. J., *Mathematical Theories of Electricity to Maxwell*, London, M.Sc. Thesis, 1949

CARRÉ, M. H., *Realists and Nominalists*, Oxford, 1946

CHERNISS, H., *Aristotle's Criticism of Presocratic Philosophy*, Baltimore, 1935

COHEN, I. B., *Benjamin Franklin's Experiments*, Cambridge, Mass., 1941

COHEN, I. B., *Franklin and Newton*, Philadelphia, 1956
COLLINGWOOD, R. G., *The Idea of Nature*, Oxford, 1945
CORNFORD, F. M., *From Religion to Philosophy*, London, 1912
— ' Mysticism and science in the Pythagorean tradition ', *Class. Quart.*, XVI, 1922, 150 ; XVII, 1923, 1
— *Laws of Motion in Ancient Thought*, Cambridge, 1931
— *Plato's Cosmology*, London, 1937
— *The Unwritten Philosophy*, Cambridge, 1950
— *Principium Sapientiae*, Cambridge, 1952
CROMBIE, A. C., *Robert Grosseteste*, Oxford, 1953
DREYER, J. L. E., *History of the Planetary Systems from Thales to Kepler*, Cambridge, 1906
DUGAS, R., *Histoire de la Mécanique*, Paris and Neuchâtel, 1950
— *La Mécanique au XVIIe Siècle*, Paris and Neuchâtel, 1954
DUHEM, P., *The Aim and Structure of Physical Theory*, trans. Wiener, Princeton, 1954
EVANS-PRITCHARD, E. E., *Witchcraft, Oracles and Magic among the Azandi*, Oxford, 1937
FRANKFORT, H. and H. A., et al., *The Intellectual Adventure of Ancient Man*, Chicago, 1946 (Eng. ed., *Before Philosophy*, 1949)
FREEMAN, K., *Ancilla to the Pre-Socratic Philosophers*, Oxford, 1948
GOMPERZ, T., *Greek Thinkers*, Vol. I trans. Magnus, Vols. II–IV trans. Berry, London, 1901–12
HALL, A. R., *Ballistics in the Seventeenth Century*, Cambridge, 1952
— *The Scientific Revolution*, London, 1954
HEATHCOTE, N. H. de V., ' Guericke's sulphur globe ', *Ann. Sci.*, VI, 1950, 293
HEIDEL, W. A., *The Heroic Age of Science*, Carnegie Institution of Washington, 1933
— *Hippocratic Medicine*, New York, 1941
JAEGER, W., *Theology of the Early Greek Philosophers*, Oxford, 1947
JAMMER, M., *Concepts of Space*, Cambridge, Mass., 1954
— *Concepts of Force*, Cambridge, Mass., 1957
JOSEPH, H. W. B., *Lectures on the Philosophy of Leibniz*, Oxford, 1949
KIRK, G. S. and RAVEN, J. E., *The Presocratic Philosophers*, Cambridge, 1957
KOYRÉ, A., ' Le gravitation universelle de Kepler à Newton ', *Archives Internationale d'Histoire des Sciences*, XXX, 1951, 638
— *Études Galiléennes, Actualités Scientifiques et Industrielles*, Paris, 1939, 852–4
— *From the Closed World to the Infinite Universe*, Baltimore, 1957
LITTLE, A. G., *Roger Bacon—Essays*, Oxford, 1914

McKie, D. and Heathcote, N. H. de V., *The Discovery of Specific and Latent Heats*, London, 1935

Millington, E. C., 'Theories of cohesion in the seventeenth century', *Ann. Sci.*, v, 1941–7, 253

— 'Studies in capillarity and cohesion in the eighteenth century', *Ann. Sci.*, v, 1941–7, 352

Onians, R. B., *The Origins of European Thought*, Cambridge, 1951

Partington, J. R., 'The origins of the atomic theory', *Ann. Sci.*, iv, 1939, 245

Partington, J. R. and McKie, D., 'Historical studies on the phlogiston theory', *Ann. Sci.*, ii, 1937, 361; iii, 1938, 1, 337; iv, 1939, 113

Paton, H. J., *Kant's Metaphysic of Experience*, London, 1936

Patterson, L. D., 'Robert Hooke and the conservation of energy', *Isis*, xxxviii, 1948, 151

— 'Hooke's gravitation theory and its influence on Newton', *Isis*, xl, 1949, 327; xli, 1950, 32

Petersson, R. T., *Sir Kenelm Digby*, London, 1956

Randall, J. H., 'Scientific method in the school of Padua', *Journ. Hist. Ideas*, i, 1940, 177

Rodier, G., *La Physique de Straton*, Paris, 1890

Russell, B., *A Critical Exposition of the Philosophy of Leibniz* (2nd ed.), London, 1937

Sambursky, S., *The Physical World of the Greeks*, London, 1956

— *Physics of the Stoics*, London, 1959

Sharp, D. E., *Franciscan Philosophy at Oxford*, Oxford, 1930

Singer, D. W., *Giordano Bruno*, New York, 1950

Skemp, J. B., *The Theory of Motion in Plato's Later Dialogues*, Cambridge, 1942

Smith, N. K., *New Studies in the Philosophy of Descartes*, London, 1952

Snow, A. J., *Matter and Gravity in Newton's Physical Philosophy*, London, 1926

Stones, G. B., 'The atomic view of matter in the fifteenth, sixteenth, and seventeenth centuries', *Isis*, x, 1928, 444

Taylor, A. E., *A Commentary on Plato's Timaeus*, Oxford, 1928

Taylor, F. S., *The Alchemists*, New York, 1949

Thorndike, L., *History of Magic and Experimental Science*, 8 vols., London, 1923–58

Todhunter, I. and Pearson, K., *A History of the Theory of Elasticity, from Galilei*, Cambridge, 1886

Toulmin, S. E., 'Criticism in the History of Science: Newton on absolute space, time and motion', *Phil. Rev.*, lxviii, 1959, 1, 203

VARTANIAN, A., *Diderot and Descartes*, Princeton, 1953
VERBEKE, G., *L'Évolution de la Doctrine du Pneuma*, Louvain, 1945
WHEWELL, W., *The History of the Inductive Sciences*, London, 1837
— *The Philosophy of the Inductive Sciences*, London, 1840
WHITTAKER, E. T., *History of the Theories of Aether and Electricity*
I : *The Classical Theories*, London, 1951
II : *The Modern Theories*, London, 1953
WHYTE, L. L., 'R. J. Boscovich S.J., F.R.S. (1711–87), and the mathematics of atomism', *Notes and Records of the Royal Society*, XIII, 1958, 46
YOST, R. M., 'Locke's rejection of the sub-microscopic', *Journ. Hist. Ideas*, XII, 1951, 111
ZELLER, E., *The Stoics, Epicureans and Sceptics*, London, 1870
ZILSEL, E., 'Copernicus and mechanics', *Journ. Hist. Ideas*, I, 1940, 113
— 'The origins of William Gilbert's scientific method', *Journ. Hist. Ideas*, II, 1941, 1

INDEX OF PROPER NAMES

Aepinus, F. U. T. 183
Agassi, J. 90, 201
Albertus Magnus 126
Albumasar 126
Alexander, H. G. 155
Alexander, P. 18
Ampère, A. M. 216f., 225, 292
Anaxagoras 39f., 47, 53
Anaximander 39, 53
Anaximenes 37, 39, 60
Aquinas, St Thomas 66, 87, 102, 126f.
Arago, F. 230
Aristarchus 131
Aristotle 26, 30, 33, 38, 42, 47, 52, 55, 57, 60–73, 75, 79f., 90, 105, 109, 126, 147, 179, 188, 291
Augustine, St 78
Averroes 127
Ayer, A. J. 11

Bacon, F. 74, 78, 85, 91–8
Bacon, R. 79f., 126
Becher, J. J. 184
Bentley, R. 150
Berkeley, G. 3, 44, 168, 181, 187
Bernoulli, Daniel 180, 190
Bernoulli, James 84
Bernoulli, John 157, 190, 225
Berthollet, A. 186
Betti, E. 220
Black, J. 185
Boerhaave, H. 184
Bohm, D. 268f., 291
Bohr, N. 261f., 291
Borelli, A. 132
Born, M. 198, 262
Boscovich, R. 163–6, 174, 200f.
Boyle, R. 59, 74f., 85, 97, 99, 115–18, 151
Braithwaite, R. B. 7, 13, 19, 20
Bridgman, P. W. 7, 197, 258
Broad, C. D. 297f.
Browne, T. 91
Bruno, G 78
Burnet, J. 41, 46

Campbell, N. R. 8, 13, 15
Carnap, R. 11f.
Carnot, N. L. S. 186
Cauchy, A. L. 195
Cavendish, H. 180, 183
Chari, C. T. K. 301
Chrysippus 76
Clarke, S. 160–3
Clausius, R. 187, 219f.
Cleghorn, W. 185
Cohen, I. B. 147
Collingwood, R. G. 101
Copernicus, N. 1, 89, 127f.
Cornford, F. M. 31, 36, 38, 41, 58
Cotes, R. 150, 187
Coulomb, C. A. 183, 209

D'Alembert, J. L. 190, 193
Dalton, J. 186
Davidson, W. 251
De Broglie, L. V. P. R. 262
Democritus 42f., 53, 62, 65
Denbigh, K. G. 301
Desaguliers, J. T. 181
Descartes, R. 10, 74, 76, 84, 101–14, 119, 121, 131, 152, 157f., 168f., 188, 291
Dicke, R. H. 278
Diderot, D. 170
Digby, K. 97
Diogenes of Apollonia 37, 52
Dirac, P. A. M. 272f., 280f., 288, 301
Duhem, P. 3, 81, 213
Dummett, A. E. 285f.

Eddington, Sir A. S. 7, 252, 256, 260
Einstein, A. 6, 139, 143, 225, 231f., 245–53, 261
Elliot, J. 188
Empedocles 40f., 47, 52f., 62
Epicurus 43, 81
Euler, L. 157, 169, 189–95
Evans-Pritchard, E. E. 34

Faraday, M. 166, 197–206, 209, 216f., 222f., 258, 281, 291f.
Fechner, G. T. 217
Feyerabend, P. K. 21, 267
Feynman, R. P. 279–86, 292, 301
Fitzgerald, G. F. 228f.
Fizeau, H. L. 154
Foucault, L. 154
Fourier, J. 186f., 193, 209, 292
Fracastoro 81
Frankfort, H., & H. A. 34f.
Franklin, B. 181, 183
Fresnel, A. 194

Galen 58f., 82, 87, 126
Galilei, Galileo 1, 44, 63, 74, 83f., 99, 101, 104, 111, 121, 126f., 189
Gassend, P. 74, 102, 117

Gauss, K. F. 210, 217, 220
Gilbert, W. 59, 79, 86–91, 94f., 100f., 127f.
Glanvill, J. 96f., 119
Gomperz, T. 53
Green, G. 187, 210
Groenewold, H. J. 281
Grosseteste, R. 78, 126
Guericke, O. von 161, 182

Hare, R. 199
Heaviside, O. 219
Heidel, W. A. 53
Heisenberg, W. 262, 266f.
Heitler, W. 274
Helmholtz, H. von 187, 208, 212, 217f., 221
Hempel, C. G. 11
Heraclitus 38, 184
Hero of Alexandria 82f.
Hertz, H. 21, 25, 212–15, 221, 224, 227
Hesiod 31, 39, 41
Hesse, M. B. 7, 18, 21, 86, 201
Higgins, B. 185
Hobbes, T. 85, 101, 104, 112–15, 133
Homer 36
Hooke, R. 118f., 127, 133f., 147
Hoskin, M. A. 163
Hull, C. L. 301
Hume, D. 168f.
Hutten, E. H. 21, 272
Huygens, Christiaan 53, 74, 84, 107f., 114, 119f., 133, 153, 157f.

Jaeger, W. 52
Jammer, M. 42, 58, 66, 68, 143
Jeffreys, H. 16
Joachim, H. H. 65
Jordan, P. 301

Kant, I. 2, 15, 166, 169–80, 201, 292
Kelvin, Lord (W. Thomson) 4, 22, 187, 198, 203, 206f., 221, 224
Kennedy, R. J. 232
Kepler, J. 2, 94f., 101, 127f., 147
Kirchhoff, G. 220
Knight, G. 181f., 188
Kohlrausch, R. 220

Lagrange, J. L. 194, 196
Landau, L. 198
Laplace, P. S. 195f., 210, 225
Larmor, J. 210, 225
Lavoisier, A. L. 185f., 188, 292
Lee, H. D. P. 85
Leibniz, G. W. 116, 157–66, 175, 182, 190, 194, 291, 302
Lesage, G. L. 224
Leslie, P. D. 181
Leucippus 38, 45, 53
Liénard, A. 219
Lifshitz, E. 198
Locke, J. 44, 121–5, 166f.
Lorentz, H. A. 208, 219, 222, 228
Lorenz, L. V. 221
Lucretius 43, 79, 81, 103

McCrea, W. H. 238, 243, 252, 255, 281
Mach, E. 3, 6, 136, 247
McKie, D. 187
Malebranche, N. 84
Margenau, H. 301
Martin, G. 172
Maupertuis, P. L. M. de 169
Maxwell, J. Clerk 5, 25, 198, 201f., 206–22, 225, 257, 260, 274, 278, 292
Mersenne, M. 113
Mesmer, A. 300
Michelson, A. A. 226
Milne, E. A. 238–45, 251f., 255
Morley, E. W. 226
Mundle, C. W. K. 302

Navier, C. L. M. H. 195
Neumann, C. 220f.
Newton, Sir Isaac 2, 15, 21, 28f., 74f., 107, 119, 122, 131, 133–57, 161, 164f., 178f., 184, 189f., 202, 222f., 232, 291f.

Ockham, William of 1f., 102, 291f.
Oersted, H. C. 216
Onians, R. B. 35, 37, 76
Oppenheim, P. 11

Parmenides 39f., 45
Partington, J. R. 186, 187
Pearson, K. 3
Peierls, R. E. 278
Pemberton, H. 152
Philo Judaeus 37
Philo of Byzantium 82
Philoponus, J. 66
Piaget, J. 290
Planck, M. 261
Plato 33, 38, 47f., 54f., 65, 68, 75
Plutarch 31, 37, 56f., 88, 127
Poisson, S. D. 183, 195f., 210
Popper, K. R. 3, 9, 16, 99
Poynting, J. H. 211
Price, H. H. 302
Ptolemy 126
Pythagorean school 39, 44f., 88, 93, 99, 101

Ramsey, F. P. 7, 13
Ravetz, J. R. 194
Reichenbach, H. 254f., 263f., 271
Rhine, J. B. 295

Rhine, L. E. 299
Riemann, B. 220, 225
Robertson, H. P. 233
Roberval, G. P. de 131f., 151
Rumford, Count (B. Thompson) 186
Russell, B. 6, 164

St Amand, John of 81
Sambursky, S. 76
Sarton, G. 98
Saurin, J. 157
Schild, A. 251
Schrödinger, E. 262f.
Schwarzschild, K. 279f., 292
Sciama, D. W. 250f.
Sellars, W. 12
Shaw, P. 184
Smythies, J. R. 302
Soal, S. G. 295, 299
Sprat, T. 97
Stahl, G. E. 184
Stokes, G. G. 187, 195, 227
Strato 82f.
Synge, J. L. 190, 232

Tetrode, H. 281, 292
Thales 38f., 57f., 90
Theophrastus 42, 57
Thompson, B. *See* Rumford
Thomson, J. J. 218
Thomson, W. *See* Kelvin
Thorndike, E. M. 232
Thouless, R. H. 302
Törnebohm, H. 253
Torricelli, E. 84, 117, 161
Toulmin, S. E. 141
Truesdell, C. A. 191, 194

Vienna Circle 7
Voltaire, F. M. A. 169
Von Neumann, J. 267f.

Waismann, F. 8
Walker, N. 99
Wassermann, G. D. 300f.
Watkins, J. W. N. 11, 293f.
Weber, W. 217, 219f., 225, 257, 292
Wellmann, M. 55
Wheeler, J. A. 270, 279–86, 292
Whitehead, A. N. 49, 251
Whittaker, E. T. 5, 219, 233, 260, 279
Wiechert, E. 219
Wiesner, B. 302
Wilson, H. A. 278
Woodger, J. H. 13

Xenophon 52

Young, T. 194
Yukawa, H. 276

Zeno 45, 289

INDEX OF SUBJECTS

absolute rotation 138f., 247
— space 138f., 147, 226, 246, 275
advanced potentials 279–86
aether 4, 84, 96, 104, 108, 114, 116f., 131, 151f., 161, 178, 194, 200, 206f., 213, 224, 226f., 274, 277f., 293
analogy 22f., 30f., 110, 123, 156, 168, 203, 206, 209, 213, 268f.
— of attraction 31, 41, 87, 99, 126f., 291
— of organism 31, 42, 72, 75, 99, 112, 118, 291
— positive and negative 24, 100
angular momentum 142, 171
animism 35f., 58, 76f., 88, 101, 111, 293
antiperistasis 55f., 67, 84f., 101
astrology 75, 80, 95f., 126, 128
atomicity 260
atomism 42–6, 81f., 99, 102f., 153f., 158, 173f., 200
attractive force 130f., 140, 145, 149f., 164f., 170, 176f., 201, 216f., 251, 278

Boyle's Law 180, 187

caloric 185f.
Cartesianism 115f., 132f., 157f.
causal anomalies 237, 261, 265, 271
cause 47, 63f., 72, 102, 133, 140f., 148f., 168f., 217, 236, 255, 272, 275, 285f., 297f.
central forces 140, 144, 155f., 164, 176, 201, 209, 217f.
centrifugal force 51f., 105, 107, 114, 117, 133, 251
chemical affinity 154, 184f.
— valency 276
cohesion 42, 84, 88, 101, 106, 114, 119, 151, 154, 164f., 177
complementarity 27, 263, 270, 281, 287
complementary variables 266, 277
continuity, principle of 158f., 163f., 171, 173
cosmological principle 241f.
Coulomb field 273f.
Coulomb's law 183, 212, 234
creations and annihilations 272f., 277

density 65, 104, 108, 174, 178f., 250, 262, 273f.
'dictionary' 13–21. *See also* interpretation
dielectric. *See* electric induction
displacement current 212, 218

effluvia 57, 88f., 100, 117
elasticity 83, 117, 120, 158, 168, 180f., 194f.
electric charge 183, 205, 208f., 212f., 225, 234, 273f.
— current 203, 205, 216f.
— induction 198f., 205, 212f., 224, 274, 278
electricity 86f., 94, 103, 107, 114, 117, 119, 121, 151, 164, 182f.
electromagnetic theory 198, 207–22, 226, 234f., 244f., 250, 274–85
electron 208, 261f., 272f.
energy 187, 209f., 258, 261, 272f., 276, 293
—, conservation of 217, 221, 224, 245, 250, 276f. *See also vis viva*
Euclidean geometry 237, 248f., 253f.
exclusion principle 273
extrasensory perception 295–303

falsifiability 8f., 17f., 98f., 116, 155, 175, 188, 215, 252f., 257, 268, 285, 292f.
Fermi-Dirac statistics 273
field 66, 91, 192, 195f., 206–12, 229, 247f., 259f., 272–82, 299f.
Fitzgerald-Lorentz contraction 228f., 235, 254
form 47, 60f., 78f., 91, 101, 105
formalist view of theories 14f., 25, 214f., 222, 270
formal tests 20, 252
four elements 40f., 62, 66, 184
four-vector 234, 276

gravitation 66, 89, 94f., 107, 112, 114, 118, 126–53, 158, 163f., 177f., 186f., 204f., 222f., 244–53, 258, 278f.
Greek science, mechanism in 51ff.

harmony 44, 90, 100
heat 54, 77f., 85, 93, 106, 112, 114, 117f., 184f.
heaviness. *See* weight
' hole ' theory 273
horror vacui 55f., 67, 82f., 92, 101, 117
hydrodynamics 111, 189–97
hypnosis 98, 300
hypothetical method 9, 88, 98, 108, 115, 119f.

hypothetico-deductive system 8–13, 28, 147f., 237, 252

impact 43, 55, 84, 103, 106, 120f., 141, 155, 163f.
impenetrability 150, 154, 165, 169, 173
impulse. *See* impact
inertia 85, 104, 106, 113, 129, 132f., 142, 159, 164, 246f., 250f.
inertial frames 231f., 246f., 253
instantaneous action 106, 113, 137, 156, 195, 198, 217, 225, 229, 237, 245, 265f., 271, 279, 284
intelligibility 27, 30, 116, 155f., 269, 285
interpretation (of formalisms) 19f., 124, 156, 212f., 238, 242, 252, 272, 274
interphenomena 264, 271f., 288, 291
inverse-square law 131, 137, 146, 157, 164, 170, 179, 183, 209, 216, 277

Kepler's laws 21, 133, 144f., 157

laws of motion (Newton's) 28, 134f., 189, 233f., 250, 268
light 77f., 91, 93, 103, 105, 113, 117f., 131, 151, 153f., 194
— velocity of 220, 226–37, 241f., 253f., 284, 286f.
lines of force 201f.
Lorentz transformation 228f., 234f., 242f., 253, 259

Mach's principle 247, 250f.
magic 32, 35, 75f., 95f.
magnetic force 137, 202f.
magnetism 46, 57f., 80f., 86f., 93f., 100, 103, 106, 114, 116, 126f., 183, 216, 234
mass 111, 136, 146, 235, 246f.
Maxwell field 274, 276
Maxwell's equations 21, 25, 207, 214f., 218f., 234, 279, 282
meson 274, 276
metaphysics 2, 9, 49, 77, 99, 116, 125, 172, 179, 290–5, 303
Michelson-Morley experiment 8, 226–32, 238, 241f., 253f.
model 14, 21–8, 101, 172, 187, 240, 250, 264, 269f., 285, 292
— fundamental 28, 49, 101, 125, 253, 285, 290, 294f.
— mathematical 23, 155f., 189–96, 229, 252
— mechanical 4, 23, 30, 52, 116, 125, 156, 194f., 206f., 213, 221
momentum, conservation of 136f., 146, 159, 211, 232, 245, 250, 279
multiplication of species 79f., 91f., 101, 130f.
myth 29–39

Neo-Platonism 77f., 133
Newtonian transformation 233f.
nuclear theory 275f.

observables 268, 270, 272, 277f.
observation statements 3–13, 124. *See also* phenomenal statements
occult qualities 100, 117, 126, 149f., 162, 173
omnipresence of God 152, 155, 161f.
operational definition 6, 146, 191, 231, 237
operationalism 6f., 148, 228, 239, 251, 256, 273

pair-production. *See* creations and annihilations
parapsychology 295–303
phenomena, 'deduction' from 2, 144, 148, 230, 235
— in Kant 15, 174f.
— in Newton 2, 15, 144f.
— in Reichenbach 263, 271, 288, 291
phenomenal statements 15f., 263
phlogiston 184f.
photon 261f., 276f.
pneuma (*spiritus*) 36f., 75f., 111
polarisation, of dielectrics. *See* electric induction
— of light 153, 194, 204, 262
positivism 1, 6f., 102, 148, 166, 187, 190, 213, 275, 291f.
positron 273f., 301
potential function 196f., 210, 219f., 234, 249
pre-acceleration 284f., 301
precognition 286, 296f.
primary qualities 30, 44, 62, 99, 103, 121f., 150, 167

radiation 77f., 91f., 184, 204f., 220, 261f., 276, 280f.
realism 1–5, 24f., 201, 206, 212, 237, 257f., 271, 275f., 287f., 292
regulative principle 175, 291
relativity theory 6, 259f.
— — general 143, 245–54, 278f.
— — special 218, 226–39, 272
repulsion 85, 91, 153f., 164, 171, 173f., 180f.
resonance 96, 114, 118
retarded potentials 221, 279–85
rigid bodies 189, 233, 237f., 253f., 284
rocket 143
Rules of Reasoning in Philosophy 144f., 230

secondary qualities 44, 103, 121f.
simplicity 16, 42, 99, 115f., 239, 247, 254f., 294f.
soul 35f., 47, 58, 66, 71, 77, 90, 101f., 118, 132, 160
sound 80, 94f., 114, 120f., 193f.
species 68, 80f., 93f., 101, 103, 105, 130f.
spin 273f.,
spirits 92, 94, 111, 140, 152, 161f. *See also pneuma*
Stoicism 37, 75f., 184
subtle matter. *See* aether
sympathetic cure of wounds 96f.
sympathy and antipathy 31, 81, 91f., 100, 103, 118, 157

telepathy 95, 98, 296f.
testability. *See* falsifiability
thought-experiment 240, 263f.
tides 94, 96, 126f., 147
time-simultaneity 231, 237, 239f.

uncertainty principle 266f., 271f., 277, 286f., 291
universal forces 254f.

vacuum 83f., 92, 110, 113, 116f., 159f., 178f., 274, 278. *See also horror vacui*
virtual interactions 276f., 286
vis viva 159, 161, 163
void. *See* vacuum *and horror vacui*
vortices 52f., 104f., 157, 170, 189

wave-equation 193f., 262, 273
wave-function 262, 274, 301
wave-motion 76, 79, 94, 101, 118f., 153f., 194
weight 42, 53, 64, 66, 89, 104, 108, 112, 132, 149, 178
witchcraft 32, 34, 75, 95

A CATALOG OF SELECTED

DOVER BOOKS

IN SCIENCE AND MATHEMATICS

History of Math

THE WORKS OF ARCHIMEDES, Archimedes (T. L. Heath, ed.). Topics include the famous problems of the ratio of the areas of a cylinder and an inscribed sphere; the measurement of a circle; the properties of conoids, spheroids, and spirals; and the quadrature of the parabola. Informative introduction. clxxxvi+326pp; supplement, 52pp. 5⅜ x 8½. 42084-1

A SHORT ACCOUNT OF THE HISTORY OF MATHEMATICS, W. W. Rouse Ball. One of clearest, most authoritative surveys from the Egyptians and Phoenicians through 19th-century figures such as Grassman, Galois, Riemann. Fourth edition. 522pp. 5⅜ x 8½. 20630-0

THE HISTORY OF THE CALCULUS AND ITS CONCEPTUAL DEVELOPMENT, Carl B. Boyer. Origins in antiquity, medieval contributions, work of Newton, Leibniz, rigorous formulation. Treatment is verbal. 346pp. 5⅜ x 8½. 60509-4

THE HISTORICAL ROOTS OF ELEMENTARY MATHEMATICS, Lucas N. H. Bunt, Phillip S. Jones, and Jack D. Bedient. Fundamental underpinnings of modern arithmetic, algebra, geometry, and number systems derived from ancient civilizations. 320pp. 5⅜ x 8½. 25563-8

A HISTORY OF MATHEMATICAL NOTATIONS, Florian Cajori. This classic study notes the first appearance of a mathematical symbol and its origin, the competition it encountered, its spread among writers in different countries, its rise to popularity, its eventual decline or ultimate survival. Original 1929 two-volume edition presented here in one volume. xxviii+820pp. 5⅜ x 8½. 67766-4

GAMES, GODS & GAMBLING: A History of Probability and Statistical Ideas, F. N. David. Episodes from the lives of Galileo, Fermat, Pascal, and others illustrate this fascinating account of the roots of mathematics. Features thought-provoking references to classics, archaeology, biography, poetry. 1962 edition. 304pp. 5⅜ x 8½. (Available in U.S. only.) 40023-9

OF MEN AND NUMBERS: The Story of the Great Mathematicians, Jane Muir. Fascinating accounts of the lives and accomplishments of history's greatest mathematical minds–Pythagoras, Descartes, Euler, Pascal, Cantor, many more. Anecdotal, illuminating. 30 diagrams. Bibliography. 256pp. 5⅜ x 8½. 28973-7

HISTORY OF MATHEMATICS, David E. Smith. Nontechnical survey from ancient Greece and Orient to late 19th century; evolution of arithmetic, geometry, trigonometry, calculating devices, algebra, the calculus. 362 illustrations. 1,355pp. 5⅜ x 8½. Two-vol. set. Vol. I: 20429-4 Vol. II: 20430-8

A CONCISE HISTORY OF MATHEMATICS, Dirk J. Struik. The best brief history of mathematics. Stresses origins and covers every major figure from ancient Near East to 19th century. 41 illustrations. 195pp. 5⅜ x 8½. 60255-9

Physics